90/

547
B.R.

20/5/99
15/ 10/99

John Brockington
B.Sc., C.Chem, M.R.S.C.

Peter Stamper
Ph.D., C.Chem, M.R.S.C.

Organic chemistry for higher education

Longman London and New York

Longman Group Limited,
Longman House,
Burnt Mill, Harlow, Essex, UK

*Published in the United States of America
by Longman Inc., New York*

© Longman Group Limited 1982

First published 1982

British Library Cataloguing in Publication Data

Brockington, John
  Organic chemistry for higher education.
  — (Longman technician series: mathematics
  and sciences sector)
  1. Chemistry, Organic
  I. Title   II. Stamper, Peter
  547'.0246   QD253

  ISBN 0-582-41230-7

**Library of Congress Cataloging in Publication
Data**

Brockington, John.
  Organic chemistry for higher education.

  (Longman technician series. Mathematics
  and sciences sector)
  Includes index.
  1. Chemistry, Organic. I. Stamper, Peter,
  1944–   II. Title. III. Series.
  QD251.2.B74   547   81-17118
  ISBN 0-582-41230-7   AACR2

Set by MHL Typesetting Ltd, Coventry
Printed in Great Britain by William Clowes (Beccles) Ltd, Beccles & London

Series sector symbol
(Mathematics and sciences)

Sector Editor:

D.R. Browning, B.Sc., F.R.I.C., A.R.I.C.S.
Principal Lecturer and Head of Chemistry, Bristol Polytechnic

Books already published in this sector of the series:

Technician mathematics Level 1 Second edition   **J.O. Bird** and **A.J.C. May**
Technician mathematics Level 2 First edition (Old syllabus U75/012, 038 and 039)   **J.O. Bird** and **A.J.C. May**
Technician mathematics Level 2 Second edition (New syllabus U80/691, 692 and 712)   **J.O. Bird** and **A.J.C. May**
Technician mathematics Level 3   **J.O. Bird** and **A.J.C. May**
Technician mathematics Levels 4 and 5   **J.O. Bird** and **A.J.C. May**
Mathematics for electrical and telecommunications technicians   *J.O. Bird* and **A.J.C. May**
Mathematics for science technicians Level 2   **J.O. Bird** and **A.J.C. May**
Mathematics for electrical technicians Level 3   **J.O. Bird** and **A.J.C. May**
Mathematics for electrical technicials Levels 4 and 5   **J.O. Bird** and **A.J.C. May**
Calculus for technicians   **J.O. Bird** and **A.J.C. May**
Statistics for technicians   **J.O. Bird** and **A.J.C. May**
Algebra for technicians   **J.O. Bird** and **A.J.C. May**
Physical sciences Level 1   **D.R. Browning** and **I. McKenzie Smith**
Engineering science for technicians Level 1   **D.R. Browning** and **I. McKenzie Smith**
Safety science for technicians   **W.J. Hackett** and **G.P. Robbins**
Fundamentals of chemistry   **J.H.J. Peet**
Further studies in chemistry   **J.H.J. Peet**
Technician chemistry Level 1   **J. Brockington** and **P.J. Stamper**
Technician physics Level 2   **E. Deeson**
Cell biology for technicians   **N.A. Thorpe**
Mathematics for scientific and technical students   **H.G. Davies** and **G.A. Hicks**
Mathematical formulae for TEC courses   **J.O. Bird** and **A.J.C. May**
Science formulae for TEC courses   **D.R. Browning**

Contents

Preface

This is one of a series of three textbooks which we have prepared primarily to meet the needs of students on the first year of TEC courses leading to higher awards in physical science and technology. It should also prove valuable to students embarking on a single honours or general honours degree course.

When compiling this book we studied numerous Level IV units which colleagues from various colleges were kind enough to supply. It was immediately apparent that there were wide variations between the syllabuses of different colleges and, furthermore, that many topics were taught at Level IV in some colleges and at Level V in others. Our ultimate selection of material was based to a large extent on the TEC Level IV Standard Unit, and also on the Level IV unit produced by the Committee of Heads of Polytechnic Chemistry Departments. We are certainly not claiming to have covered every topic in these two units. For example, the determination of melting and boiling temperatures has been omitted on the grounds that it should have been adequately taught at an earlier stage. Certain topics, notably Grignard reagents, posed a particular problem. Eventually we decided to include these compounds, but to leave other organometallic compounds to Level V. If readers disagree strongly with our choice of subject matter we should be very glad to hear from them.

We have not hesitated to depart from these units where we feel that there are sound chemical reasons for doing so. For instance, we have adopted a chemical bond approach, for the reasons given in Chapter 1, rather than a functional group approach. Also, we have avoided any mention of the dubious and obsolescent concept of hybridisation, and have considered instead the contribution which each atomic orbital makes to a molecular orbital. For this treatment we are indebted to *Frontier Orbitals and Organic Chemical Reactions* by Ian Fleming (John Wiley and Son Ltd), 1976. Throughout the book we have used IUPAC names and given data in SI units. We have checked the former against *Nomenclature of Organic Chemistry* (Pergamon), 1980, and taken the latter from *Chemistry Data Book* by J. G. Stark and H. G. Wallace (John Murray).

In conclusion we should like to express our thanks to all those who have worked so hard to make this book a success. We are particularly grateful to David Browning of Bristol Polytechnic, who has edited the manuscript, and to the Longman's Editor and Designer respectively, who have attended to the production of the book with such care.

John Brockington and Peter Stamper
May 1981

Acknowledgements

We are grateful to the following for permission to reproduce copyright material:

Shell UK Ltd. for our Fig. 6.1; Longman Group Ltd. for our Fig. 11.1.

Chapter 1

Chemical bonds in organic compounds

The properties of organic compounds are governed by the chemical bonds which are present in their molecules. Every type of covalent bond has its own particular reactions. The C—C bond, for instance, behaves in essentially the same way in alkanes, alcohols, aldehydes, ketones, carboxylic acids and all other kinds of organic compounds, although in some cases the behaviour may be modified by other bonds in the molecules. The chemical reactions of any compound are therefore the sum of the reactions of its component bonds. For example, ethanal,

$CH_3C\underset{H}{\overset{O}{<}}$, displays C—C bond chemistry plus C—H bond chemistry

plus C=O bond chemistry. Throughout this book we shall consider organic chemistry on a chemical bond basis, and we must begin by discussing the nature of covalent bonds.

A covalent bond cannot always be formed between two neighbouring atoms; certain conditions must be satisfied. Each atom must have one electron in its outermost atomic orbital (AO), and these electrons must be spinning in opposite directions. Depending on the various forces of attraction and repulsion between them, two such atoms may approach each other sufficiently closely for their outer orbitals to overlap. The overlapping causes the two atomic orbitals to combine to form two *molecular orbitals,* known as a *bonding orbital* and an *anti-bonding orbital.* The latter, which is relatively high in energy, remains unoccupied while the bonding molecular orbital, commonly called simply the 'molecular orbital' (MO), holds both the electrons. In a molecular orbital, as in a fully occupied atomic orbital, the two electrons have opposite spins.

Electron density is not uniform throughout a molecular orbital, but is greatest along the internuclear axis. The high concentration of (negative) electrons between the two (positive) atomic nuclei serves to bind the nuclei together. In this situation the atoms are said to be linked by a 'covalent bond'.

Three types of MO are recognised; in decreasing order of strength they are named as σ (sigma), π (pi) and δ (delta). σ MOs may result from the overlap of s, p or d atomic orbitals, π MOs from overlapping p or d orbitals, and δ MOs from d orbitals only. In this book we are concerned only with σ and π MOs. Single covalent bonds consist of σ MOs, while multiple covalent bonds, i.e. double and triple bonds, involve both σ and π MOs.

1.1 Single covalent bonds

σ MOs can be studied on the basis of the AOs from which they are constructed. Numerous combinations are possible, but we shall restrict discussion to σ MOs which result from the overlap of: (*a*) two s AOs; (*b*) two p AOs; and (*c*) one s and one p AO.

Overlap of two s AOs, as illustrated by the H_2 molecule

A hydrogen atom has the electronic configuration $1s^1$. As two such atoms join together to give a hydrogen molecule the 1s orbitals overlap to give a σ MO, designated as σ_{1s}. Because the hydrogen molecule has two such electrons in this MO its configuration is given as σ_{1s}^2.

The 1s AO is spherically symmetrical, but when two such orbitals overlap the MO which is formed has a boundary surface that is ellipsoid in shape (Fig. 1.1). Thus, the shape of the MO is not derived just by partially superimposing the two AOs.

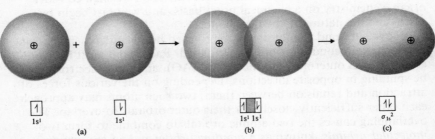

Fig. 1.1 Formation of the hydrogen molecule. (a) The approach of two hydrogen atoms, possessing electrons of opposite spins. (b) Overlap of atomic orbitals. (c) The resultant molecular orbital.

A hydrogen molecule is more stable than two separate hydrogen atoms. Essentially, stability is due to the high electron density which exists between the two atomic nuclei (Fig. 1.2).

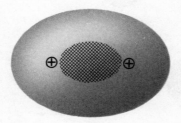

Fig. 1.2 The region of a hydrogen molecule where the probability of locating the two electrons is at a maximum.

The two electrons in this region reduce repulsion between the two nuclei. Furthermore, each positive nucleus is attracted to the cloud of two electrons. Thus, the two hydrogen atoms are firmly held in close proximity to each other.

Bonding is accompanied by a reduction in both potential and kinetic energy. That some sort of energy is lost is clear from the great amount of heat which is evolved during the combination.

Overlap of two p AOs, as illustrated by the F_2 molecule

Fluorine has the electronic configuration $1s^2\ 2s^2\ 2p_x^2\ 2p_y^2\ 2p_z^1$. When two fluorine atoms join together to give a fluorine molecule it is the unpaired $2p_z$ electrons of each that are involved. The $2p_z$ electrons are referred to as *bonding electrons*, and all the others as *non-bonding electrons*.

The overlap of p orbitals could conceivably occur in two ways (Fig. 1.3). Wherever possible — and this applies to the formation of the fluorine molecule — there is a head-on approach of p orbitals because in this way there is a greater degree of overlap and the system attains maximum stability. The shape of the resultant MO, like that of the MO in the hydrogen molecule, stems from the fact that the electron density is greatest between the two atomic nuclei. The overlapping lobes of the AOs are thus increased in size, while the outer lobes are correspondingly decreased.

The F—F σ orbital formed in this way is properly written as σ_{2p_z}, but may be abbreviated to σ_z or $z\sigma$. Each fluorine atom in the molecule has two 1s, two 2s, two $2p_x$ and two $2p_y$ electrons that are not involved in bonding. Those in the outer shell, i.e. the 2s, $2p_x$ and $2p_y$ electrons, are said to constitute *lone pairs* of electrons, while those which belong to the inner shell, i.e. the 1s electrons, are referred to as *core electrons*. Lone pairs of electrons are important because they can be utilised in forming coordinate bonds.

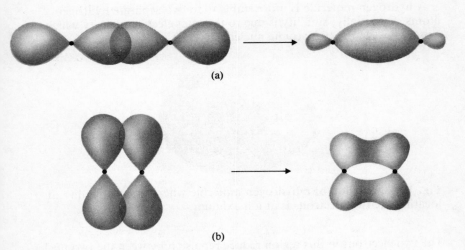

Fig. 1.3 (a) Head-on overlap of p orbitals, leading to a σ MO.
(b) Sideways overlap of p orbitals, leading to a π MO.

In certain molecules with multiple bonds, e.g. $CH_2{=}CH_2$ and
$CH{\equiv}CH$, p orbitals can overlap only in a sideways manner to give
rather weak π MOs (§1.2).

Overlap of one s and one p AO, as illustrated by the HF molecule

If the AOs contributing to the MO are of different types, the MO is
not symmetrical but has a shape that is derived from those of the
constituent AOs. Thus, in the hydrogen fluoride molecule, where the
overlapping AOs are 1s (from H) and $2p_z$ (from F), the shape is as
shown in Fig. 1.4.

Fig. 1.4 Formation of the HF molecule.

Overlap involving s or p AOs, available through promotion, as illustrated by the CH_4 molecule

It follows from the previous discussion that the covalency of an element,
i.e. the number of covalent bonds which are formed by an atom of that
element, should be equal to the number of unpaired electrons in the

atom. A glance at the electronic configuration of carbon, however, shows that this is not always so.

Electronic configuration			Number of unpaired electrons	Covalency
1s	2s	2p		
		x y z		
⇅	⇅	↑ ↑ ☐	2	4

The explanation lies in the fact that this electronic configuration relates to the ground state (i.e. the most stable state) of the carbon atom, but promotion of a 2s electron to the vacant 2p orbital can occur as other atoms approach for combination:

Electronic configuration			Number of unpaired electrons
1s	2s	2p	
		x y z	
⇅	↑	↑ ↑ ↑	4

The number of unpaired electrons now corresponds to the number of covalent bonds which is actually formed. In any carbon compound, therefore, we may expect at least two types of MO to be formed, one involving the singly filled 2s orbital and one involving a singly filled 2p orbital.

In a molecule of methane, CH_4, the four hydrogen atoms, labelled H_a, H_b, H_c and H_d in Fig. 1.5, are distributed tetrahedrally around the carbon atom. This serves to minimise repulsion between them. In this position the 2s AO of the carbon atom overlaps the 1s AO of each hydrogen atom to form a σ MO (Fig. 1.5(a)). Three other σ MOs are formed by the interaction of the three singly filled 2p orbitals with the 1s AOs of the hydrogen atoms (Fig. 1.5(b)). A total of four MOs is thus formed to accommodate eight bonding electrons, four from carbon and one from each of the four hydrogen atoms.

To summarise, in a methane molecule, *there is no single MO which is responsible for a C—H bond*. Each C—H bond consists of a combination of four MOs formed by the interactions shown in Fig. 1.5. The combination of orbitals is the same for all four bonds. For this reason the bonds are identical to one another, as indicated by their equal lengths (0.109 nm).

Although the C—H bonds in methane consist of a combination of MOs, they must not be regarded as multiple bonds. With eight electrons involved in bonding, each bond comprises only two electrons and is therefore a single covalent bond.

6

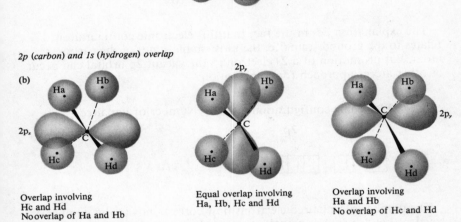

2s (carbon) and 1s (hydrogen) overlap

(a)

Equal overlap involving
Ha, Hb, Hc and Hd

2s

2p (carbon) and 1s (hydrogen) overlap

(b)

2p$_x$

Overlap involving
Hc and Hd
No overlap of Ha and Hb

Equal overlap involving
Ha, Hb, Hc and Hd

2p$_y$

Overlap involving
Ha and Hb
No overlap of Hc and Hd

2p$_y$

Fig. 1.5 Formation of the CH_4 molecule.

1.2 Multiple covalent bonds

Multiple bonds, such as the carbon–carbon double bond in ethene and
the carbon–carbon triple bond in ethyne, involve both σ and π MOs.
σ MOs have been discussed above; π MOs may result from overlapping
p or d AOs, but we shall consider only those which arise from the
sideways overlap of p orbitals (Fig. 1.3).

The ethene molecule

The two carbon atoms in a molecule of ethene cannot be held together
merely by a σ bond; there must be a π bond as well. This is suggested
by a number of experimental facts, e.g.

(i) the high strength of the bond. The C=C bond has a bond dissociation enthalpy of 612 kJ mol⁻¹, whereas that of the C—C bond is
only 348 kJ mol⁻¹;

(ii) the shortness of the bond: 0.134 nm, compared with 0.154 nm for
the C—C bond;

(iii) restricted rotation about a C=C bond. Compounds such as

1,2-dichloroethene exist in two isomeric forms, designated *cis* and *trans* (§2.2). The carbon−carbon double bond locks the molecule in position, and prevents one isomer from changing into the other.

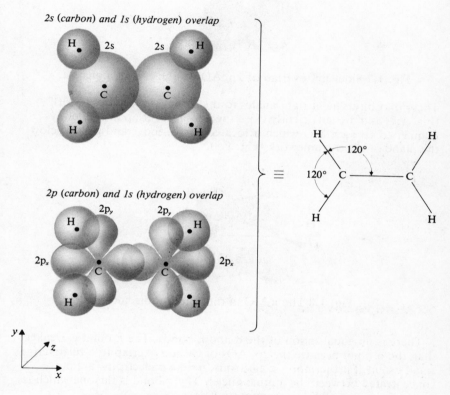

Fig. 1.6 The σ-framework of the ethene molecule.

The ethene molecule possesses a planar *σ-framework* (Fig. 1.6) in which the carbon and hydrogen atoms are held together by σ bonds, rather like the carbon and hydrogen atoms in a molecule of methane. Thus, the C—C single bond is due partly to the overlap of 2s AOs and partly to the overlap of, let us say, $2p_x$ AOs. Each C—H bond is due partly to the 2s AO of the carbon atom overlapping the 1s AO of the hydrogen atom, and partly to the $2p_x$ and $2p_y$ AOs of the carbon atom overlapping the same 1s AO of the hydrogen atom.

Each of the carbon atoms still has a $2p_z$ AO available for bonding.

8

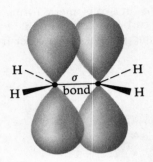

Fig. 1.7 Sideways overlap of $2p_z$ AOs in the ethene molecule.

These p_z orbitals lie at right angles to the plane of the σ-framework (Fig. 1.7) and are sufficiently close together to overlap in a sideways manner to give a π MO which is located partly above and partly below the plane of the σ-framework (Fig. 1.8).

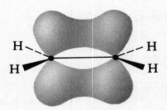

Fig. 1.8 The π MO in the ethene molecule.

There is no equalisation of the σ and π bonds. The π bond is weaker than the σ bond because the $2p_z$ AOs of carbon overlap to a relatively small extent. Furthermore, π electrons, unlike σ electrons, are not concentrated between the atomic nuclei. The π bond is the one which is disrupted during addition reactions (§7.4).

The ethyne molecule

The carbon–carbon bond in ethyne is exceptionally short (0.120 nm) and exceptionally strong (bond dissociation enthalpy = 837 kJ mol⁻¹), because the two carbon atoms are held together by a triple bond, consisting of one σ and two π bonds.

The molecule has a linear σ-framework, H—C—C—H, the formation of which involves the 2s and, let us say, the $2p_x$ AOs of the carbon atoms. This leaves the $2p_y$ and $2p_z$ AOs of each carbon atom free for further bonding. They remain directed at right angles to each other and to the line of the C—C σ bond (Fig. 1.9).

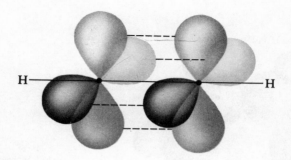

Fig. 1.9 Sideways overlap of 2p AOs in the ethyne molecule. (For the sake of clarity the distance between the carbon atoms has been exaggerated. Dotted lines indicate where overlap occurs.)

The two 2p AOs are close enough together for them both to overlap, so that two π MOs are formed. They are stronger than the π MO in ethene, and are so close to each other that they interact to give a cylindrical electron cloud around the axis of the C—C σ bond (Fig. 1.10).

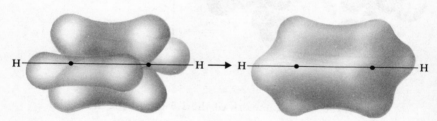

Fig. 1.10 Interaction of π MOs in the ethyne molecule.

1.3 Delocalised multiple bonds

Hydrocarbons with two C=C bonds are called *dienes*; those with three such bonds are *trienes*, and so on. Most compounds of this sort have properties which are entirely predictable from those of ethene, but compounds which are *conjugated*, i.e. which possess alternating single and double bonds, have some totally unexpected properties. This suggests that they possess a structural feature which is absent in ethene.

The 1,3-butadiene molecule, CH_2=CH—CH=CH_2

The σ-framework of the 1,3-butadiene molecule (Fig. 1.11) is similar to that of the ethene molecule in that it is planar and involves the 2s and, say, the $2p_x$ and $2p_y$ AOs of the carbon atoms.

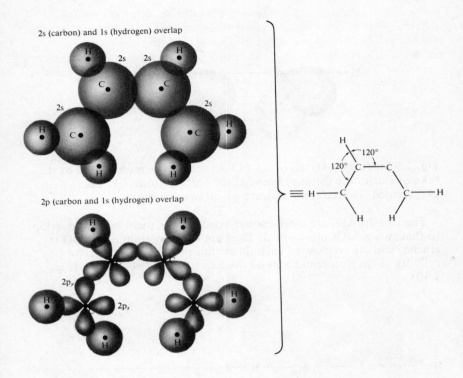

2s (carbon) and 1s (hydrogen) overlap

2p (carbon and 1s (hydrogen) overlap

Fig. 1.11 The σ-framework of the 1,3-butadiene molecule.

Each carbon atom (C_1, C_2, C_3 and C_4) has its $2p_z$ AO available for sideways overlap. Figure 1.12 shows these AOs, and makes it clear that *they can interact with one another to an equal extent.*

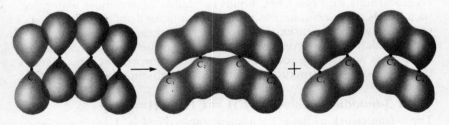

Fig. 1.12 Formation of delocalised π MOs in the 1,3-butadiene molecule.

The four $2p_z$ AOs combine to give four π MOs which are described as *delocalised*, because they are spread over the whole of the conjugated system. Two of these delocalised π MOs are bonding and two are anti-bonding. The former are shown in Fig. 1.12, while the latter are unoccupied. The net result is that while double bonding is most marked between C_1 and C_2, and between C_3 and C_4, the bond between C_2 and C_3 also has a certain amount of double bond character.

The benzene molecule

The bonding in benzene is similar to that in 1,3-butadiene, the only difference being that all the carbon–carbon bonds in the molecule are exactly the same as one another. Because of this, benzene is not regarded as a conjugated hydrocarbon.

The benzene molecule is planar, with a C—C—C bond angle of $120°$. The σ-framework of the molecule (Fig. 1.13) resembles that of ethene and involves the 2s and, say, the $2p_x$ and $2p_y$ AOs of the carbon atoms.

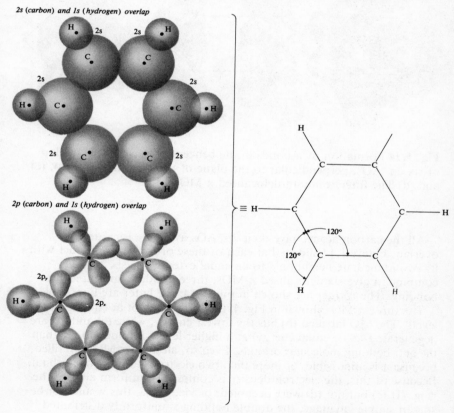

Fig. 1.13 The σ-framework of the benzene molecule.

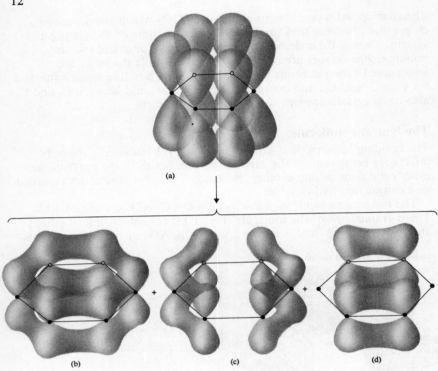

(a)

(b)   (c)   (d)

Fig. 1.14 Formation of π bonds in the benzene molecule. (a) Overlap of six 2p$_z$ AOs, perpendicular to the plane of the σ-framework. (b), (c) and (d) The three resultant delocalised π MOs.

All the carbon atoms have their 2p$_z$ AOs available for sideways overlap. Figure 1.14 shows that each of these orbitals can interact with its two immediate neighbours to an equal extent. The six 2p$_z$ AOs combine to give six delocalised π MOs, three bonding and three anti-bonding. The former are shown in Fig. 1.14; the latter are unoccupied.

The three π MOs shown in Fig. 1.14(b)–(d) are not at equal energy levels. The MO labelled (b) has the lowest energy, while (c) and (d) are degenerate (i.e. of equal energy) at a higher level, although lower than the anti-bonding molecular orbitals. Even so, all three π MOs are filled, because it is impossible for more than two electrons to enter one orbital. Because of this, the electron density is completely uniform around the ring. (If (c) but not (d) were occupied, or vice versa, this would not be so.) In simple language, the double bonding is uniformly distributed around the ring. It is misleading to represent benzene by the formula

⬡, as this suggests that the $2p_z$ AOs interact in pairs to give

three *localised* π MOs. The best representation is ⬡, in which the circle symbolises the delocalised π orbital.

Delocalisation of π electrons takes place in any situation where it can lead to a reduction in electrical potential energy and hence to improved stability. We can determine the difference between the potential energy of, say, benzene, and the potential energy that benzene would have if the molecules possessed three localised π orbitals. Consider the hydrogenation of cyclohexene:

⬡ + H₂ = ⬡

cyclohexene   cyclohexane

$\Delta H = -119.6$ kJ mol⁻¹

If there were three such bonds in the benzene molecule the enthalpy of hydrogenation would be $3 \times (-119.6) = -358.8$ kJ mol⁻¹. The observed value is only -208.4 kJ mol⁻¹, giving a difference of -150.4 kJ mol⁻¹. The gain of stability conferred on benzene by delocalisation can therefore be put at 150.4 kJ mol⁻¹ (Fig. 1.15). This value is referred to as the *stabilisation energy* (or stabilisation enthalpy) of benzene. It is often called the *delocalisation energy*, although strictly this is incorrect. The latter term refers purely to the gain of stability due to the delocalisation of electrons, while the former is a net value of a number of effects.

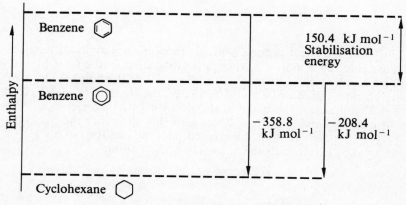

Fig. 1.15 The stabilisation energy of benzene.

1.4 Electronic effects in molecules

Whenever two identical atoms are joined together by a covalent bond (e.g. Cl—Cl), there is no *permanent* displacement of the bonding electrons towards one atomic nucleus or the other. However, if two different atoms are covalently bonded together (e.g. H—Cl), the bonding electrons will almost certainly be displaced towards one atom or the other. In this example, they are displaced towards the chlorine atom, which thereby acquires a partial negative charge (δ^-): the hydrogen atom is left with a corresponding partial positive charge (δ^+). The covalent bond in this situation is said to be *polarised*.

A measure of the polarity of a compound is provided by its *dipole moment* (μ), which is the turning moment that its molecules possess, by virtue of their charge separation, when the compound is placed between oppositely charged electrodes. The SI unit for expressing dipole moments is the coulomb metre (C m), although a non-SI unit, the debye (D), is often used instead.

There are two quite distinct effects which can cause the permanent polarisation of covalent bonds. They are known in organic chemistry as the *inductive effect* (I) and the *mesomeric effect* (M), and we shall consider each in turn.

Inductive effect (induction effect)

Inductive effects in molecules are of two types. An alkyl group (e.g. CH_3 or C_2H_5) has a *positive inductive effect* ($+I$) because, in comparison with a hydrogen atom, it releases electrons towards a carbon atom to which it is attached. A halogen atom, in contrast, has a *negative inductive effect* ($-I$) because, compared with a hydrogen atom, it draws electrons away from an adjoining carbon atom. Both $+I$ and $-I$ effects are symbolised by drawing arrows on the covalent bonds concerned:

$$CH_3 \rightarrow C\diagdown^{\diagup} \qquad\qquad Cl \leftarrow C\diagdown^{\diagup}$$

The inductive effect is due essentially to electronegativity differences. Because chlorine is more electronegative than carbon, the C—Cl bond is polar covalent in character, i.e. the electrons of the bond are located relatively near the chlorine atom. Similarly, the electron releasing effect of a methyl group is attributed to the electronegativity difference between carbon and hydrogen. Because of this difference the electron density on the carbon atom is increased, and the electrons of a CH_3—C bond are repelled towards the adjoining carbon atom.

$$C \overset{\bullet}{\underset{\times}{\cdot}} C \overset{H}{\underset{H}{\overset{\times}{\underset{\times}{\cdot}}}} H \qquad\qquad C \overset{\cdot}{\underset{\times}{}} Cl$$

The electronegativity difference between carbon and a halogen is much greater than that between carbon and hydrogen. In consequence, the $-I$ effect of a halogen atom is much greater than the $+I$ effect of an alkyl group.

Both $+I$ and $-I$ effects are transmitted along a hydrocarbon chain, but become progressively weaker with increasing distance from the source of the effect.

Mesomeric effect (conjugative effect)

The mesomeric effect always involves multiple bonds, and stems from the readiness with which the π orbitals of such bonds can be disturbed so that they interact with other orbitals on adjoining atoms. It is possible for a π orbital to interact either with another π orbital, or with a neighbouring p orbital holding a lone pair of electrons. Mesomeric effects of the latter kind have a major bearing on the chemistry of certain aromatic compounds (e.g. chlorobenzene) and carbonyl compounds (e.g. carboxylic acids and their derivatives), and we shall restrict discussion to cases of this sort.

Mesomeric effect in chlorobenzene (Fig. 1.16)

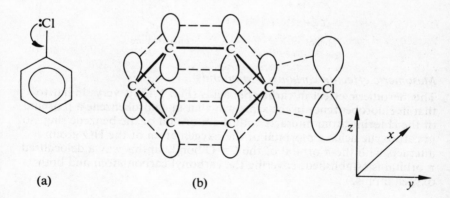

Fig. 1.16 Mesomeric effect in chlorobenzene.

Chlorine has the following electronic configuration: $1s^2\ 2s^2\ 2p^6\ 3s^2\ 3p_x^2\ 3p_y^2\ 3p_z^1$. The singly occupied 3p orbital forms a σ bond with one of the carbon atoms of the benzene ring, while the other 3p orbitals hold lone pairs of electrons. One of these 3p orbitals lies in the same plane as the σ-framework of the benzene ring, but the other is perpendicular to it and interacts to a certain extent with the $2p_z$ orbital of the carbon atom (Fig. 1.16(b)), thus introducing a certain amount of double bond character into the carbon–chlorine bond.

Because the 2p$_z$ orbital of the carbon atom contributes to the delocalised π orbital of the aromatic ring, we can argue that the p orbital of the chlorine atom combines to some extent with the π orbital of the ring. In this way the chlorine lone pair tends to increase the electron density of the benzene ring, and we can say that the lone pair is 'drawn towards the ring'. However, it is most important to realise that in this particular case, because of the high electronegativity of chlorine, the inductive effect operates in the opposite direction.

The mesomeric effect in chlorobenzene does not destroy the stable aromatic character of the benzene ring system, but it does mean that the C—Cl bond in chlorobenzene is shorter and stronger than the corresponding bond in an alkyl chloride (0.169 nm, compared with 0.177 nm).

Other aromatic compounds which display the mesomeric effect include phenol and phenylamine:

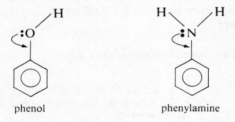

phenol   phenylamine

Mesomeric effect in carbonyl compounds

The mesomeric effect in carboxylic acids (Fig. 1.17) is very similar to that in chlorobenzene. In the same way that in chlorobenzene a p orbital of the chlorine atom interacts with the π orbital of the benzene ring, so in carboxylic acids a p orbital of the oxygen atom of the HO group interacts with the π orbital of the C=O bond. In this way a delocalised π orbital is established, covering the carbonyl carbon atom and both oxygen atoms.

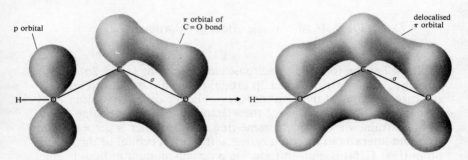

Fig. 1.17 Mesomeric effect in the carboxyl group.

The 2p orbital on the hydroxyl oxygen atom is occupied by a lone pair of electrons, and for most purposes it is sufficient to argue that this lone pair is drawn towards the carbonyl carbon atom. The effect is symbolised in the following way:

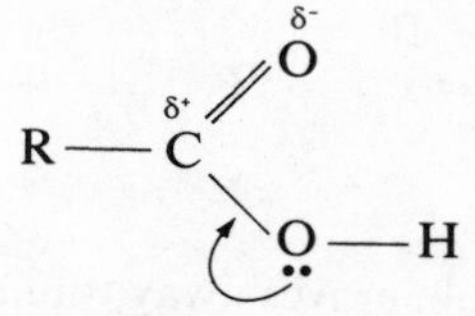

The mesomeric effect is also present in carboxylic acid derivatives, such as esters, amides and acid chlorides:

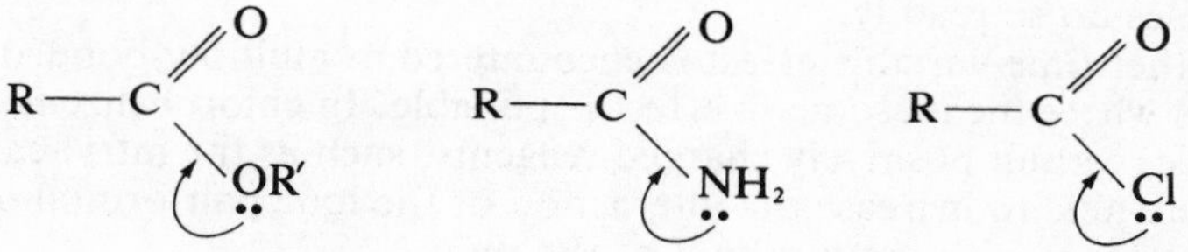

The mesomeric effect is at a maximum in carboxylate anions, i.e. ions which are formed from carboxylic acids by the loss of a proton. A carboxylate ion is often represented by the formula

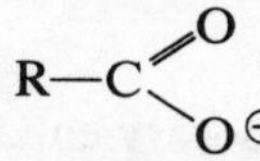

However, the loss of a (positive) proton from an acid molecule permits an increase in the mesomeric effect, i.e. a lone pair of electrons on the singly bonded oxygen atom can be drawn more strongly towards the carbonyl carbon atom. As a result, both carbon–oxygen bonds become identical, with a length of 0.127 nm and a character which is intermediate between that of a single bond and that of a double bond. Negative charge, instead of residing on one oxygen atom only, is distributed among both oxygen atoms and the carbon atom, so that the ion is best represented as follows. The O—C—O bond angle is 124°.

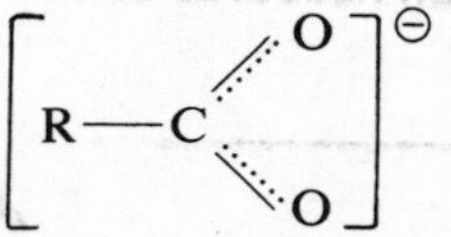

Temporary effects

So far, the only polarising effects which we have discussed have been permanent ones. However, it is possible for a covalent bond to become temporarily polarised by the approach of a polar molecule or an ion. In bromine, for example, the Br—Br bond can be polarised by a neighbouring molecule of ethanol, in which the inductive effect is responsible for a permanent polarisation:

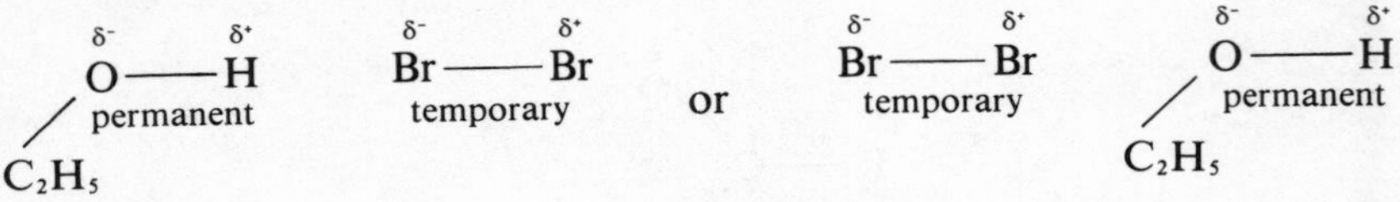

$$
\underset{\substack{| \\ C_2H_5}}{\overset{\delta^-}{O}}\!\!-\!\!\overset{\delta^+}{H}
\quad\quad
\overset{\delta^-}{Br}\!\!-\!\!\overset{\delta^+}{Br}
\quad\quad \text{or} \quad\quad
\overset{\delta^-}{Br}\!\!-\!\!\overset{\delta^+}{Br}
\quad\quad
\underset{\substack{| \\ C_2H_5}}{\overset{\delta^-}{O}}\!\!-\!\!\overset{\delta^+}{H}
$$

permanent   temporary   temporary   permanent

Once the ethanol molecule moves away, the bromine molecule reverts to its original state. A temporary inductive effect of this sort, called an *inductomeric effect*, can lead to a considerable increase in the rate of a chemical reaction. For example, in certain circumstances unpolarised bromine molecules will not react with alkenes, although polarised molecules do so readily.

Another time-variable effect is encountered in multiply-bonded compounds where the mesomeric effect is possible. In chlorobenzene, for example, certain positively charged reagents, such as the nitryl cation, NO_2^+, are able to increase the interaction of the lone pair orbital on the chlorine atom with the π orbital of the ring:

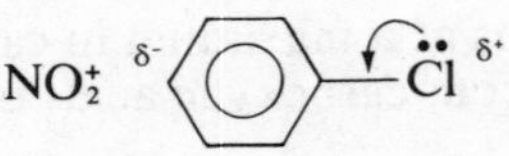

$$
NO_2^+ \quad \overset{\delta^-}{\bigcirc}\!\!-\!\!\overset{\displaystyle\cdot\cdot}{\underset{\displaystyle\cdot\cdot}{Cl}}{}^{\delta^+}
$$

This temporary enhancement of the mesomeric effect is termed an *electromeric effect*.

Chapter 2

Isomerism

Different compounds with the same molecular formula are known as *isomers*. The phenomenon, called *isomerism*, is commonly encountered among organic compounds and inorganic complexes. In this book we shall discuss only the former.

There are two main types of isomerism, namely *constitutional isomerism* and *stereoisomerism*. Constitutional isomers possess fundamentally different molecular structures, while stereoisomers have similar constitutions but differ in their molecular geometry, i.e. in the spatial arrangements of their atoms or groups.

2.1 Constitutional isomerism (structural isomerism)

Carbon compounds display four varieties of constitutional isomerism, which are as follows.

Skeletal isomerism

Otherwise known as *chain isomerism* or *nuclear isomerism*, this is concerned with the structure of the hydrocarbon skeleton. Thus, butane and 2-methylpropane are skeletal isomers:

$$CH_3CH_2CH_2CH_3$$

butane

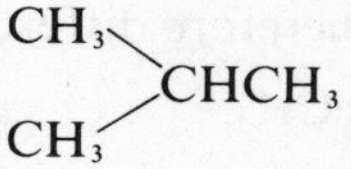

$$CH_3 \atop CH_3 \Big\rangle CHCH_3$$

2-methylpropane

In general, skeletal isomers have similar physical properties, although a branched chain compound invariably has a lower boiling temperature than its unbranched isomer, e.g. the boiling temperatures of butane and 2-methylpropane are -0.5 °C (272.7 K) and -11.7 °C (261.5 K) respectively. Branched chain molecules are relatively easily separated from one another, partly because they are compact and do not become entangled, and partly because the atoms at the centres of such molecules are remote from atoms of other molecules and cannot contribute to van der Waals' forces of attraction.

Skeletal isomers differ very little in the nature of their chemical reactions, although in some cases there may be a considerable variation in reactivity. 2-Methylpropane, for example, is much more reactive than butane.

Position isomerism

This occurs whenever a functional group occupies different positions on the same hydrocarbon skeleton, e.g.

$$CH_3CH_2CH_2OH \qquad \text{and} \qquad CH_3CH(OH)CH_3$$

1-propanol   2-propanol

$$CH_3CH(OH)CH_2CH_2CH_3 \qquad \text{and} \qquad CH_3CH_2CH(OH)CH_2CH_3$$

2-pentanol   3-pentanol

Also 1,2- (*ortho*), 1,3- (*meta*) and 1,4- (*para*) isomers, e.g.

2-methylphenol   3-methylphenol   4-methylphenol

The position of a functional group often has little bearing on physical and chemical properties. Thus, 2-pentanol and 3-pentanol have almost identical properties. There is, however, considerably less similarity between 1-propanol and 2-propanol. The former is a primary alcohol and on oxidation gives an aldehyde and ultimately a carboxylic acid, while the latter is a secondary alcohol and on oxidation gives a ketone.

Functional group isomerism

As the term implies, functional group isomers have different functional groups and therefore different physical and chemical properties, e.g.

$$CH_3-O-CH_3 \qquad \text{and} \qquad CH_3CH_2OH$$

methoxymethane   ethanol

also

$$CH_3CH_2COOH, \quad HCOOCH_2CH_3 \quad and \quad CH_3COOCH_3$$

propanoic acid   ethyl methanoate   methyl ethanoate

Metamerism

Isomers which have different hydrocarbon radicals attached to a particular atom are called *metamers*. The atom concerned may be oxygen, e.g.

$$CH_3-O-CH_2CH_2CH_3 \quad CH_3-O-CH(CH_3)_2 \quad CH_3CH_2-O-CH_2CH_3$$

1-methoxypropane   2-methoxypropane   ethoxyethane

or nitrogen, e.g.

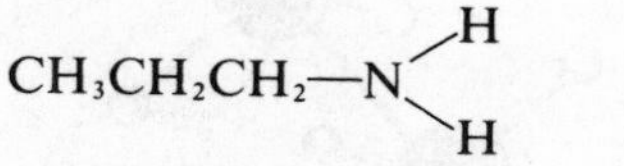

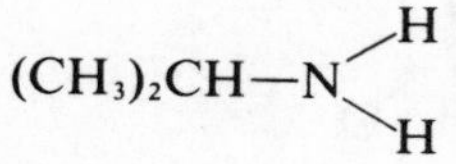

$$CH_3CH_2CH_2-N\begin{smallmatrix}H\\H\end{smallmatrix} \qquad (CH_3)_2CH-N\begin{smallmatrix}H\\H\end{smallmatrix}$$ primary amines

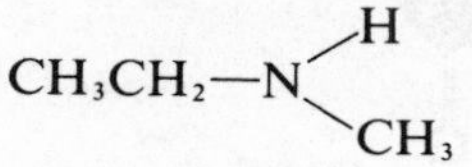

$$CH_3CH_2-N\begin{smallmatrix}H\\CH_3\end{smallmatrix}$$ secondary amine

$$CH_3-N\begin{smallmatrix}CH_3\\CH_3\end{smallmatrix}$$ tertiary amine

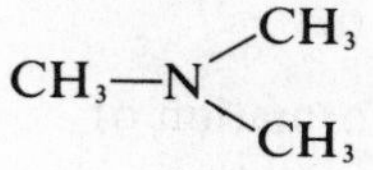

or it may be the carbon atom of a carbonyl group, $\diagdown C{=}O$, e.g.

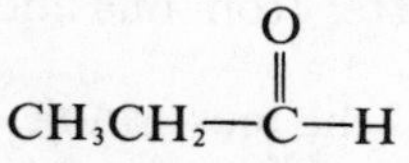

$$CH_3CH_2-\overset{\overset{\textstyle O}{\|}}{C}-H \qquad\qquad CH_3-\overset{\overset{\textstyle O}{\|}}{C}-CH_3$$

propanal (an aldehyde) 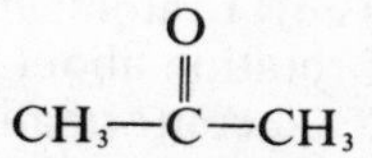   propanone (a ketone)

Certain metamers, such as the ethers quoted above, are very similar to one another, but others may differ considerably. In particular, there are significant differences between primary, secondary and tertiary amines, and also between aldehydes and ketones.

2.2 Stereoisomerism

Three forms of stereoisomerism are recognised, namely *conformational isomerism, geometrical isomerism* and *optical isomerism*. Organic stereoisomers always have the same hydrocarbon skeleton and the same functional groups.

Conformational isomerism

All molecules can absorb heat energy from their surroundings. Some of

22

this energy can cause pairs of atoms, linked together by single covalent bonds, to rotate about their internuclear axes. Because of this, the atoms in a molecule of ethane can adopt a great many positions relative to one another without violating the H—C—H and H—C—C bond angles of 109.5° required by the Sidgwick—Powell theory. Two of these possible arrangements are symmetrical; they are referred to as the *eclipsed conformation* (Fig. 2.1(a)) and the *staggered conformation* (Fig. 2.1(b)). As the carbon atoms rotate about the C—C bond, an ethane molecule can pass through an infinite number of other conformations between these two extremes.

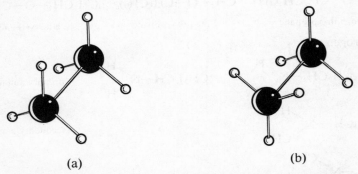

(a)   (b)

Fig. 2.1 (a) Eclipsed conformation, (b) staggered conformation of ethane.

The *conformations* of a molecule are defined by IUPAC as the various arrangements of its atoms in space that differ from one another only as the result of rotation about single bonds.

The staggered conformation of ethane is the most stable. In this arrangement the hydrogen atoms are as far away from one another as possible; consequently, the forces of repulsion between the electrons of the C—H bonds and between the hydrogen nuclei are at a minimum. Such forces are known as *non-bonding repulsions*.

The eclipsed conformation, where the hydrogen atoms of the two methyl groups are closest together, is the least stable. Nevertheless, the potential energy difference between the staggered and eclipsed conformations is only 12.5 kJ mol^{-1}. Such a small amount of energy is readily available from the surroundings. As a result there is ready interconversion of the two forms, even at low temperatures, and we say that there is *free rotation about a carbon—carbon single bond*.

Despite this free rotation, molecules of ethane show a preference for the staggered conformation. In other words, at any instant of time, most of the molecules are likely to possess this particular conformation. Because there is only one low energy conformation, ethane does not exhibit stereoisomerism.

Representation of conformations

The conformation of molecules, although best studied with three-dimensional models, can be represented on paper by either *sawhorse* or *Newman projections*.

Sawhorse projections

These are drawn in essentially the same way as Fig. 2.1, the only difference being that carbon atoms are not shown, while other atoms are represented by chemical symbols rather than spheres:

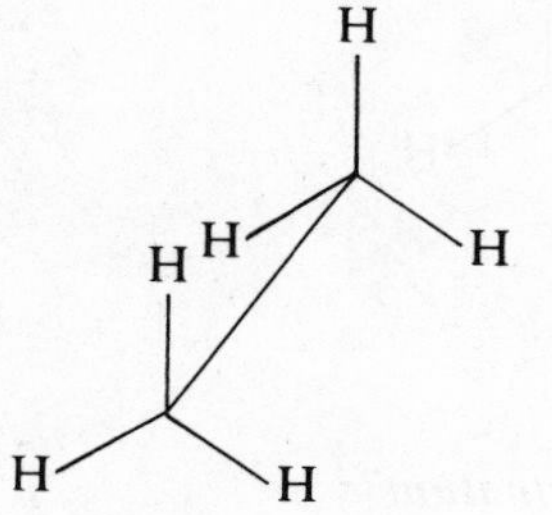

eclipsed conformation 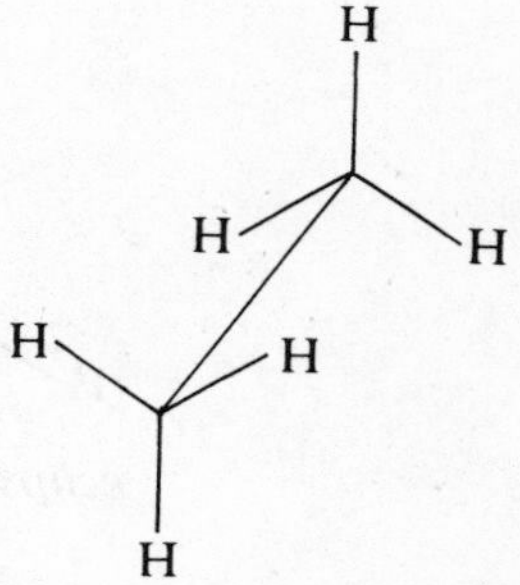   *staggered conformation*

When sketching sawhorse projections, we always imagine that we are looking down on the molecule from the right.

Newman projections

According to this convention, we represent the carbon atoms by a circle (and other atoms by symbols), and imagine that we are looking along the molecule from slightly above and to the right of the internuclear axis of the carbon atoms. The carbon atoms (i.e. the circle) partly block our view of distant bonds.

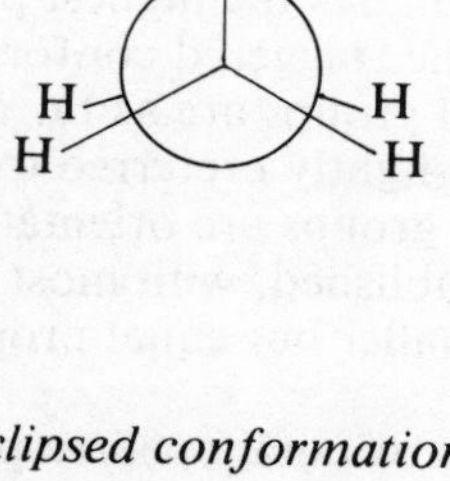

eclipsed conformation 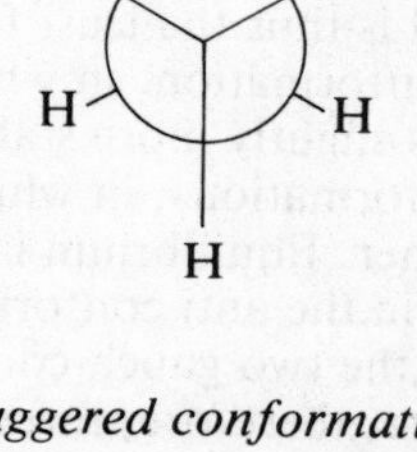   *staggered conformation*

Conformers

If we study propane in the same way that we have just studied ethane, we again find that there is no stereoisomerism. For the higher alkanes, however, this is not the case. Butane can show one eclipsed conformation, and *three* staggered conformations. We can easily draw sawhorse projections of all these by modifying the ethane conformations shown above so that one hydrogen atom at each end of the molecule is replaced by a CH_3 group.

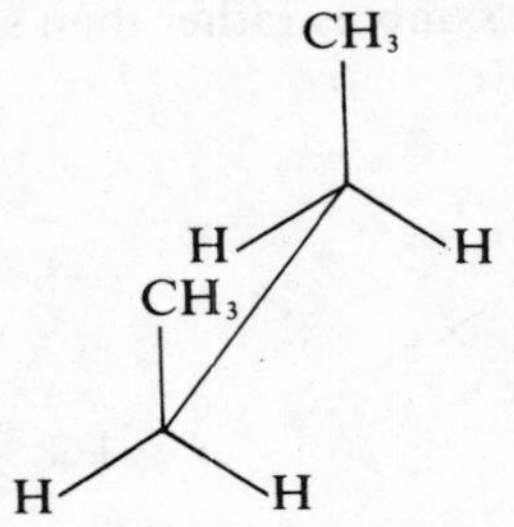

eclipsed conformation

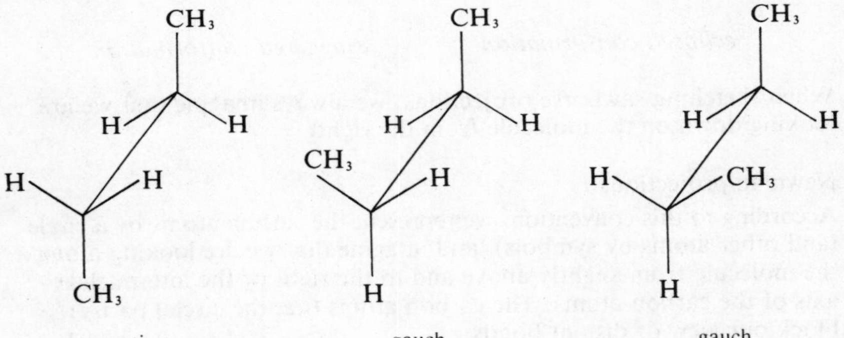

anti   gauch   gauch

staggered conformations

As with ethane, the eclipsed conformation has the highest potential energy and is thus the least favoured. Of the staggered conformations, the *anti* conformation, in which the methyl groups are as far apart as possible, is slightly more stable and hence slightly preferred to the two *gauch* conformations, in which the methyl groups are orientated at 60° to each other. Equilibrium is therefore established, with most butane molecules in the anti conformation and smaller but equal proportions in each of the two gauch conformations.

There are, therefore, three stereoisomers of butane, corresponding to the three stable conformations that the molecule can adopt. They are referred to as *conformational isomers* or *conformers* of butane. The term 'conformer' is defined by IUPAC as a molecule in a conformation into which the atoms return spontaneously after small displacements.

Conformations of cyclohexane

There are two ways in which the carbon atoms of a cyclohexane ring can be arranged with retention of C—C—C bond angles of 109.5°. (Deviations from this angle cause strains within the molecule and are thus unlikely.) The two arrangements are easily shown by molecular models and represent the *boat* and *chair* forms of cyclohexane:

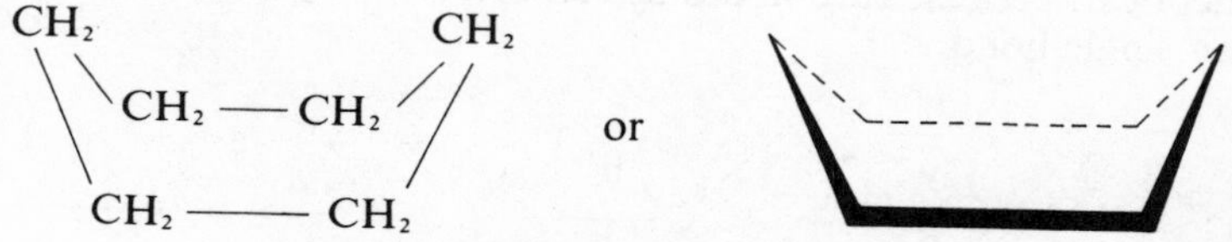

boat conformation

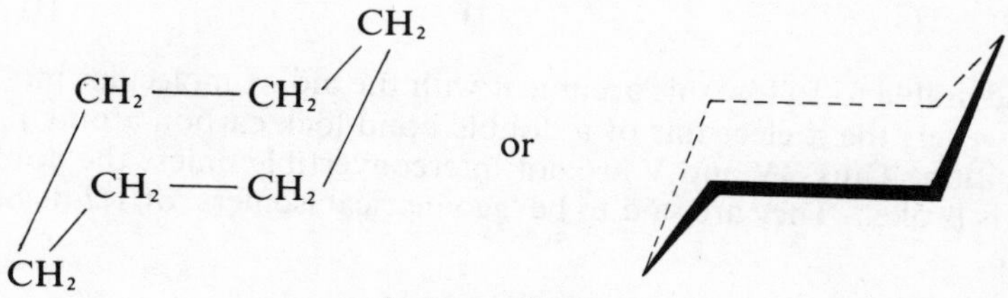

chair conformation

The chair form is considerably the more stable because, in this form, all adjoining pairs of hydrogen atoms are staggered. In contrast, in the boat form, four of these pairs are eclipsed, and there is also repulsion between a pair of hydrogen atoms at the 'boat-ends'. (Models are an essential aid to understanding.)

The change from one form to the other involves a twisting motion, during which the molecule assumes an infinite number of intermediate conformations. Two of these are relatively stable, and are termed *twist conformations*:

twist conformations of cyclohexane

These twist conformations, together with the chair conformation, represent three conformers of cyclohexane. Equilibrium is established, with the vast majority of cyclohexane molecules (about 9 999/10 000) in the chair form and the remainder equally distributed between the two twist forms. It is important to realise that the high energy boat form is *not* a conformational isomer of cyclohexane.

Geometrical isomerism

Geometrical isomerism results when carbon atoms are unable to rotate freely about the bonds that join them together. It cannot occur in saturated compounds, unless they are cyclic. Thus, forms I, II and III are identical, because carbon atoms can rotate freely about a carbon–carbon single bond.

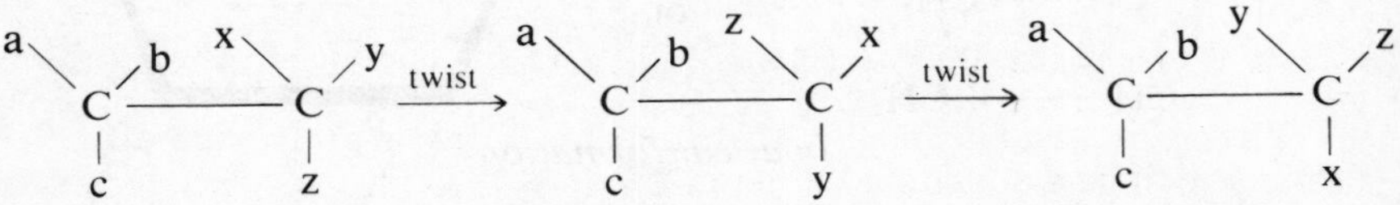

It is advisable to follow this argument with the aid of molecular models.

However, the π electrons of a double bond lock carbon atoms firmly in position. Thus, IV and V are not interconvertible unless the double bond is broken. They are said to be 'geometrical isomers' of 1,2-dichloro-ethene.

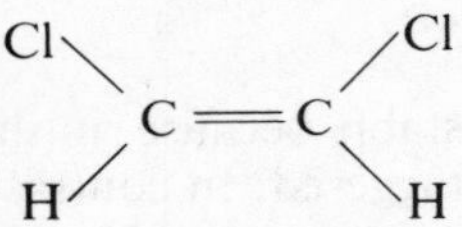

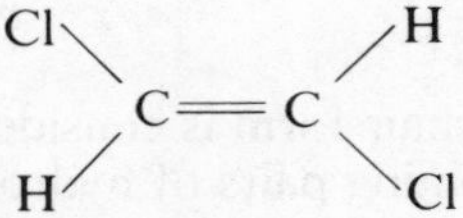

Although a carbon–carbon double bond is the commonest cause of restricted rotation, a cyclic structure can also be responsible. Thus, VI and VII, which represent 1,4-dimethylcyclohexane molecules, are also geometrical isomers.

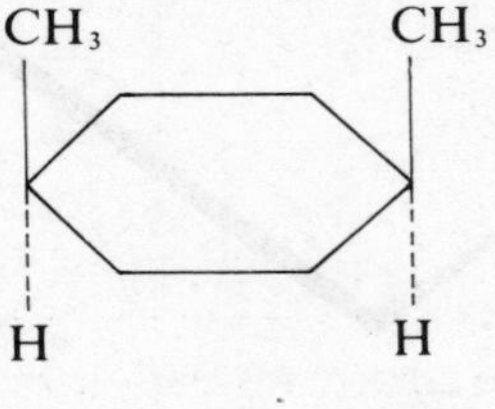

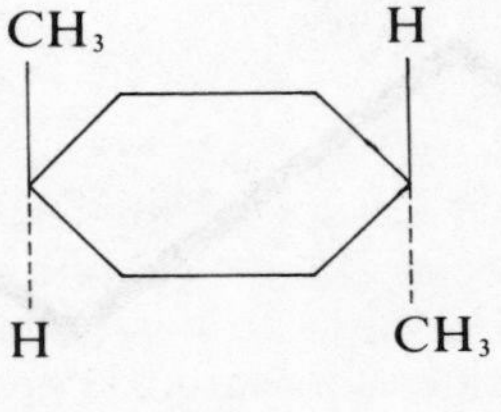

Geometrical isomers are distinguished by the prefixes *cis* and *trans*: *cis* (Latin — 'hither') relates to the form in which the two similar atoms or groups lie on the same side of the ring or double bond, e.g. to forms IV and VI, while *trans* (Latin — 'across') denotes the isomer in which they are on opposite sides, e.g. forms V and VII. Because of this notation, geometrical isomerism is sometimes referred to as *cis—trans isomerism*.

One of the commonest examples of geometrical isomerism relates to maleic acid (*cis*-butenedioic acid) and fumaric acid (*trans*-butenedioic acid):

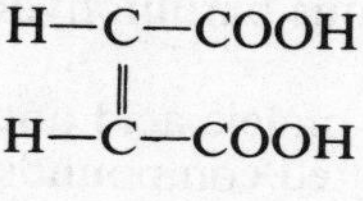

H—C—COOH
‖
H—C—COOH

maleic acid

H—C—COOH
‖
HOOC—C—H

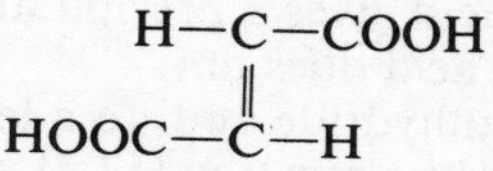

fumaric acid

Physically and chemically, they are very different from each other. Maleic acid has a melting temperature of 130 °C (403 K) with decomposition; fumaric acid melts at 286 °C (559 K) with decomposition. Maleic acid is highly soluble in water (78.8 g per 100 cm^3 at 25 °C (298 K)), whereas fumaric acid is only slightly soluble (0.7 g per 100 cm^3 at the same temperature). Maleic acid is a considerably weaker acid than fumaric acid, with pK_1 and pK_2 values of 1.92 and 6.23 respectively as opposed to 3.02 and 4.38. The reason is that although one proton can readily be lost from a molecule of maleic acid, the other is held as a member of a ring by hydrogen bonding:

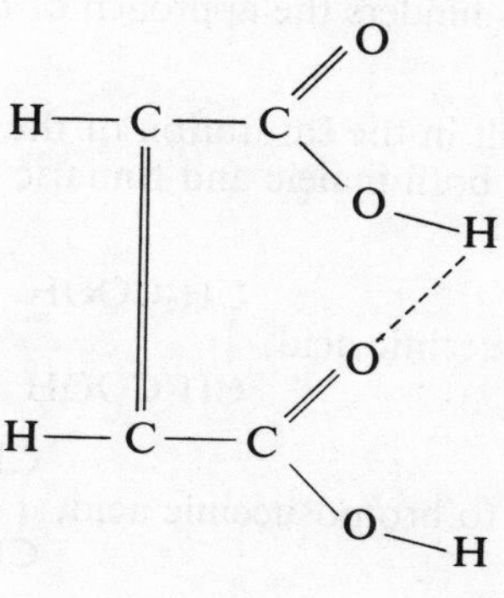

No such restriction applies to fumaric acid.

On heating, either alone or with ethanoic anhydride, maleic acid readily gives maleic anhydride, but fumaric acid cannot form fumaric anhydride as its carboxyl groups are too far apart. Instead, it rearranges to give maleic acid, which then dehydrates to maleic anhydride:

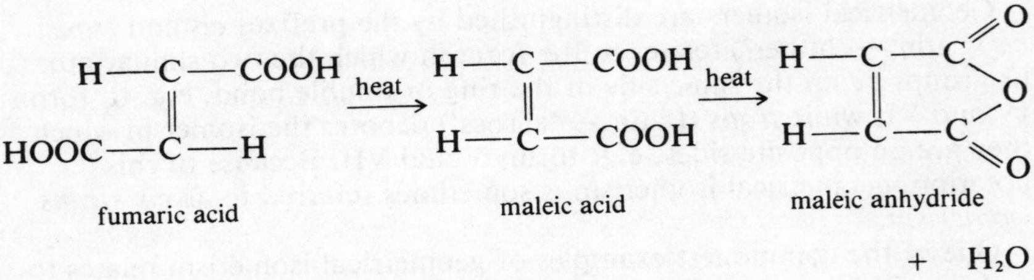

$$\text{fumaric acid} \xrightarrow{\text{heat}} \text{maleic acid} \xrightarrow{\text{heat}} \text{maleic anhydride}$$

$$+ \quad H_2O$$

Other chemical differences include the following.

(i) Maleic acid gives a precipitate with aqueous barium hydroxide, but fumaric acid does not.

(ii) Maleic anhydride and, to a lesser extent, maleic acid undergo the *Diels–Alder reaction* (§7.5) with conjugated compounds such as 1,3-butadiene:

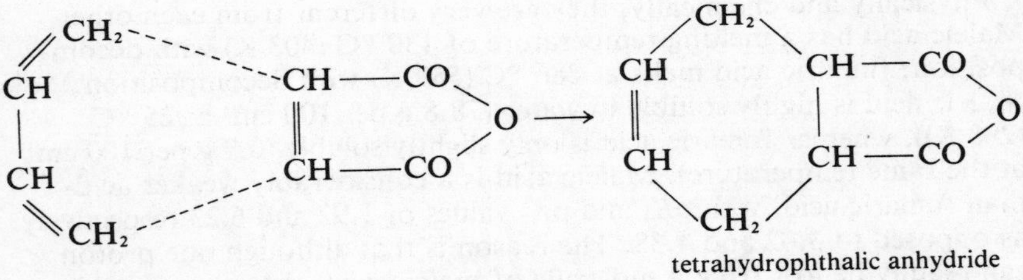

tetrahydrophthalic anhydride

Fumaric acid does not take part in such reactions, because one of the carboxyl groups hinders the approach of the butadiene molecule to the double bond.

Reactions which result in the saturation of the double bond lead to the same product from both maleic and fumaric acid. Examples are as follows:

(i) hydrogenation to succinic acid, $\begin{array}{l} CH_2COOH \\ | \\ CH_2COOH \end{array}$

(ii) hydrobromination to bromosuccinic acid, $\begin{array}{l} CHBrCOOH \\ | \\ CH_2COOH \end{array}$

The configuration of geometrical isomers is usually established by physical methods, especially infrared spectroscopy. X-ray analysis, by yielding interatomic distances, is also valuable. It shows, for instance, that in one form of 1,2-dichloroethene the inter-chlorine distance is 0.35 nm, whereas in the other isomer it is 0.41 nm. Thus, the former must be *cis* and the latter *trans*:

$$H—C—Cl$$
$$\parallel$$
$$H—C—Cl$$

Cl—Cl = 0.35 nm

$$H—C—Cl$$
$$\parallel$$
$$Cl—C—H$$

Cl—Cl = 0.41 nm

Dipole moments may also be helpful. All *cis* isomers have a dipole moment, because electronic displacements in their bonds are in the same direction and reinforce each other, but simple *trans* isomers do not, because their electron shifts are in opposite directions and cancel out.

Optical isomerism

This is due to a lack of symmetry within a molecule. The absence of symmetry in an article is relatively uncommon, for most things possess a *plane of symmetry* which divides the article into two symmetrical halves. A person or a hat, for example, has a plane of symmetry (Fig. 2.2).

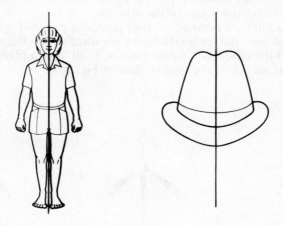

Fig. 2.2 Planes of symmetry.

If any object has a plane of symmetry there is only one form of it, in the sense that it is identical to its mirror image. The mirror image of a hat is exactly the same as the original hat; if we were to make the 'mirror image hat' it would not only look the same as the original but the two could be superimposed on each other without any difficulty (Fig. 2.3(a)). They could be neatly stacked together.

However, if an object does not have a plane of symmetry it should be possible to make a mirror image version that is quite different from the original. A left-hand glove, for instance, does not possess a plane of symmetry, and its mirror image is not another left-hand glove but a

Object Mirror Image   Object Mirror Image

Object and mirror image are identical   Object and mirror image are not identical

Fig. 2.3 Objects and their mirror images.

right-hand glove. The two gloves are not identical, for they cannot be superimposed. If we were to lay one on top of the other the fingers and thumbs would clash (Fig. 2.3).

A glove or any other object lacking a plane of symmetry is said to possess *chirality*, pronounced ki-rality, which means 'handedness'. (Greek: *cheir* = the hand) Any *chiral* object exists in two forms, one of which is the mirror image of the other.

Most molecules are *achiral*, i.e. they possess a plane of symmetry, so that they and their mirror image forms are identical. Tetrahedral molecules of the type Ca_3b, Ca_2b_2 and Ca_2bc all have a plane of symmetry, as shown by the broken lines in Fig. 2.4.

Fig. 2.4 Planes of symmetry in the tetrahedral molecules Ca_3b, Ca_2b_2 and Ca_2bc. In all these drawings the atoms or groups on the left and right of the central carbon atom extend out of the plane of the paper, while the atom or group below the carbon atom extends into the plane of the paper. The atom or group shown at the top lies directly above the carbon atom and in the plane of the paper.

In every case the mirror image (Fig. 2.5) is superimposable on the original molecule, proving that the two are identical. There can be no isomerism.

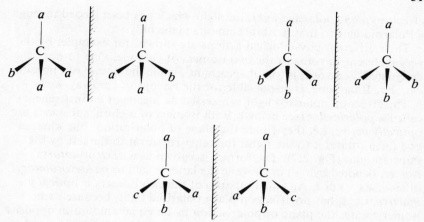

Fig. 2.5 Object to mirror image relationships of compounds Ca_3b, Ca_2b_2 and Ca_2bc.

Some molecules, however, do not have a plane of symmetry. For any such chiral molecule there can be obtained a closely related yet distinctive molecule that is its mirror image. The commonest, but by no means only, cause of asymmetry is the presence of an *asymmetric carbon atom*, i.e. a carbon atom which is joined to four different atoms or groups of atoms, as in a molecule $Cabcd$ (Fig. 2.6).

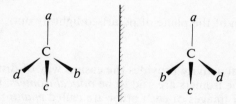

Fig. 2.6 Optical isomers of a compound $Cabcd$.

That these two forms are not identical is easily proved by constructing molecular models and showing that it is impossible to superimpose them without any of the groups clashing.

Thus, any compound which consists of chiral molecules must be isomeric with another compound composed of molecules of opposite chirality. Because of their structural similarity, it is hardly surprising that any two such isomers have very similar properties. These include identical melting temperatures, boiling temperatures, solubilities, refractive indices and chemical properties. The isomers differ in two respects only, namely in their physiological effects (if any) and their effect on *plane polarised light*, i.e. light which has been passed through a Polaroid lens so that it vibrates in one plane only.

Their different physiological actions are shown, for example, by the very different flavours of the two isomers of carvone. One form, designated S-carvone, is 'oil of spearmint', while the isomer of opposite chirality, R-carvone, is responsible for the flavour of caraway seeds.

The effect on polarised light is assessed by means of an instrument called a *polarimeter* (see below). Both isomers of a chiral substance are *optically active*, i.e. they rotate the plane of polarisation, but whereas one form rotates it to the right, the other rotates it to the left by the same amount (Fig. 2.7). The former is known as a *dextrorotatory isomer*, denoted by (+) or *d*, while the latter is said to be *laevorotatory*; otherwise (−) or *l*. An equal mixture of the two isomers is *optically inactive*, i.e. it has no effect on plane polarised light, because each isomer rotates the plane of polarisation by an equal amount in opposite directions. The mixture is referred to as a *racemic modification*, and may be denoted by (±) or *dl*.

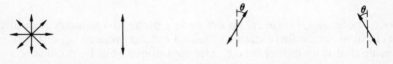

Ordinary light vibrating in all planes

Polarised light vibrating in a vertical plane

Plane of vibration after passage through a dextrorotatory isomer

Plane of vibration after passage through a laevorotatory isomer under the same conditions

Fig. 2.7 Rotation of the plane of polarised light by optically active isomers.

Because optical activity provides the easiest way of distinguishing between them, the isomers are said to be *optical isomers*. Optical isomers which are mirror images of each other are called *enantiomers* or *enantiomorphs*.

It must be clearly understood that polarised light is in no way the *cause* of optical isomerism. Such isomerism exists in complete independence of any lighting conditions, and is purely the result of molecular asymmetry. Optical isomerism may exist in addition to any of the other types of isomerism. For example, 2-butanol, $CH_3CH(OH)CH_2CH_3$ (C represents an asymmetric carbon atom), exhibits optical isomerism, but is also structurally isomeric with 1-butanol, 2-methyl-1-propanol and 2-methyl-2-propanol.

Conditions for chirality

In the preceding discussion, for the sake of simplicity, we have used the lack of a plane of symmetry as a condition for chirality. Strictly, however, we should consider the absence of a plane of symmetry or a *centre of symmetry.*

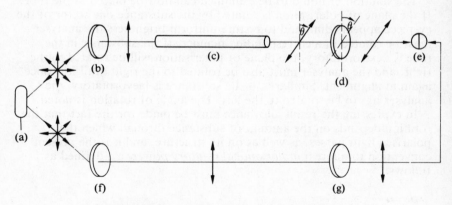

Certain molecules, e.g. dimethyldiketopiperazine, do not have a plane of symmetry, yet possess a centre of symmetry as represented by the dot in the above formula. If a line is drawn from any group in the molecule to the centre of symmetry, and then extended an equal distance beyond this point, it meets the mirror image of the original group. The molecule thus possesses a certain amount of symmetry, and is super-imposable on its mirror image. For this reason dimethyldiketopiperazine does not exhibit optical isomerism.

The polarimeter

Fig. 2.8 The optical system of a polarimeter.

A polarimeter enables a compound to be examined in solution or as a neat liquid. The instrument (Fig. 2.8) consists essentially of the following components.

(i)   A monochromatic *light source* (a), usually a sodium lamp.

(ii) A *polariser* (b), to convert ordinary light into plane polarised light. The polariser consists of a Polaroid sheet or a Nicol prism, and is mounted in a fixed position. (A *Nicol prism* is made from two pieces of calcite, a form of calcium carbonate, cemented together with Canada balsam.)

(iii) A *tube* (c) for holding the substance under examination.

(iv) Another Polaroid sheet or Nicol prism (d), called the *analyser*. This one is free to be rotated, and the angle through which it is turned can be read from a graduated scale. The analyser, like the polariser, transmits light vibrating in one plane only. When this plane coincides with that of the light leaving the polariser, there is maximum light transmission. Rotation of the analyser through 90° from this position leads to a condition known as 'crossed polars', in which minimum light is transmitted.

(v) An *eyepiece* (e), which is usually divided into two sectors. One sector receives light which has passed through the polariser, tube and analyser, while the other receives light which has passed through two more Polaroid sheets (f and g) so aligned that there is maximum light transmission. This sector of the eyepiece is always bright.

Before the instrument is used its zero point is checked. With the sample tube empty and the polariser and analyser in alignment, corresponding to a zero reading on the scale, the whole of the eyepiece should be uniformly bright. Rotation of the analyser should cause one sector to become dim and reach minimum brightness when the angle is 90°.

The solution or liquid to be examined can now be placed in the tube. If the plane of polarisation is rotated by the substance one sector of the eyepiece appears dim, and to restore uniform brightness the analyser must be rotated in a corresponding manner. If the substance in the tube is dextrorotatory, the plane of polarisation will be rotated to the right, and the analyser must also be rotated to the right until it is once again in alignment. Similarly, if the substance is laevorotatory, the analyser has to be rotated to the left. The angle of rotation is noted.

In expressing the result allowance must be made for the fact that the rotation depends on the amount of substance through which the polarised light passes, as well as on its structure, and it is the present convention to quote a *molar optical rotatory power*, α_n, defined as follows:

$$\alpha_n = \frac{\alpha}{cl}$$

where α = observed angle of optical rotation, in radians,
c = concentration of the solution, in mol m^{-3},
l = length of the tube containing the solution, in metres.
The units of α_n are therefore rad m^2 mol^{-1}.

Compounds exhibiting optical isomerism

One of the best known compounds with an asymmetric carbon atom is

lactic acid (2-hydroxypropanoic acid), $CH_3\overset{*}{C}H(OH)COOH$. One enantiomer will cause dextrorotation and the other laevorotation, although which is which is hard to say. (We shall return to this aspect later.) (+)-Lactic acid can be isolated from muscle tissue, and (−)-lactic acid can be obtained by the fermentation of sucrose. Both have a melting temperature of 26 °C (299 K). The (±)-form has a sharp melting temperature of 16.8 °C (290.0 K). The absence of a melting range indicates that this is a racemic compound rather than a racemic mixture (see below). It is obtained naturally from sour milk, and can also be prepared synthetically by the following routes:

$$CH_3C\overset{O}{\underset{H}{\diagup}} + HCN \rightarrow CH_3CH(OH)CN \xrightarrow{\text{hydrolysis}} CH_3CH(OH)COOH$$

ethanal   2-hydroxypropanenitrile

or

$$CH_3CH_2COOH + Br_2 \rightarrow CH_3CHBrCOOH \xrightarrow{\text{hydrolysis}} CH_3CH(OH)COOH$$

propanoic acid   2-bromopropanoic acid

It is seldom that just one enantiomer of a compound is produced by synthesis unless conditions are deliberately arranged so as to encourage this. Consider the latter synthesis in more detail:

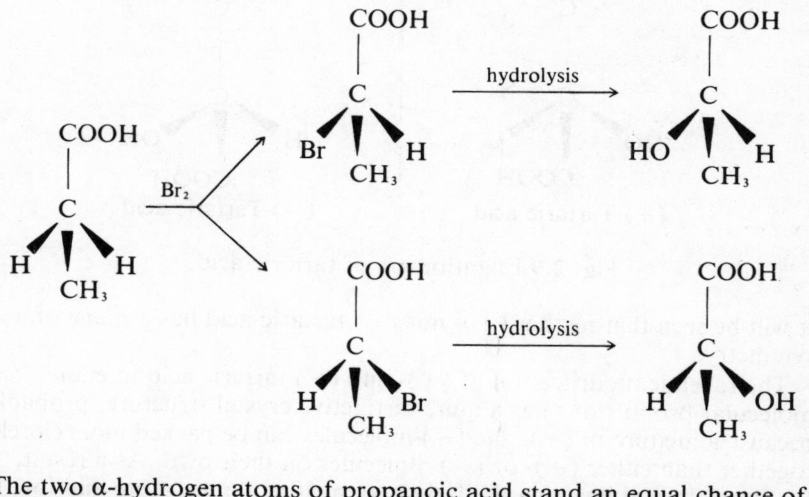

The two α-hydrogen atoms of propanoic acid stand an equal chance of becoming substituted by bromine atoms, so that both forms of 2-bromopropanoic acid are produced in equal amounts. In other words, (±)-2-bromopropanoic acid is formed, and on hydrolysis this yields (±)-lactic acid.

Although an asymmetric carbon atom is often the cause of optical activity, the presence of two such atoms, paradoxically, may result in no activity at all. This situation can arise when the asymmetric carbon atoms are identically substituted, as in tartaric acid (2,3-dihydroxy-butanedioic acid), HOOC—$\overset{*}{C}$H(OH)$\overset{*}{C}$H(OH)COOH.

A tartaric acid molecule is in effect built up from two —CH(OH)COOH units. Two types of unit are possible, one dextrorotatory and the other (the mirror image) laevorotatory:

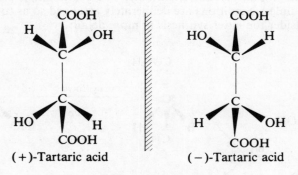

A molecule of (+)-tartaric acid consists of two (+)-units. (Rotation of polarised light is double that of a single unit.) (−)-Tartaric acid, the enantiomer, consists of two (−)-units (Fig. 2.9).

Fig. 2.9 Enantiomers of tartaric acid.

It will be seen that neither (+)- nor (−)-tartaric acid has a plane of symmetry.

The racemic modification of (+)- and (−)-tartaric acid in equimolecular proportions has a quite distinctive crystal structure, probably because a mixture of (+)- and (−)-molecules can be packed more closely together than either (+)- or (−)-molecules on their own. As a result, (±)-tartaric acid has a particularly high melting temperature and low water solubility (Table 2.1).

A third isomer of tartaric acid has molecules composed of one (+)-unit and one (−)-unit. Each unit rotates the plane of polarised light by an equal amount but in opposite directions, so that the net effect is nil. The compound, called *meso*-tartaric acid, is said to be optically inactive by internal compensation.

Table 2.1 Physical characteristics of the three optical isomers of tartaric acid, and the racemic modification

	(+)-Acid	(−)-Acid	*meso*-Acid	(±)-Acid
melting temperature/ °C (K)	170 (443)	170 (443)	140 (413)	205 (478)
water solubility/g per 100 g at 15°C (288 K)	139	139	125	20.6

There is only one form of *meso*-tartaric acid, for it is identical to its mirror image. (If the mirror image shown in Fig. 2.10 is rotated through 180° it can be superimposed on the original.) Furthermore, X−Y (Fig. 2.10) is a plane of symmetry.

meso-Compounds are defined as those which contain asymmetric carbon atoms, but whose molecules are identical to their mirror images. As may be seen from Table 2.1, *meso*-compounds have their own physical characteristics.

meso-Tartaric acid and (+)- or (−)-tartaric acid cannot be said to be enantiomers as they are not mirror images of each other. Rather, they are described as *diastereomers* or *diastereoisomers*. This term can be applied to any pair of optical isomers which do not bear an object to mirror image relationship.

If a compound has two differently substituted asymmetric carbon atoms there are, ignoring racemic modifications, four possible optical isomers. A compound with three C* atoms has 2^3, i.e. 8 isomers. For a general case, a structure with n C* atoms has 2^n isomers. This applies only if the C* atoms are all differently substituted. If any are similarly substituted, as with tartaric acid, the number is lower due to the existence of *meso*-forms.

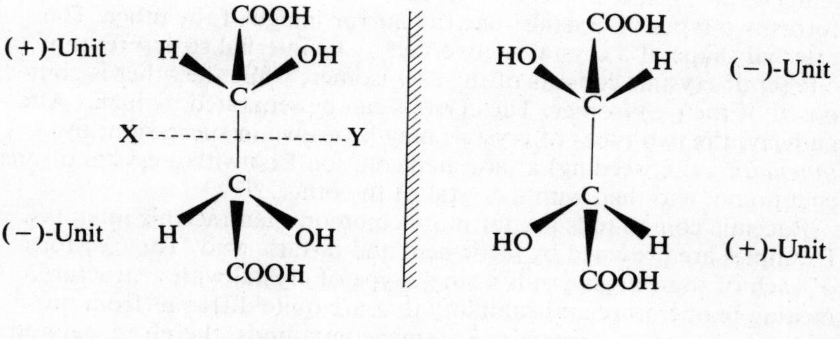

Fig. 2.10 *meso*-Tartaric acid.

Racemic modifications

Racemic modifications are formed either by synthesis, as shown above, or by *racemisation*, i.e. the conversion of one or both enantiomers to an equimolecular mixture of the two. The change may be effected by heat, light, or by dissolving in a solvent. For example, if (+)-tartaric acid is heated with a little water, (±)-tartaric acid will form. The change is believed to occur via a planar configuration (Fig. 2.11).

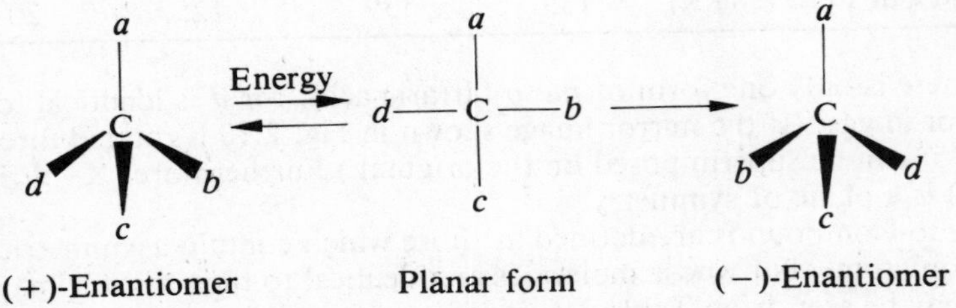

(+)-Enantiomer   Planar form   (−)-Enantiomer

Fig. 2.11 Racemisation of a compound *Cabcd*.

When, say, a (+)-enantiomer is heated, energy is absorbed and the molecule vibrates more and more strongly. Eventually a planar configuration is momentarily achieved. When the molecule returns from the planar state to the tetrahedral state it stands an equal chance of going to the (+)- or (−)-form, and so a racemic modification is obtained.

The separation of a racemic modification into its enantiomers is a process known as *resolution*. The technique employed depends on whether the racemic modification is a *racemic mixture* of dextrorotatory and laevorotatory crystals, or a *racemic compound*, with equal amounts of (+)- and (−)-isomers in the one type of crystal.

Compounds which give racemic mixtures are rare. One such is ammonium sodium tartrate, as discovered by Louis Pasteur in 1848. At temperatures below 27 °C (300 K) the racemic modification of this salt forms two types of crystals, one the mirror image of the other. The external shape of a crystal is governed by its internal structure; thus, one set of crystals consists of the (+)-isomer, while the other is composed of the (−)-isomer. The crystals can be separated by hand. Alternatively, the two types of crystals may be grown in succession, by *inoculating* (i.e. seeding) a saturated solution first with a crystal of one enantiomer and then with a crystal of the other.

Racemic compounds are far more common than racemic mixtures. Examples are provided by lactic acid and tartaric acid. The (±)-form of each of these acids yields a single type of crystal with a structure, melting temperature and solubility that are quite different from those of the (+)- and (−)-isomers. Racemic compounds, therefore, cannot be separated by mechanical means or by the inoculation of a solution, and must be resolved by conversion into diastereomers. Diastereomers, unlike enantiomers, differ from one another in their physical properties, and may be separated by such techniques as crystallisation or chromatography. For instance, a racemic compound of an acid can be resolved by reaction with an optically active base, such as brucine or strychnine:

$$
\begin{array}{ccccccc}
 & & & & (+)\text{-A} & & (-)\text{-A} \\
 & & & & | & & | \\
(\pm)\text{-A} & + & 2(-)\text{-B} & = & (-)\text{-B} & + & (-)\text{-B} \\
\text{acid — racemic} & & \text{base — laevo} & & \text{salt — two diastereomers} & &
\end{array}
$$

After separation, the diastereomers are hydrolysed to regenerate the free acids.

Bases can be resolved in a similar fashion with an optically active acid. Tartaric acid is of limited use for the purpose, owing to the ease with which it racemises on heating, and camphor-10-sulphonic acid is often preferred. Again, the salts that are formed are hydrolysed after separation to yield the original substance in $(+)$- and $(-)$-forms.

For certain racemic modifications a biochemical method of separation can be employed. Pasteur in 1858 devised such a method for obtaining $(-)$-ammonium tartrate from the racemic form. He added the mould *Penicillium glaucum*, which destroyed $(+)$-ammonium tartrate faster than the $(-)$-enantiomer. Eventually, when the $(+)$-form had disappeared, the other could be isolated.

Representation of configuration

The three-dimensional arrangement of atoms in a molecule is known as its *configuration*. On the printed page, configuration is represented by a convention such as that employed in Figs. 2.4, 2.5, etc, in which perspective is achieved through the use of tapered bonds. In hand-written work it is more convenient to use Fischer *projection formulae*, as explained by Fig. 2.12.

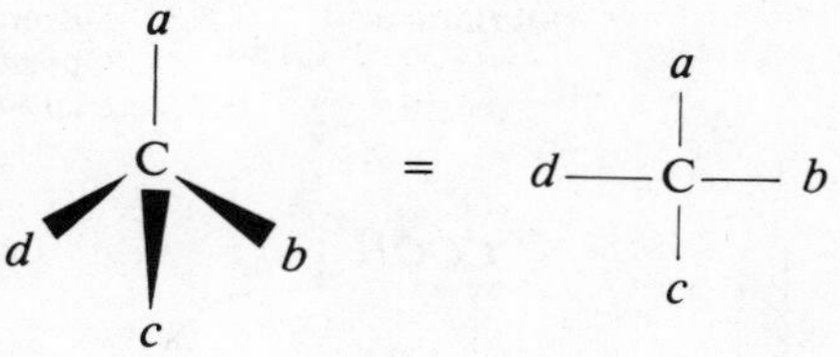

(a)   (b)

Fig. 2.12 (a) Tetrahedral structure of a molecule $Cabcd$. (b) Fischer projection formula of the same molecule.

The following notes are for guidance in writing and using Fischer projection formulae.

(i) Such formulae are used only for molecules which possess at least one asymmetric carbon atom.

(ii) The main carbon chain is drawn vertically, with the lowest numbered carbon atom at the top. For example, the two enantiomers of 2-butanol are written as follows:

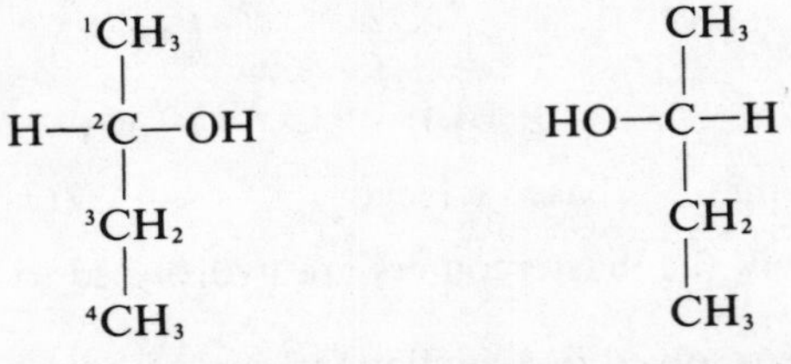

$$^1CH_3 \qquad\qquad CH_3$$
$$H-^2C-OH \qquad HO-C-H$$
$$^3CH_2 \qquad\qquad CH_2$$
$$^4CH_3 \qquad\qquad CH_3$$

(iii) The horizontal lines to an asymmetric carbon atom represent bonds that project towards us, out of the plane of the paper, while vertical lines represent bonds that extend away from us, behind the plane of the paper.

(iv) To decide whether the mirror image of a certain structure is superimposable on the original, the mirror image must be rotated through 180° in the plane of the paper. If it then corresponds to the original, object and mirror image are identical; otherwise they are enantiomers. Reconsider (+)- and *meso*-tartaric acid:

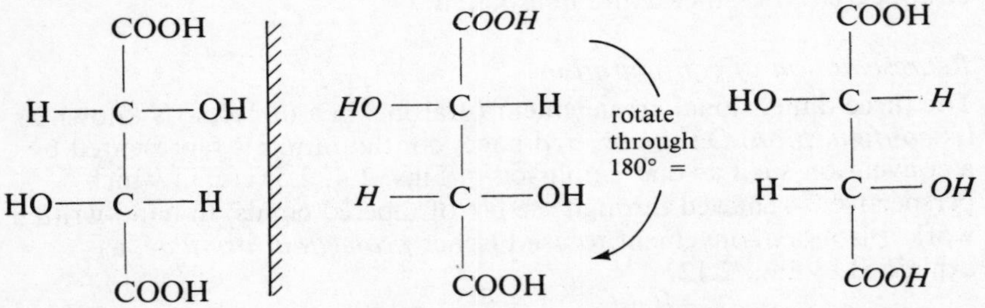

$$\begin{array}{ccc} COOH & COOH & COOH \\ H-C-OH & HO-C-H & HO-C-H \\ HO-C-H & H-C-OH & H-C-OH \\ COOH & COOH & COOH \end{array}$$

rotate through 180° =

(+)-tartaric acid   (−)-tartaric acid

Different from the original. Object and mirror image are enantiomers

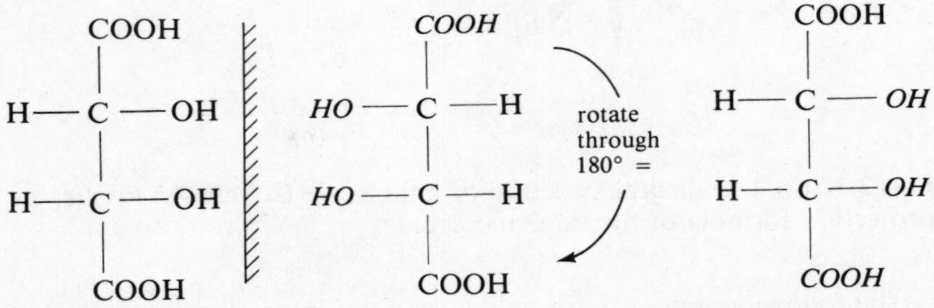

$$\begin{array}{ccc} COOH & COOH & COOH \\ H-C-OH & HO-C-H & H-C-OH \\ H-C-OH & HO-C-H & H-C-OH \\ COOH & COOH & COOH \end{array}$$

rotate through 180° =

meso-tartaric acid

Same as the original. Object and mirror image are identical

Designation of configuration

Although projection formulae are simple to draw, they suffer from the obvious limitation that we cannot use them in conversation unless we happen to have a piece of paper with us. Clearly, we need to be able to name an optical isomer so as to indicate its configuration. It would be very convenient if we could do this by referring to its optical activity, but unfortunately there is no simple connection between the two. In fact, the complete name of an optical isomer specifies both its configuration and its optical activity.

Two systems are in use for denoting the configuration of enantiomers, namely the R and S system, introduced by Cahn, Ingold and Prelog in 1964, and the Rosanoff convention, which dates from 1906.

R and S system

The four atoms or groups of atoms attached to the asymmetric carbon atom are first arranged in order of priority. If the substituent atoms are all different from one another, as they are in 1-bromo-1-chloroethane, $\overset{*}{C}HBrClCH_3$, they are merely arranged in decreasing order of atomic number, thus:

Group	Br	Cl	CH₃	H
Atomic number of substituent atom	35	17	6	1
Priority	1	2	3	4

If two or more groups possess similar substituent atoms, as in lactic acid, $CH_3\overset{*}{C}H(OH)COOH$, priority is decided on the basis of the atomic numbers of the atoms attached to the substituent atom, thus:

Group	COOH	CH₃	HO	H
Atomic number of substituent atom	6	6	8	1
Atomic numbers of atoms attached to the substituent atom	8,8	1,1,1	1	
Priority	2	3	1	4

The molecule is then orientated so that the group of lowest priority (usually hydrogen) is directed away from us, while the other three groups appear in ascending or descending order of priority. Two configurations therefore result (Fig. 2.13).

If progress from groups 1 to 3 is clockwise, the configuration is designated by the letter R, from the Latin word *rectus* = right, while if such movement is anticlockwise the letter S is used, standing for *sinister* = left.

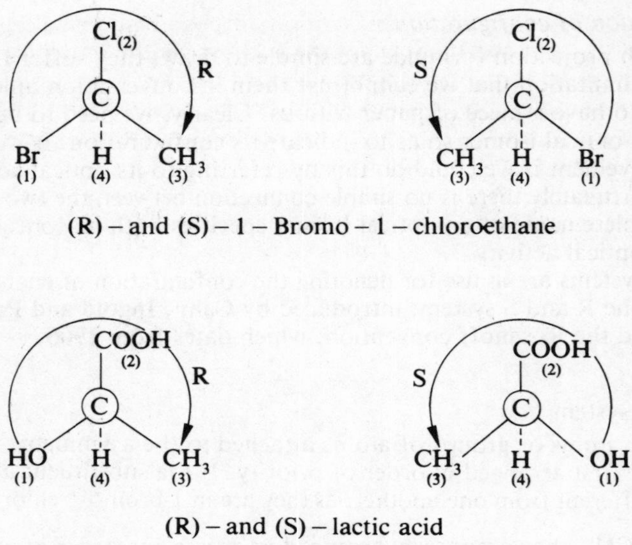

(R) – and (S) – 1 – Bromo – 1 – chloroethane

(R) – and (S) – lactic acid

Fig. 2.13 R and S notation of optical isomers.

D and L system

According to the *Rosanoff convention*, optically active compounds are designated D or L (note the use of small capital letters) depending on whether they are structurally related to D- or L-glyceraldehyde.

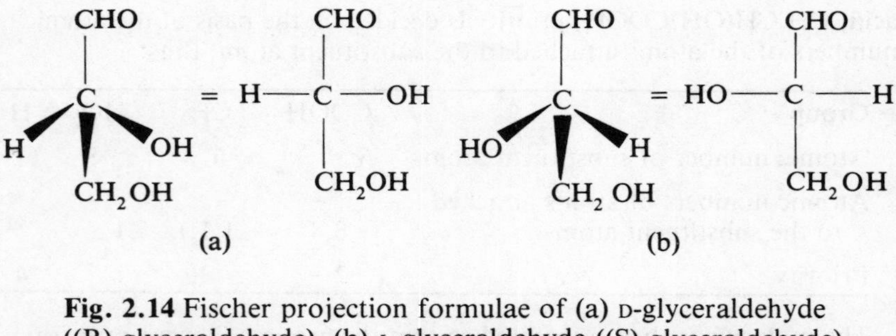

Fig. 2.14 Fischer projection formulae of (a) D-glyceraldehyde ((R)-glyceraldehyde), (b) L-glyceraldehyde ((S)-glyceraldehyde).

The two enantiomers of glyceraldehyde are shown in Fig. 2.14. In a molecule of D-glyceraldehyde, as represented by the Fischer projection formula, the HO group lies to the *right* of the asymmetric carbon atom, whereas in the L isomer it lies to the *left*. It should be clearly understood that the letters D and L relate only to configuration, and do not represent an effect on plane polarised light.

Many optically active organic compounds can be derived from glyceraldehyde by syntheses which do not involve the breaking of bonds to the asymmetric carbon atom, and which do not, therefore, involve a change in configuration. Compounds that are derived from D-glyceraldehyde are designated D-compounds, while those from L-glyceraldehyde are L-compounds.

The convention is widely used in naming carbohydrates and amino-acids. For carbohydrates, such as glucose, the letter D or L relates to the position of the HO group on the lowest asymmetric carbon atom of the projection formula, thus:

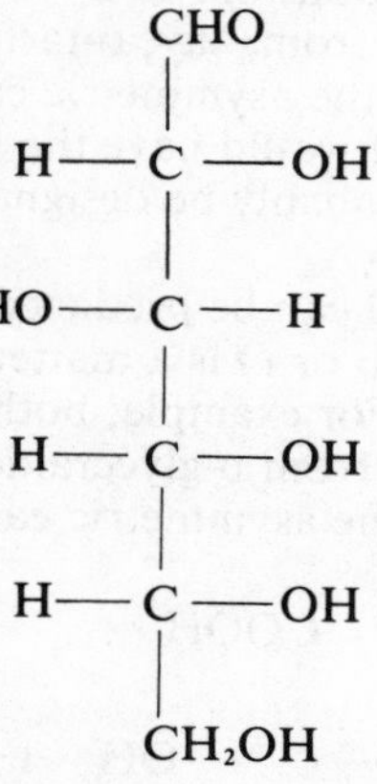

$$
\begin{array}{cc}
\text{CHO} & \text{CHO} \\
| & | \\
\text{H——C——OH} & \text{H——C——OH} \\
| & | \\
\text{HO——C——H} & \text{HO——C——H} \\
| & | \\
\text{H——C——OH} & \text{H——C——OH} \\
| & | \\
\text{H——C——OH} & \text{HO——C——H} \\
| & | \\
\text{CH}_2\text{OH} & \text{CH}_2\text{OH} \\
\text{D-glucose} & \text{L-glucose}
\end{array}
$$

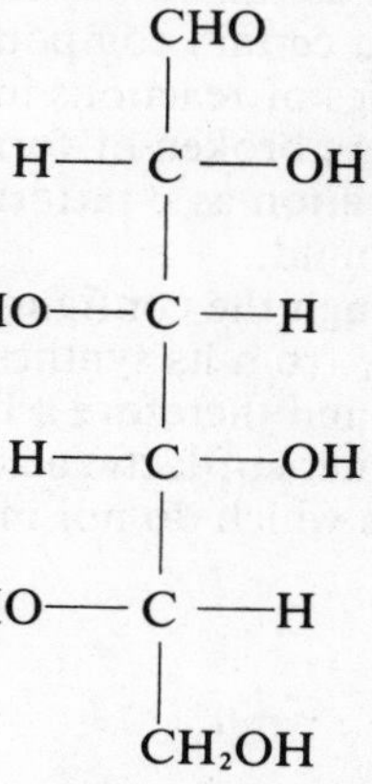

For amino-acids, the position of the NH_2 group is specified, thus:

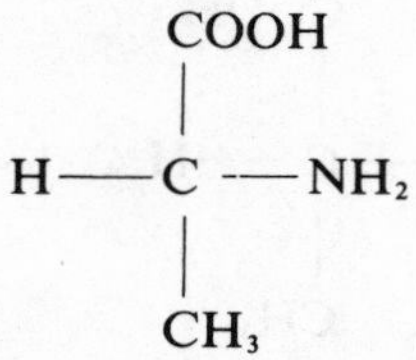

$$
\begin{array}{cc}
\text{COOH} & \text{COOH} \\
| & | \\
\text{H——C——NH}_2 & \text{H}_2\text{N——C——H} \\
| & | \\
\text{CH}_3 & \text{CH}_3 \\
\text{D-alanine} & \text{L-alanine}
\end{array}
$$

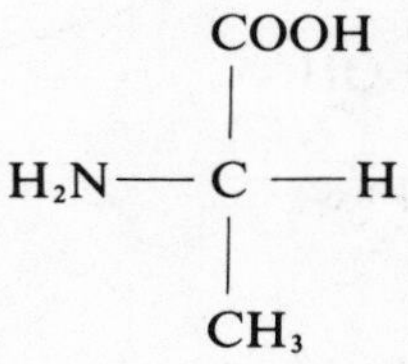

Although this notation is chemically sound, it suffers from one or two disadvantages. Apart from the obvious one, that confusion may arise between D and d, or between L and l, there is the more serious objection that certain hydroxyamino-acids and related carbohydrates possess similar configurations, yet are differently designated. In theory the D and L system could become obsolete, but in practice it is likely to be retained, especially for carbohydrates.

Establishment of configuration

Deciding which of a pair of structures relates to the (+)-form of a compound and which to the (−)-form proved impossible until 1949. In that year, however, J.M. Bijvoet discovered, by a special X-ray analytical technique, that (+)-tartaric acid had the structure shown in Fig. 2.9. Because of its structural relationship to L-glyceraldehyde, the isomer can be named in full as L-(+)-tartaric acid. The laevorotatory enantiomer is D-(−)-tartaric acid. (There are, of course, no such compounds as D-(+)-tartaric acid and L-(−)-tartaric acid.)

With this knowledge, the configurations of all other optically active compounds can be worked out from their relationships to D- or L-tartaric acid. If a certain compound were synthesised from, say, D-tartaric acid by a series of reactions in which no bonds to the asymmetric carbon atom were broken or formed, that compound would have the same configuration as D-tartaric acid and would probably be designated as a D-compound.

Although the configuration of a compound can be predicted with certainty from its synthesis, the designation (D or L) is a matter of convention and therefore a little unpredictable. For example, both enantiomers of lactic acid can be synthesised from D-glyceraldehyde by methods which do not involve the bonds to the asymmetric carbon atom:

$$
\begin{array}{ccc}
& \text{COOH} & \\
& | & \\
\text{H} - \text{C} - \text{OH} & & (-)\text{-lactic acid} \\
& | & \\
& \text{CH}_3 &
\end{array}
$$

$$
\begin{array}{ccc}
& \text{CHO} & \\
& | & \\
\text{H} - \text{C} - \text{OH} & & \\
& | & \\
& \text{CH}_2\text{OH} &
\end{array}
$$

D-glyceraldehyde

$$
\begin{array}{ccc}
& \text{COOH} & \\
& | & \\
\text{HO} - \text{C} - \text{H} & & (+)\text{-lactic acid} \\
& | & \\
& \text{CH}_3 &
\end{array}
$$

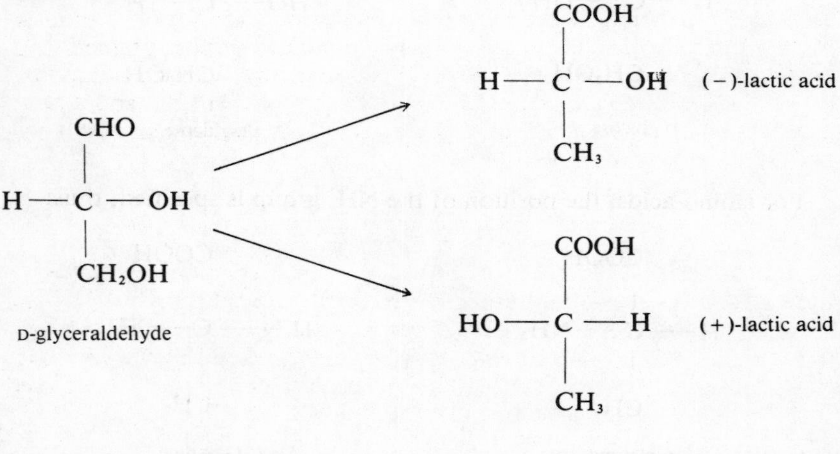

The details of the syntheses have been omitted, but they indicate the configurations of (+)- and (−)-lactic acids beyond all reasonable doubt. Since both acids are derived from D-glyceraldehyde either of them could logically be named as D-lactic acid, but by convention this name is given to the (−)-form, in which the HO group lies to the right of the asymmetric carbon atom, as it does in D-glyceraldehyde.

Chapter 3

Strengths of organic acids and bases

One of the first definitions of the terms 'acid' and 'base' was proposed in 1887 by S. Arrhenius. Although somewhat restricted in scope, it enables much of the aqueous chemistry of acids and bases to be explained, and is still of current importance. Arrhenius defined an *acid* as a compound which, in aqueous solution, produces hydrogen ions, $H^+(aq)$, and a *base* as a substance which gives rise to hydroxide ions, HO^-. Neutralisation, on this theory, amounts to:

$$H^+(aq) + HO^-(aq) = H_2O(l)$$

An acid which produces one hydrogen ion from each molecule of acid that ionises is described as *monoprotic* or *monobasic*. If two or more hydrogen ions are produced per molecule of acid the terms *diprotic* (dibasic), *triprotic* (tribasic) and so forth are used. If one molecule of a monoprotic acid is neutralised by one molecule of a base then the base is described as *monoacidic*. If two or more molecules of acid are required for neutralisation the base is said to be diacidic, triacidic, etc.

3.1 Dissociation constants and pK values

Acids

Acids such as nitric acid and hydrochloric acid give a high concentration of hydrogen ions in aqueous solution (relative to the concentration of the un-ionised acid) and are referred to as *strong acids*. Those, such as ethanoic acid and hydrogen cyanide, which give only a low concentration of hydrogen ions (relative to that of the acid) are called *weak acids*.

The differences arise from the extent to which acid molecules ionise in aqueous solution. In a dilute solution of hydrochloric acid ionisation is virtually complete, whereas in a dilute solution of ethanoic acid only a few per cent of ethanoic acid molecules form ions. The terms 'weak' and 'strong' are relative and the range of acid strengths is wide. A quantitative expression of the strength of an acid can be obtained in terms of its dissociation constant, as follows.

In an aqueous solution of a weak monoprotic acid, HA, a dynamic equilibrium is established between un-ionised molecules and hydrated ions:

$$HA + H_2O \rightleftharpoons H^+(aq) + A^-(aq)$$

Application of the equilibrium law of mass action gives:

$$\frac{[H^+(aq)] \, [A^-(aq)]}{[HA] \, [H_2O]} = \text{constant at a given temperature}$$

The concentration of water in dilute solutions differs little from that in pure water and may be taken as a constant. For example, the concentration of H_2O in pure water is 55.54 mol dm^{-3}, whereas in hydrochloric acid containing 0.1 mol of HCl per dm^3 the concentration of H_2O is 55.37 mol dm^{-3}. The above equations can thus be simplified:

$$HA \rightleftharpoons H^+ + A^-$$

$$\therefore \quad \frac{[H^+] \, [A^-]}{[HA]} = \text{constant } (K_a)$$

where K_a is the *dissociation constant* or *ionisation constant* of the acid, HA. Values of K_a provide a *direct* indication of acid strength, i.e. the greater the dissociation constant the stronger is the acid. Some typical values for relatively weak acids at 25 °C (298 K) are quoted in Table 3.1.

K_a as defined above does not apply to strong acids. The reason is that at high ionic concentrations the oppositely charged ions are relatively close together and strongly attract one another. For example, in hydrochloric acid many H$^+$ ions are surrounded by Cl$^-$ ions, and many Cl$^-$ ions by H$^+$ ions, in effect reducing the concentration of free H$^+$ and free Cl$^-$ ions in the solution. The concentration of free H$^+$ ions, known as the *activity* or *effective concentration* of H$^+$ ions, is thus lower than the molar concentration, and likewise for the Cl$^-$ ions. Strictly, the expression for K_a requires the use of activities, and if molar concentrations are used instead K_a is not constant for strong acids. For weak acids, where ionic concentrations are low, interionic attraction is relatively unimportant and molar concentration is approximately equal to activity.

Discussions of acid strengths in terms of negative indices can be avoided by the use of pK_a *values*. By definition,

$$pK_a = - \lg K_a$$

Table 3.1 K_a and pK_a values of some common acids at 25°C (298 K)

Acid	K_a/mol dm^{-3}	pK_a	
CCl$_3$COOH	2.24×10^{-1}	0.65	
CHCl$_2$COOH	5.13×10^{-2}	1.29	
H$_3$PO$_4$	7.08×10^{-3}	2.15 (pK_1)	
CH$_2$ClCOOH	1.38×10^{-3}	2.86	
HNO$_2$	4.57×10^{-4}	3.34	
HCOOH	1.78×10^{-4}	3.75	
CH$_2$ClCH$_2$COOH	7.94×10^{-5}	4.10	decreasing acid strength
C$_6$H$_5$COOH	6.31×10^{-5}	4.20	
CH$_3$COOH	1.74×10^{-5}	4.76	
HClO	3.72×10^{-8}	7.43	
HBrO	2.00×10^{-9}	8.70	
HCN	3.98×10^{-10}	9.40	
C$_6$H$_5$OH	1.00×10^{-10}	10.0	
HIO	3.02×10^{-11}	10.52	
H$_2$O	1.81×10^{-16}	15.74	
C$_2$H$_5$OH	$\sim 10^{-16}$	~ 16	

Thus, for ethanoic acid,

$$K_a = 1.74 \times 10^{-5} \text{ mol dm}^{-3}$$

and

$$pK_a = -\lg (1.74 \times 10^{-5})$$

$$= -(0.24 - 5) = 4.76$$

pK_a values for weak acids are all positive. Because the sign is changed on going from K_a to pK_a, the latter is *inversely* related to acid strength, i.e. the smaller a pK_a value the stronger is the acid (Table 3.1).

For a diprotic acid, the degree of ionisation (α) for the second stage of ionisation is less than that for the first, owing to the difficulty of removing a proton from a negatively charged ion. For a triprotic acid, the degree of ionisation for the third stage is always very small. Thus, if an acid has a basicity greater than one, each stage of the ionisation has its own dissociation constant and its own pK value. For such acids we no longer use the notation pK_a, but pK_1, pK_2, etc.

Bases

Substances which give rise to a high concentration of hydroxide ions in aqueous solution (relative to the concentration of the substance itself) are referred to as *strong bases*. They comprise ionic compounds, e.g. sodium hydroxide, which are completely dissociated in solution. In contrast, *weak bases* give only a low concentration of hydroxide ions in solution. They are covalent compounds which ionise to a very limited extent, e.g.

$$NH_3 + H_2O \rightleftharpoons NH_4^+(aq) + HO^-(aq)$$

$\alpha = 0.013$ for a 0.1 M solution at 25 °C (298 K) (α is defined as the fraction of molecules that are ionised at equilibrium).

By application of the equilibrium law,

$$\frac{[NH_4^+(aq)][HO^-(aq)]}{[NH_3][H_2O]} = \text{constant at a given temperature}$$

For the reason given above this can be simplified to:

$$\frac{[NH_4^+][HO^-]}{[NH_3]} = \text{constant } (K_b)$$

where K_b stands for the *dissociation constant* or *ionisation constant* of the base. pK_b is defined analogously to pK_a. Some values are given in Table 3.2.

Table 3.2 K_b and pK_b values of some common bases at 25°C (298 K)

Base	K_b/mol dm^{-3}	pK_b	
$(CH_3)_2NH$	5.25×10^{-4}	3.28	decreasing base strength
CH_3NH_2	4.37×10^{-4}	3.36	
$(CH_3)_3N$	6.31×10^{-5}	4.20	
$C_6H_5CH_2NH_2$	2.34×10^{-5}	4.63	
NH_3	1.78×10^{-5}	4.75	
NH_2NH_2	8.51×10^{-7}	6.07	
NH_2OH	6.61×10^{-9}	8.18	
$C_6H_5NH_2$	4.17×10^{-10}	9.38	
CH_3CONH_2	7.9×10^{-16}	15.1	

3.2 Factors affecting the strengths of organic acids in aqueous solution

All organic compounds whose molecules contain one or more HO groups are to some extent acidic. Examples are alcohols, phenols, carboxylic acids and sulphonic acids. In aqueous solution they ionise thus:

$$X{-}O{-}H \rightleftharpoons XO^-(aq) + H^+(aq)$$

where X represents the group of atoms to which the hydroxyl group is bonded. Before discussing how acid strength depends on the nature of X, we shall consider the enthalpy changes accompanying the steps involved in ionisation. These steps are illustrated below in a Born−Haber cycle.

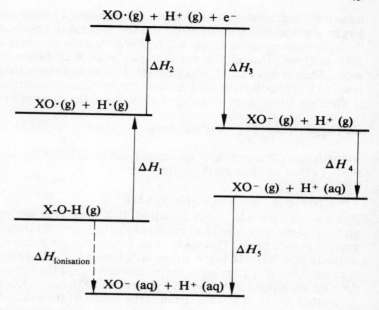

ΔH_1 is the bond dissociation enthalpy of the O—H bond.

ΔH_2 is the ionisation enthalpy of hydrogen.

ΔH_3 is the electron affinity of the acid radical, i.e. the enthalpy evolved when an electron is added to an isolated radical.

ΔH_4 is the hydration enthalpy of a proton.

ΔH_5 is the hydration enthalpy of the oxoanion, XO^-.

The sum $\Delta H_1 + \Delta H_2 + \Delta H_3 + \Delta H_4 + \Delta H_5 =$ the enthalpy of ionisation of the acid ($\Delta H_{\text{ionisation}}$).

In general, the more exothermic the ionisation, the stronger is the acid. For all acids, the enthalpy changes ΔH_2 and ΔH_4 are constant. Therefore, as ΔH_1 decreases and/or ΔH_3 and ΔH_5 increase, $\Delta H_{\text{ionisation}}$ becomes increasingly exothermic and the ionisation becomes more extensive. Conversely, as ΔH_1 increases and/or ΔH_3 and ΔH_5 decrease then acid strength diminishes.

Note Reaction feasibility, i.e. the possibility of a reaction taking place, is dictated strictly by the free energy change, ΔG. However, ΔH provides a reasonable guide when comparing reactions of the same type. In general, the more exothermic a reaction, the more likely it is to proceed.

The effect of X on bond dissociation enthalpy (ΔH_1)

Let us consider the $XO\cdot$ radical. If X reduces the electron density on the oxygen atom it will be more difficult for the oxygen atom to utilise its unpaired electron in reforming the O—H bond. In other words, if X has an electron withdrawing effect the O—H bond will be weaker, and its bond dissociation enthalpy will be lower, than in the absence of such an effect. This applies, for example, when X contains a halogen atom. Such atoms exert a $-I$ effect (§1.4), and as a result the O—H bond in a halocarboxylic acid is weaker than in an unsubstituted acid.

The opposite arguments apply if X increases the electron density on the oxygen atom. For example, the O—H bond in R—CCOOH (with R groups) is strengthened compared with the same bond in CH_3COOH because of the $+I$ effect of three alkyl groups.

The effect of X on electron affinity (ΔH_3)

Again we will consider the XO· radical. If X lowers the electron density on the oxygen atom it reduces the repulsion between an isolated electron and the oxygen atom. This leads to an *increase* in the electron affinity of the radical XO·. Halogen atoms in X have an effect of this sort.

In contrast, if X increases the electron density on the oxygen atom the electron affinity *decreases*. Alkyl groups in X result in this situation.

Another factor to be considered is the ability of the group X to delocalise the negative charge of the oxoanion, XO^-, that is formed. Delocalisation of electrons always leads to a lowering in energy (cf. benzene), and in this case increases the electron affinity as the following scheme indicates.

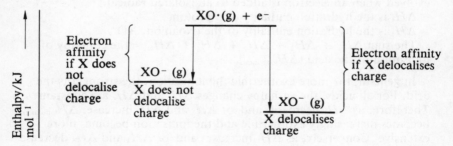

Delocalisation occurs to some extent in halocarboxylate ions. For example, the charge on a CCl_3COO^- ion is not confined to the oxygen atoms, but is shared between them and the chlorine atoms. However, the effect is more pronounced in sulphonate ions, RSO_3^-, where the charge is delocalised over the three oxygen atoms bonded to the central sulphur atom. As delocalisation increases, the enthalpy of the oxoanion XO^- is lowered and the electron affinity of the radical XO· increases.

The effect of X on enthalpy of hydration (ΔH_5)

Oxoanions become hydrated by the formation of hydrogen bonds between water molecules and oxygen atoms bonded to the central atom. The more oxygen atoms are present, the more water molecules can be hydrogen bonded and the greater is the hydration enthalpy. Thus, the hydration enthalpy of a sulphonate ion, RSO_3^-, is greater than that of a carboxylate ion, $RCOO^-$.

Summary Ionisation, and hence acid strength, is increased by atoms or groups which withdraw electrons. Such atoms or groups:

(i)   decrease electron density on the hydroxyl oxygen atom;
(ii) delocalise the charge on the oxoanion;
(iii) promote hydration of the oxoanion.

In contrast, ionisation is hindered and acid strength is decreased by groups which release electrons. Such groups increase electron density on the hydroxyl oxygen atom, and are often ineffective at delocalising the charge on the oxoanion or hydrating it.

For most acids the dominant factors that govern acid strength are ΔH_1 and ΔH_5. We shall now use these principles to discuss qualitatively the acid strengths of some well known organic compounds.

Alcohols and phenols

Alcohols are less acidic than water because alkyl groups release electrons to the oxygen atom ($+\,I$ effect):

$$R\!\rightarrow\!O\!-\!H$$

Phenols are more acidic than water because the benzene ring effectively withdraws a lone pair of electrons from the oxygen atom (mesomeric effect).

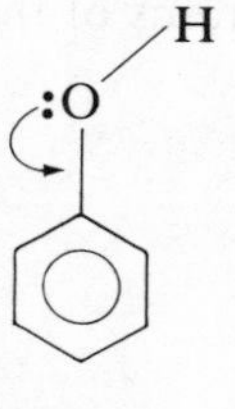

phenol molecule 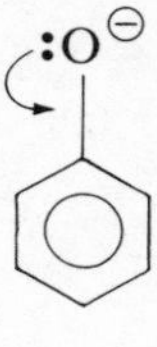   phenoxide ion

The effect is present in the phenol molecule and the phenoxide ion and in both cases promotes an increase in acidity. In the molecule, the effect results in a weakening of the O—H bond, and in the ion it allows the negative charge to become delocalised over the benzene ring. (No such delocalisation can occur in an alkoxide ion.) The latter is particularly important, because the delocalisation of charge results in a considerable increase in stability.

Carboxylic acids

Carboxylic acids are more acidic than water. This is because they possess the carbonyl group, C=O, which contains a highly electronegative oxygen atom. Consequently, in comparison with a hydrogen atom, the carbonyl group draws electrons away from the hydroxyl oxygen atom.

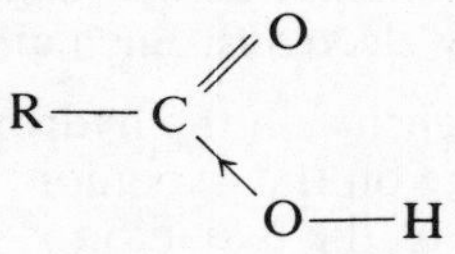

When discussing the relative strengths of different carboxylic acids we can ignore this effect because it is common to all of them. All we need to consider is whether the group R withdraws or repels electrons from the carboxyl group. Such effects are transmitted through the molecule to the hydroxyl oxygen atom, i.e.

R withdraws electrons   R repels electrons

If R withdraws electrons it increases acid strength, because it weakens the O—H bond and also stabilises the carboxylate ion by removing some of the negative charge on the oxygen atoms of the ion:

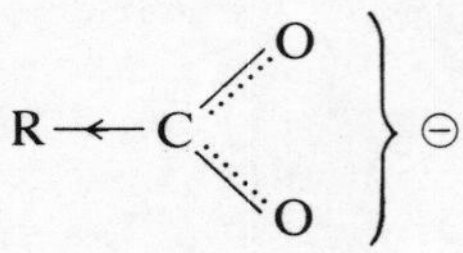

carboxylate ion

In general, the greater the electron withdrawal effect of R, the stronger is the acid. Conversely, if R repels electrons, acid strength decreases.

It is interesting to compare the strengths of carboxylic acids against methanoic acid, HCOOH, where the hydrogen atom of the C—H bond has no significant effect on the electron density of the hydroxyl oxygen atom. Because of the $+I$ effect of the methyl group, ethanoic acid is weaker than methanoic acid. In contrast, chloroethanoic acid,

CH$_2$ClCOOH, is a stronger acid than methanoic acid, due to the $-$I effect of the chlorine atom. Dichloroethanoic acid, CHCl$_2$COOH, is stronger than chloroethanoic acid because of the greater electron withdrawing capacity of its two chlorine atoms. By similar reasoning trichloroethanoic acid, CCl$_3$COOH, is stronger still. 3-Chloropropanoic acid, CH$_2$ClCH$_2$COOH, is weaker than chloroethanoic acid because the chlorine atom is relatively remote from the hydroxyl group. Bromine substituted acids are weaker than chloro-acids, and iodo-acids are even weaker (although still stronger than unsubstituted acids). This is in accordance with the electronegativity order of the halogens: Cl > Br > I.

Benzoic acid, C$_6$H$_5$COOH, is a weaker acid than methanoic acid because of the presence of a mesomeric effect. The π electrons of the benzene ring are drawn towards the (somewhat positive) carbonyl carbon atom, so that they interact with the π electrons of the C=O bond.

benzoic acid

4-nitrobenzoic acid

The effect is sometimes called a *negative mesomeric effect* ($-$M), and it causes the benzene ring to behave as an electron source. In chlorobenzene, phenol and phenylamine the mesomeric effect operates in the opposite direction (§1.4) and is termed a *positive mesomeric effect* ($+$M). This causes the benzene ring to act as an electron sink.

Nitrobenzoic acids are stronger than benzoic acid because a nitro group, in any position on the ring, withdraws electrons from it. This is another example of a $-$M effect.

Sulphonic acids

Carboxylic acids in general are weak acids, but sulphonic acids are considerably more acidic. Benzenesulphonic acid, C$_6$H$_5$SO$_3$H, for example, is approximately as strong as hydrochloric acid. Two factors contribute to their strength. First, there are three oxygen atoms over which the negative charge of the sulphonate ion is delocalised, compared with only two in a carboxylate ion. The electron affinity of a sulphonate radical is therefore greater than that of a carboxylate radical. Second,

54

the sulphonate ion has a higher hydration enthalpy than the carboxylate ion. Consequently, $\Delta H_{ionisation}$ is greater for sulphonic acids than for carboxylic acids.

3.3 Factors affecting the strengths of organic bases in aqueous solution

In this section we shall restrict discussion of organic bases to amines. Such compounds are derivatives of ammonia, and on ionisation in aqueous solution they form hydroxide ions and substituted ammonium ions:

$$R_3N + H_2O \rightleftharpoons R_3NH^+(aq) + HO^-(aq)$$

The degree of ionisation, and hence the base strength, depends on the nature of R. As with acids the process of ionisation involves a number of steps, each with its own enthalpy change, as illustrated by the following Born–Haber cycle.

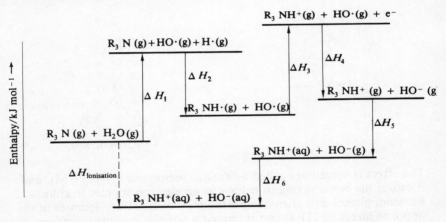

ΔH_1 is the bond dissociation enthalpy of the O—H bond in the water molecule.

ΔH_2 is the negative bond dissociation enthalpy of the N—H bond in the $R_3NH\cdot$ radical, i.e. the enthalpy evolved when R_3N forms a covalent bond with a hydrogen atom.

ΔH_3 is the ionisation enthalpy of the $R_3NH\cdot$ radical.

ΔH_4 is the electron affinity of the hydroxyl radical.

ΔH_5 is the hydration enthalpy of the R_3NH^+ ion.

ΔH_6 is the hydration enthalpy of the hydroxide ion.

Thus the sum $\Delta H_1 + \Delta H_2 + \Delta H_3 + \Delta H_4 + \Delta H_5 + \Delta H_6 =$ the enthalpy of ionisation of the base ($\Delta H_{ionisation}$).

In general, as for acids, the more exothermic the ionisation the stronger is the base. For any base the enthalpy changes ΔH_1, ΔH_4 and

ΔH_6 are constant, and as ΔH_2 and ΔH_5 increase and/or ΔH_3 decreases $\Delta H_{\text{ionisation}}$ becomes increasingly exothermic and the base strength increases.

The effect of R on bond dissociation enthalpy (ΔH_2)

If R increases the electron density on the nitrogen atom it will be easier for the nitrogen atom to utilise its electrons in forming an N—H bond. Thus if R releases electrons the N—H bond will be stronger and its bond dissociation enthalpy will be greater than in the absence of such an effect. The opposite argument applies if the nitrogen atom is bonded to an electron withdrawing group.

The effect of R on ionisation enthalpy (ΔH_3)

If R increases the electron density on the nitrogen atom then electron loss from the $R_3NH\cdot$ radical to form the R_3NH^+ cation is easier than if the effect is absent. Therefore the ionisation enthalpy (ΔH_3) decreases if R releases electrons, and conversely the ionisation enthalpy increases if R withdraws electrons.

The effect of R on hydration enthalpy (ΔH_5)

The cation R_3NH^+ becomes hydrated by the formation of hydrogen bonds between water molecules and hydrogen atoms bonded to the nitrogen atom. The extent to which hydration occurs, and hence the magnitude of the hydration enthalpy (ΔH_5), therefore increases as the number of hydrogen atoms bonded to the nitrogen atom increases.

Summary Groups which increase the electron density on the nitrogen atom promote base strength, while groups which lower electron density reduce base strength. Base strength is also enhanced if the cation formed can become extensively hydrated.

We shall now use these principles to compare qualitatively the base strengths of some simple amines against that of ammonia. We shall see that in many cases the dominant factors controlling base strength are ΔH_2 and ΔH_3, although ΔH_5 is important for tertiary amines.

Aliphatic amines

The $+I$ effect of the methyl group results in methylamine, CH_3NH_2, being a stronger base than ammonia. Dimethylamine, $(CH_3)_2NH$, possesses two electron releasing methyl groups and is a stronger base than methylamine. In general, base strength varies in the order

$$NH_3 < RNH_2 \text{ (primary amine)} < R_2NH \text{ (secondary amine)}$$

provided that both amines contain the same alkyl group.

Extending this argument to tertiary amines, we might expect trimethylamine, $(CH_3)_3N$, to be an even stronger base because there are three methyl groups. Trimethylamine, however, is a *weaker* base than dimethylamine. The reason is that the cation derived from the tertiary amine, $(CH_3)_3NH^+$, is less extensively hydrated and therefore less stabilised than that from the secondary amine, $(CH_3)_2NH_2^+$. This is because there are fewer hydrogen atoms to which water molecules can become hydrogen bonded:

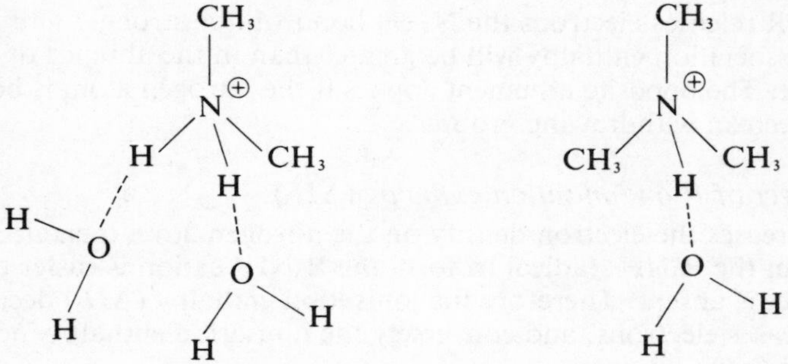

This effect is sufficient to account for the lower base strength of tertiary amines compared with secondary amines.

To summarise, as successive alkyl groups are introduced into ammonia, the base strengths of the amines depend on the balance between the inductive effect and the extent to which the cations are stabilised by hydration:

$$NH_3 \qquad RNH_2 \qquad R_2NH \qquad R_3N$$

+ I effect of R increases base strength

$\longrightarrow$

decreasing stabilisation of cation by hydration reduces base strength

Aromatic amines

Phenylamine (aniline), $C_6H_5NH_2$, is a much weaker base than ammonia because the benzene ring effectively withdraws electrons from the nitrogen atom and hence reduces its electron density (mesomeric effect). Diphenylamine, $(C_6H_5)_2NH$, is even weaker than phenylamine, because there are two benzene rings reducing the electron density on the nitrogen atom, and triphenylamine, $(C_6H_5)_3N$, is an exceedingly weak base.

When the nitrogen atom is situated in a side chain, as in the case of benzylamine, $C_6H_5CH_2NH_2$, the electron withdrawing effect of the ring is largely isolated from the nitrogen atom by the intervening saturated carbon atom. Consequently, the base strength of benzylamine is roughly the same as that of ammonia and much greater than that of phenylamine.

3.4 Lewis acids and bases

The Arrhenius theory of acids and bases applies only to compounds in aqueous solution, and is too restrictive to be useful elsewhere. For example, the dehydration of alcohols to alkenes is catalysed by Arrhenius acids, such as sulphuric acid and phosphoric acid, and by certain other substances, such as aluminium oxide and zinc chloride, which cannot be regarded as acids on the Arrhenius theory. The reaction proceeds in essentially the same way in all cases, and we therefore need to redefine the term 'acid' so as to include all species capable of catalysing the change.

Sulphuric acid, phosphoric acid and other Arrhenius acids bring about the dehydration of alcohols because they are *protic acids*, i.e. they liberate protons, and it is these protons which act as the real catalyst for the reaction. In the first stage of the mechanism, a proton accepts a lone pair of electrons from a molecule of the alcohol in the formation of a coordinate bond:

$$
\begin{array}{c}
R \\
\diagdown \\
\diagup \\
H
\end{array}
O\colon \;+\; H^+ \;=\;
\begin{array}{c}
R \\
\diagdown \\
\diagup \\
H
\end{array}
O \longrightarrow H^+
$$

(The rest of the mechanism is discussed later in the book.) Aluminium oxide, zinc chloride and certain other non-hydrogen containing compounds catalyse the reaction because they, too, can accept a lone pair of electrons from the alcohol. In this context, therefore, it becomes logical to define an acid as *any species which is capable of accepting a lone pair of electrons*, a definition first proposed by G.N. Lewis in 1923.

The H_2SO_4 molecule is not an acid on the Lewis theory, for it cannot accept a lone pair of electrons. (It behaves as an acid only because it readily releases a proton; the latter is the true Lewis acid.) The Lewis theory is thus not 'all embracing' as is sometimes claimed; it is merely different from other theories and more appropriate in certain situations. It is common practice in organic chemistry to divide acids into two groups: (i) protic acids, e.g. H_2SO_4; and (ii) Lewis acids, e.g. Al_2O_3.

A similar argument applies to bases. For example, organic carbonyl compounds, such as aldehydes and ketones, react with a considerable number of reagents. Some of them, e.g. ammonia, hydrazine and hydroxylamine, are bases on the Arrhenius theory, but others, e.g. alcohols, are not. However, all these reagents behave in fundamentally the same way, by donating a lone pair of electrons to the carbonyl compound, e.g.

$$H_3N: \quad + \quad \diagdown C = O \quad = \quad H_3N \longrightarrow C = O$$

$$\begin{matrix} R \diagdown \\ H \diagup \end{matrix} O: \quad + \quad \diagdown C = O \quad = \quad \begin{matrix} R \diagdown \\ H \diagup \end{matrix} O \longrightarrow C = O$$

It is therefore sensible, in this and other similar contexts, to regard all such substances as bases, and to define a base, as did G.N. Lewis, as *a chemical species which is capable of donating a lone pair of electrons.*

Chapter 4

Principles of organic reaction mechanisms

4.1 Free radicals, electrophiles and nucleophiles

Substrates and reagents

In most organic reactions there are two reactants (i.e. reacting substances), one of which is referred to as the *substrate* and the other as the *reagent*. A reagent is defined as an attacking species, and a substrate as a species which is undergoing attack. The concept is somewhat artificial because in the liquid or gaseous phase molecules of *both* reactants are in motion, and chemical reactions occur when the molecules collide with one another. If an organic reaction involves an organic and an inorganic reactant, the former is regarded as the substrate and the latter as the reagent, e.g.

$$C_2H_5Br + NaOH = \underbrace{C_2H_5OH + NaBr}$$

substrate　　reagent　　　　　　products

If a reaction involves two organic species, convention alone dictates which is the substrate and which the reagent. For instance, in an esterification reaction, it is customary to state that the acid is attacked by the alcohol, e.g.

$$CH_3COOH + C_2H_5OH = \underbrace{CH_3COOC_2H_5 + H_2O}$$

substrate　　reagent　　　　　　products

Reagents are divided into three categories, namely:

(i)   free radicals;
(ii) electrophiles;
(iii) nucleophiles.

Free radicals

A *free radical* is defined as an atom or group of atoms which possesses one or more unpaired electrons. A simple example is a chlorine atom. Because there is an odd number of electrons (17) in this atom, one of them must be unpaired. The unpaired electron can be symbolised by a dot written to the right of the formula. Thus, instead of referring to a chlorine atom, Cl, we can, for emphasis, refer to a chlorine free radical, Cl·. Other common examples are alkyl free radicals, such as the methyl free radical, CH_3·.

It is important not to confuse the terms 'free radical' and 'radical'. The latter is employed in chemical nomenclature to denote a group of atoms which occurs repeatedly in various compounds.

Free radicals are formed whenever a covalent bond is broken symmetrically. Consider, for example, a covalent bond in a diatomic molecule A$\overset{.}{\underset{\times}{-}}$B. (The dot and cross represent the shared pair of electrons that constitutes the covalent bond.) If the bond is broken in such a way that one electron joins atom A while the other joins atom B, then two free radicals are formed:

$$A\overset{.}{\underset{\times}{\frown}}B \rightarrow A\cdot + B\cdot$$

The curved half-arrows are used to denote the movement of single electrons.

This type of bond fission is known as *homolytic fission* or *homolysis*. The prefix *homo* means 'the same', and indicates that in this type of fission the two fragments are similar to each other in the sense that they are both free radicals. Homolytic fission is the exact opposite of covalent bond formation.

Chlorine free radicals are formed when the element chlorine is exposed to ultraviolet light or heated to temperatures above 400 °C (673 K):

$$Cl\overset{.}{\underset{\times}{\frown}}Cl \rightarrow 2Cl\cdot$$

Homolytic fission does not always result in identical fragments. The decomposition of iodine chloride, for example, yields two different free radicals:

$$I\overset{.}{\underset{\times}{\frown}}Cl \rightarrow I\cdot + Cl\cdot$$

The reaction is nevertheless regarded as homolytic fission, because it conforms to the principle stated above.

The breaking of a covalent bond does not always involve homolytic fission, because it is possible for *both* the bonding electrons to migrate to one atom:

$$A \overset{\frown}{\underset{x}{\bullet}} B \rightarrow A^+ + B^-$$

A curved arrow with a complete arrowhead is used in mechanistic studies to denote the movement of a pair of electrons.

This type of bond fission, which is the opposite of coordinate bond formation, is known as *heterolytic fission* or *heterolysis*. The prefix *hetero*, meaning 'different', is used to signify that in this type of fission the two fragments differ from each other in that they are ions of opposite charge.

Electrophiles

An *electrophile* or *electrophilic reagent* is defined as one which is electron seeking, i.e. one which seeks out an electron-rich centre in the substrate for its attack. For a species to act as an electrophile it must satisfy three basic requirements.

(i) It must contain an atom possessing a positive electrical charge. For maximum attraction to the electron-rich centre of the substrate an electrophile should carry a full positive charge, i.e. it should be ionic, but in practice most electrophiles are polar molecules in which there is only a partial positive charge (δ^+).

(ii) It must be able to accept a lone pair of electrons. An electrophile reacts by forming a coordinate bond with the substrate, and it is important that the electrophile has a vacant orbital available at a relatively low energy level to accommodate the incoming pair of electrons.

(iii) It must be able to form a strong bond with a carbon atom, otherwise there cannot be a stable reaction product.

An example of a cationic electrophile is the nitryl cation, NO_2^+. Most common cations, e.g. Na^+, Ca^{2+}, Al^{3+} and NH_4^+, are not electrophiles because they do not satisfy conditions (ii) or (iii). Examples of molecular electrophiles are HBr (i.e. $\overset{\delta^+}{H}$—$\overset{\delta^-}{Br}$) and H_2SO_4.

Nucleophiles

A *nucleophile* or *nucleophilic reagent* is defined as one which seeks out, for its attack, a part of the substrate that is positively charged, cf. an atomic nucleus. The essential requirements of a nucleophile are as follows.

(i) It must possess a negative electrical charge, − or δ^-.

(ii) It must be able to donate a lone pair of electrons. A nucleophile, like an electrophile, acts by forming a coordinate bond with the substrate, but in this case the reagent is the donor, not the acceptor.

(iii) It must be able to form a strong bond with a carbon atom, for the reason given above.

Nucleophiles are more numerous than electrophiles and include most common anions, e.g. HO^-, CH_3COO^-, $C_2H_5O^-$ and CN^-. Examples of molecular nucleophiles are H_2O, NH_3 and CH_3NH_2.

Nucleophiles (and, to a lesser extent, electrophiles) vary considerably in their reactivity. The most reactive, which are said to be 'good entering groups', have the following characteristics:

(i) a full negative charge, as opposed to a partial charge;

(ii) the ability to be readily polarised by the positive charge on the substrate. 'Polarisation', in this context, relates to the distortion of electron shells. For example, the ions Cl^- and I^- have related electronic configurations ($Cl^- = 2,8,8$; $I^- = 2,8,18,18,8$), and we

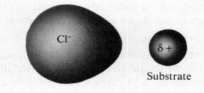

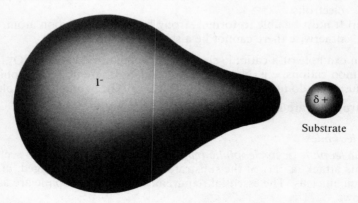

Fig. 4.1 Comparison of Cl^- and I^- ions as nucleophiles.

might expect them to be similar in their behaviour as nucleophiles. In fact I⁻ is much the better, because the ready distortion of its large electron cloud towards the positive centre greatly facilitates coordination (Fig. 4.1).

4.2 Classification of organic reactions

In the study of organic reaction mechanisms, only four basic types of reaction are recognised, namely substitution, addition, elimination and rearrangement. Other terms which are in common use to classify chemical reactions, e.g. oxidation–reduction or double decomposition, are not employed because they are not needed. For instance, the chlorination of methane,

$$CH_4 + Cl_2 = CH_3Cl + HCl$$

is an oxidation–reduction reaction, but can equally well be described as a substitution reaction.

Substitution reactions (symbol S)

Substitution means 'replacement', and in such a reaction an atom or group of atoms belonging to the substrate is replaced by another atom or group of atoms from the reagent. An example is the chlorination of methane, quoted above.

Substitution reactions are classified according to the nature of the reagent, as follows.

(i) Free radical substitution (S_R). Such reactions are shown by alkanes and alkylaromatic hydrocarbons. Examples are the chlorination of methane, and of methylbenzene in the side chain:

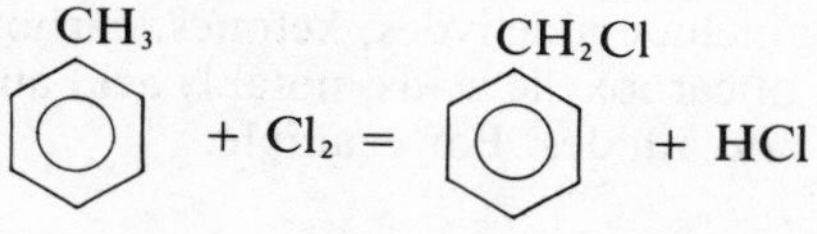

(ii) Electrophilic substitution (S_E). Shown by benzene and all other aromatic compounds, e.g.

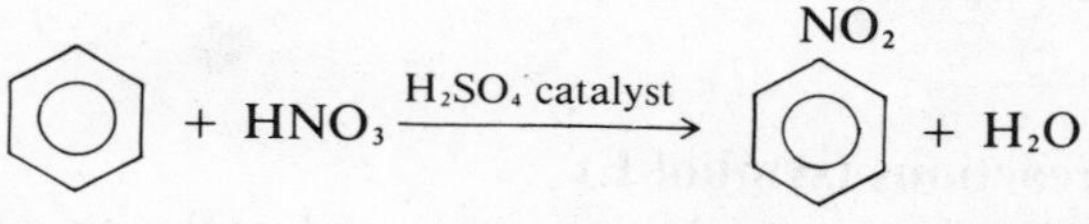

(iii) Nucleophilic substitution (S_N). Shown by alkyl halides and, to a lesser extent, by alcohols, e.g.

$$C_2H_5Br + NaOH(aq) = C_2H_5OH + NaBr$$
$$C_2H_5OH + HBr = C_2H_5Br + H_2O$$

Addition reactions (symbol A)

Such reactions are restricted to unsaturated compounds, i.e. those possessing double or triple bonds. The reagent adds across the multiple bond of the substrate, to give one substance only. This is known as an *adduct*, i.e. addition product. For example,

$$CH_2{=}CH_2 + Br_2 = CH_2BrCH_2Br$$

substrate   reagent   adduct

Addition reactions are classified, like substitutions, according to the type of reagent.

(i) Free radical addition (A_R). Shown by alkenes, alkynes and benzene in the presence of ultraviolet light. Examples are the bromination of ethene in ultraviolet light (see above) and of benzene under similar conditions:

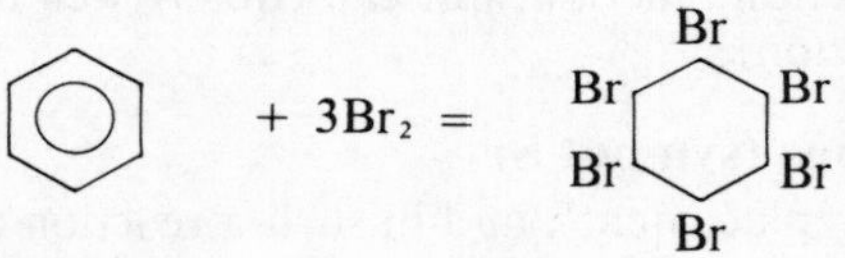

(ii) Electrophilic addition (A_E). Shown by alkenes and alkynes in the absence of ultraviolet light, e.g.

$$CH_2{=}CH_2 + Br_2 \xrightarrow[\text{in ethanol}]{\text{in solution}} CH_2BrCH_2Br$$

(iii) Nucleophilic addition (A_N). Shown principally by carbonyl compounds, i.e. those possessing a carbon−oxygen double bond. Carbonyl compounds include aldehydes, ketones, carboxylic acids, and derivatives of carboxylic acids, notably acid anhydrides, acid chlorides, esters and amides. For example,

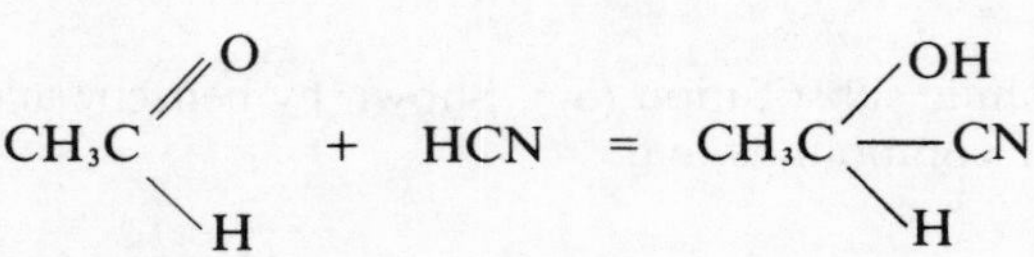

Elimination reactions (symbol E)

Elimination is the reverse of addition, and involves the removal of a small molecule from the substrate to produce an unsaturated compound. The two commonest types of elimination reaction are as follows.

(i) Dehydration, i.e. the elimination of H_2O. An acid catalyst is usually required, e.g.

$$CH_3CH_2OH \xrightarrow[\text{H}_2\text{SO}_4 \text{ catalyst}]{\text{concentrated}} CH_2{=}CH_2 + H_2O$$

(ii) Dehydrohalogenation, i.e. the elimination of HCl, HBr or HI. The reaction is brought about by means of a strong base, e.g.

$$CH_3CH_2Br + KOH = CH_2{=}CH_2 + KBr + H_2O$$

Rearrangement reactions (symbol R)

Such reactions, in which compounds are converted into their isomers, are relatively uncommon among simple organic substances. When rearrangements do occur they may involve either the hydrocarbon skeleton of the molecule or a functional group. Certain rearrangements occur spontaneously, while others require the presence of a catalyst. Examples are as follows:

$$CH_3CH_2CH_2CH_3 \xrightarrow{\text{AlBr}_3 \text{ catalyst}} \begin{array}{c} CH_3 \\ {\diagdown} \\ CH_3 {\diagup} \end{array}\!\!CHCH_3$$

butane   2-methylpropane

$$CH_3CH_2NO \xrightarrow{\text{no catalyst}} CH_3CH{=}NOH$$

nitrosoethane   ethanal oxime

Tautomerism

It can happen that an organic compound A rearranges of its own accord into an isomer B, which in turn can change back into A.

$$A \rightleftharpoons B$$

At any given temperature a dynamic equilibrium is established, at which the rate of the forward change is equal to that of the reverse change. The name given to this spontaneous reversible rearrangement is *tautomerism*, and the isomers A and B are referred to as *tautomers*.

The proportions of A and B in the equilibrium mixture depend on temperature, in accordance with the usual rules governing reversible reactions. Thus, we can argue that the change is exothermic in one direction and endothermic in the other. If we raise the temperature, in other words if we supply heat, we assist the reaction which requires the heat, i.e. we promote the endothermic change more than the exothermic one, and equilibrium is disturbed accordingly. Conversely, if we lower the temperature, we preferentially favour the exothermic reaction.

If, to the mixture, a reagent is added which chemically attacks, say, A, then equilibrium is disturbed. To restore equilibrium, in accordance with the Equilibrium Law of Mass Action, some B changes to A. If enough reagent is added, all B changes to A and the tautomeric mixture behaves as though it consisted of pure A. Similarly, the introduction of any reagent which reacts with B causes the mixture to behave as though it were composed of pure B.

The commonest type of tautomerism is *keto–enol tautomerism*, in which a ketone (or aldehyde) exists in equilibrium with an *enol*, i.e. an unsaturated alcohol possessing the $-CH=\overset{\underset{\displaystyle |}{\textstyle OH}}{C}-$ grouping. The simplest enol is ethenol (vinyl alcohol), $CH_2=CHOH$; thus, the simplest example of keto–enol tautomerism involves this compound and ethanal:

$$CH_2=\overset{\underset{\displaystyle |}{\textstyle OH}}{CH} \rightleftharpoons \left[CH_2=\overset{\underset{\displaystyle |}{\textstyle O}}{CH} \right]^- + H^+ \rightleftharpoons CH_3C\begin{smallmatrix} \diagup O \\ \diagdown H \end{smallmatrix}$$

Equilibrium lies almost entirely to the right-hand side. The reason is that the change in both directions involves the migration of a proton, H^+. When the proton joins the $\left[CH_2=\overset{\underset{\displaystyle |}{\textstyle O}}{CH} \right]^-$ ion, it has a greater tendency to join the CH_2 carbon atom to give CH_3CHO than the oxygen atom to give $CH_2=CHOH$. This is because CH_3CHO is a weaker acid than $CH_2=CHOH$, and weak acids are generally formed at the expense of stronger ones. **It is a general principle of keto–enol tautomerism that equilibrium favours the keto form.**

A well known example of keto–enol tautomerism concerns ethyl 3-oxobutanoate (ethyl acetoacetate), which, as we shall see, is an important reagent in organic syntheses. At room temperature the tautomeric mixture consists of approximately 8 per cent of the enol form (I) and 92 per cent of the keto form (II):

$$CH_3\overset{\underset{\displaystyle |}{\textstyle OH}}{C}=CHCOOC_2H_5 \rightleftharpoons CH_3\overset{\underset{\displaystyle \|}{\textstyle O}}{C}CH_2COOC_2H_5$$

$$\text{I} \qquad\qquad\qquad\qquad \text{II}$$

Determination of the composition of a tautomeric mixture by chemical analysis is not easy, because any reaction which removes one of the tautomers causes a disturbance of equilibrium and gives a spuriously high result. There are two possible approaches. One is to reduce the temperature before analysis so as to 'freeze equilibrium', i.e. to lower the rate of rearrangement. Thus, at 0 °C (273 K) the enol form of ethyl 3-oxobutanoate can be estimated by titration with an ethanolic solution of bromine:

$$CH_3\overset{\underset{\displaystyle |}{\textstyle OH}}{C}=CHCOOC_2H_5 + Br_2 = CH_3\overset{\underset{\displaystyle |}{\textstyle OH}}{C}BrCHBrCOOC_2H_5$$

If the titration is performed rapidly, there is not enough time for a significant amount of the keto form to change to the enol form.

A better solution to the problem lies in using a physical method, in which there is no danger of disturbing equilibrium. Ultraviolet spectroscopy is generally used at present. Another well known method is based on the use of an Abbé refractometer to measure the refractive index of the mixture. The refractive index of each tautomer is also found, either by calculation or by direct measurement on the pure keto and enol forms. (These can be obtained from the mixture at −78 °C (195 K). Provided they are in a high state of purity, they can be warmed up to room temperature without rearrangement.) If a linear relationship is assumed between refractive index and composition, the composition of the equilibrium mixture can easily be determined.

Another common type of tautomerism concerns simple sugars. An example is D-glucose, which exists as a mixture of linear and cyclic forms:

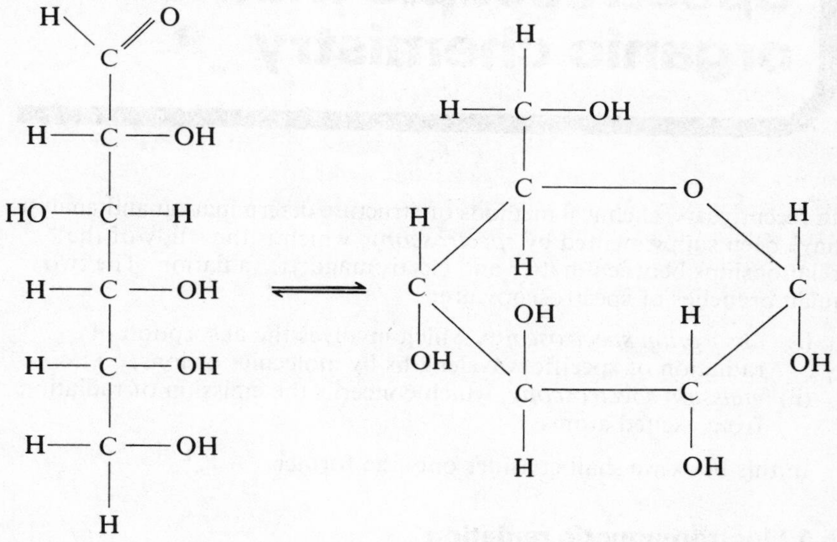

The equilibrium mixture contains approximately 99 per cent of the cyclic form. Despite this, if a reagent is added which attacks the aldehyde group ($-C\overset{\displaystyle O}{\underset{\displaystyle H}{}}$), glucose behaves as though it were entirely linear.

In this discussion we have used the term 'tautomerism' in a broad sense to describe the ready interconversion of any pair of isomers. In recent years, however, its use has become increasingly limited to pairs (or groups) of essentially similar structures which are interconverted by the migration of a hydrogen atom. On this narrower definition keto−enol interchange is still regarded as tautomerism, but the equilibrium between linear and cyclic sugars is not.

Chapter 5

Spectroscopic methods in organic chemistry

In recent years, chemical methods of structure determination and analysis have been supplemented by *spectroscopy*, which is the study of the relationships between matter and electromagnetic radiation. The two main branches of spectroscopy are:

(i) *absorption spectroscopy*, which involves the absorption of radiation of specific wavelengths by molecules or ions;

(ii) *emission spectroscopy*, which concerns the emission of radiation from excited atoms.

In this book we shall consider only the former.

5.1 Electromagnetic radiation

J.C. Maxwell (1881) showed that light travels at a constant velocity (2.998×10^8 m s^{-1} *in vacuo*), and has, associated with it, electric and magnetic fields which show a constant periodic variation similar to that of a sine wave (Fig. 5.1). Thus, we often refer to light as *electromagnetic radiation* or simply *radiation*.

Light is characterised by its *wavelength* (λ) (Fig. 5.1), and its *frequency* (ν), which is defined as the number of cycles or waves passing a fixed point in 1 s. These are related to the velocity of light (c) by the equation:

$$c = \nu\lambda$$

Frequency is expressed in hertz, Hz, where 1 Hz is equal to one cycle

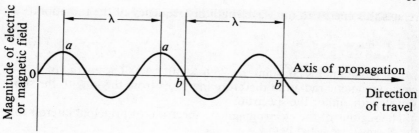

Fig. 5.1 Representation of the oscillating electric or magnetic field of light. *Wavelength* (λ) is the distance between adjacent corresponding points as shown. *a* to *a* (i.e. peak to peak) or *b* to *b* represents one *cycle*.

per second. Wavelength is measured in metres or nanometres (1 nm = 10^{-9} m).

The colour of visible light varies with its wavelength. For example, red light and violet light have wavelengths of approximately 750 and 400 nm respectively. White light comprises radiation of all wavelengths between these two limits. This is shown by splitting white light into its components (i.e. its spectrum) with a prism or a diffraction grating; radiation of all wavelengths in the 750−400 nm range is observed, each producing a different colour sensation to the eye. Electromagnetic radiation outside the 750−400 nm range also exists but is invisible. The various regions of the complete electromagnetic spectrum (Fig. 5.2) are given names, although the boundaries between one region and another are not rigidly defined.

Electromagnetic radiation is a form of energy. *Planck's equation*

				Visible			
Radio waves	Television waves	Micro waves	Infrared		Ultraviolet	X-rays	γ-rays

λ /m	10^2	1	10^{-2}	10^{-4}	10^{-6}	10^{-8}	10^{-10}	10^{-12}
ν /MHz	3	3×10^2	3×10^4	3×10^6	3×10^8	3×10^{10}	3×10^{12}	3×10^{14}

(1MHz = 10^6 Hz)

Fig. 5.2 The complete electromagnetic spectrum, with names of the various regions.

relates this energy to the wavelength or frequency of the radiation:

$$E = h\nu = \frac{hc}{\lambda}$$

where h is Planck's constant, $6.625\,6 \times 10^{-34}$ J s. The energy of electromagnetic radiation therefore increases from the long to the short wavelength end of the spectrum.

The regions of the electromagnetic spectrum of principal interest to the chemist are listed below.

(i) *The ultraviolet, visible and infrared regions*. These regions are the ones most widely used in organic chemistry.

(ii) *Radio waves*. Nuclear magnetic resonance spectroscopy (n.m.r.) utilises the fact that the nuclei of certain atoms, notably $^1_1$H, $^{13}_6$C, $^{31}_{15}$P and $^{19}_9$F, possess a weak magnetic moment. The method involves the recording of the frequency of radio waves absorbed by a solution of a compound when placed in a magnetic field. The results indicate the positions of atoms relative to their neighbours in the molecule.

(iii) *The X-ray region*. X-Ray diffraction enables the positions of atoms or ions in a crystalline solid to be determined.

We shall now consider spectroscopy in these regions (except for the X-ray region) in more detail.

5.2 Ultraviolet and visible spectroscopy

Let us consider two electronic energy levels, A and B, of an atom. Suppose that the lower level, A, is occupied (i.e. contains at least one electron), while the higher level, B, is vacant (Fig. 5.3).

If ΔE_1 represents the difference in energy between the two levels, the atom will absorb radiation of wavelength λ_1 in promoting an electron from level A to level B to produce an excited atom. The relationship between ΔE_1 and λ_1 is given by Planck's equation, i.e.

$$\Delta E_1 = \frac{hc}{\lambda_1}$$

In practice more than one vacant higher energy level is available, and radiation of various wavelengths is absorbed in promoting an electron to different levels. For example, an electron may be promoted from A to C (Fig. 5.3), but because ΔE_2 is greater than ΔE_1 the wavelength (λ_2) of the absorbed radiation is shorter than λ_1.

The difference in energy between the lower level and the various higher levels varies from atoms of one element to those of another. Therefore, the range of wavelengths absorbed, called the *absorption spectrum*, is different for each element. The absorbed radiations usually fall in the ultraviolet or visible part of the spectrum, and we often refer to the *ultraviolet spectrum* or the *visible spectrum* of an element, depending

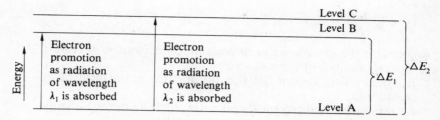

Fig. 5.3 Representation of three energy levels, A, B and C, in an atom.

on which of the two regions is involved. The study of such spectra is called *ultraviolet spectroscopy* or *visible spectroscopy* respectively.

Molecules are similar to atoms in that they also possess discrete energy levels. Thus, the absorption of ultraviolet or visible radiation results in the promotion of an electron from a relatively low energy level to vacant levels of higher energy, and the wavelengths of absorbed radiation constitute the ultraviolet spectrum or the visible spectrum of the molecule.

The relative energy levels of a molecule, and hence the absorption spectrum obtained, are dictated by its structure; in particular, by the types of bonds that are present. In general, only the electrons associated with the π bonds in a molecule are excited by radiation in the ultraviolet and visible regions. This is particularly the case in π conjugated systems. The more extensive the conjugation, the *smaller* is the difference between the lower and higher levels and the *longer* is the wavelength of the absorbed radiation. The ultraviolet or visible spectrum therefore provides information about the unsaturated part of a molecule.

Applications of ultraviolet and visible spectroscopy

Analysis of unsaturated compounds

Ultraviolet and visible spectroscopy are widely used for the quantitative analysis of unsaturated compounds in solution, e.g. trace amounts of drugs or vitamins extracted from body fluids and tissues. Analysis is achieved by the application of two laws relating to the absorption of radiation by matter. J. Beer's law (1852) states that the fraction of the incident light absorbed by a compound dissolved in a transparent solvent (i.e. a solvent which does not absorb in the same range as the compound) depends only on the concentration of the compound in the light path. J.H. Lambert's law (1760) was originally related to solids but applies equally well to solutions. The law states that equal fractions of the incident light are absorbed by successive layers of equal thickness of the absorbing compound.

The combined laws, known as the *Beer–Lambert law*, expressed mathematically for solutions, is as follows:

$$D_i = \lg \frac{I_0}{I} = \varepsilon \times c \times d$$

where: D_i is the *absorbance* or *internal transmission density* (formerly 'optical density') of the solution,

I_0 is the incident radiant flux (i.e. the intensity of the incident radiation),

I is the transmitted radiant flux (i.e. the intensity of the radiation after completely passing through the sample),

c is the concentration of the solution,

d is the path length of the solution,

ε is the absorption coefficient (formerly 'extinction coefficient') of the absorbing substance. For a given compound, at a fixed wavelength, the coefficient is constant. If the concentration is expressed in mol dm^{-3}, then ε is called the *molar absorption coefficient*. Values of ε for many compounds at stated wavelengths are tabulated in data books.

It follows from the Beer–Lambert law that, at a given wavelength and a fixed path length, absorbance is directly proportional to concentration. Thus, when a spectrophotometer is used to determine concentration, the cells (which contain the solutions) must all be of the same

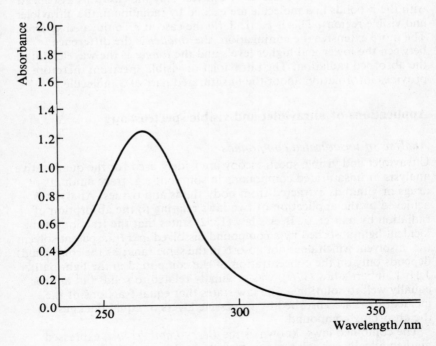

Fig. 5.4 Ultraviolet spectrum of nitrobenzene in ethanol.

thickness. The instrument provides ultraviolet or visible spectra, i.e. graphs of absorbance against wavelength (Fig. 5.4), for standard solutions of the compound concerned. A graph, called a 'calibration graph' is then plotted of absorbance against concentration. (The values for absorbance, which are read off from the spectra, should all relate to a particular wavelength where absorbance is at a maximum.) The concentration of any other solution of the compound can then be found by recording its absorbance and referring to the calibration graph.

Alternatively, for some compounds, the value of ε at a given wavelength can be obtained from tables. Concentration can then be found by applying the Beer—Lambert law.

Detection of impurities

Ultraviolet and visible spectroscopy can be used for detecting unsaturated impurities in saturated compounds. For example, the presence of benzene in cyclohexane or ethanol can be detected by its ultraviolet spectrum, because neither cyclohexane nor ethanol absorbs ultraviolet radiation of the same wavelengths as benzene.

5.3 Infrared spectroscopy

Two atoms joined together by a covalent bond behave in many ways like two spheres joined by a spring. In the same way that the spheres vibrate along the axis of the spring with a natural frequency which depends on the mass of the spheres and the strength of the spring, so the atoms in a molecule are always vibrating along the axis of the bond, even at temperatures approaching absolute zero. This *stretching frequency* or *vibrational frequency* is characteristic of the bond and the masses of the atoms. In a molecule of hydrogen chloride, for example, the stretching frequency is 8.97×10^{13} Hz. When the molecule is exposed to radiation of precisely this frequency, absorption occurs and the amplitude of vibration (i.e. the distance moved by the atoms during a vibration) is increased. This is referred to as an increase in the *vibrational energy* of the molecule.

A spring also has a natural *bending frequency*, which is lower than its stretching frequency. In the same way, bending vibrations or *deformations* also occur in molecules containing three or more atoms, and as a result radiation may again be absorbed by a process similar to that described above.

The stretching and bending frequencies of covalent bonds correspond to radiation in the infrared region of the spectrum. Therefore, a molecule will absorb infrared radiation of specific wavelengths depending on the types of bond which are present and the masses of the atoms that are bonded together. Because each type of bond in a molecule, e.g. C—C, C≡C, O—H, etc, possesses its own stretching frequency and bending frequency, the frequencies of absorbed radiation, i.e. the *infrared* (absorption) *spectrum* of the molecule, provide an insight into the bonds present.

Absorption spectra are usually recorded using a double beam spectro-photometer, an instrument that essentially records the extent to which a sample absorbs radiation at various wavelengths. The spectra (Fig. 5.5) are commonly presented as plots of percentage transmittance (i.e. $I/I_0 \times 100$) against wavelength or *wavenumber*. Wavenumber, the reciprocal of wavelength expressed in cm, has units of cm^{-1}.

As shown by the examples in Fig. 5.5, an infrared spectrum comprises one or more *peaks* or *absorption bands* at wavelengths of radiation where the sample absorbs. The line obtained on the spectrum where no sample absorption occurs is called the *base line*. The interpretation of a spectrum involves the measurement of the wavelength at the maximum of each peak, and possibly the absorbance or percentage transmittance.

Applications of infrared spectroscopy

Detection of bonds and groups

The precise stretching frequency of a particular bond in a molecule depends upon the neighbouring groups or atoms and the state of the sample. Stretching frequencies in the solid state often differ from those obtained from a solution, where the nature of the solvent can affect the frequencies. In many cases, however, a particular type of bond is found to vibrate and hence absorb at approximately the same frequency in a wide range of different molecules. For example, the carbon–oxygen

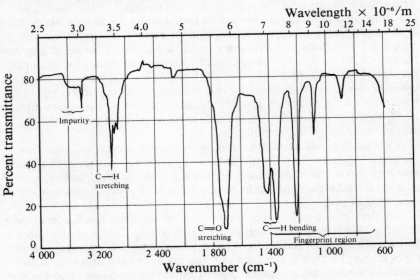

(a)

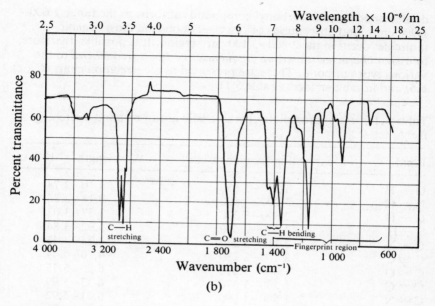

(b)

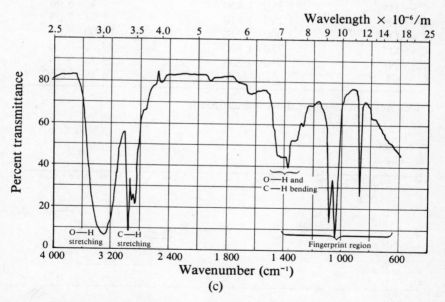

(c)

Fig. 5.5 The infrared spectra and assignments of the principal peaks for (a) propanone, (b) butanone, and (c) ethanol.

double bond in most carbonyl compounds absorbs in the range 1 600–1 900 cm⁻¹, and absorptions of the C—H bonds in most organic molecules occur in the 2 800–3 100 cm⁻¹ region. It is possible, therefore, to assign *characteristic group frequencies* or *bond frequencies* to the various types of bond. These are presented as a correlation chart (Fig. 5.6) and in tabular form (Table 5.1).

Table 5.1 Some approximate characteristic group or bond stretching frequencies

Bond	Stretching frequency/cm⁻¹	Wavelength $\times$ 10⁻⁶/m
O — H	3 230–3 650	3.10–2.74
N — H	3 200–3 500	3.13–2.86
C — H	2 800–3 300	3.57–3.03
S —H	2 550–2 600	3.92–3.84
C≡N	2 110–2 260	4.74–4.43
C≡C	2 100–2 250	4.76–4.44
C=O	1 600–1 870	6.25–5.35
C=C	1 580–1 670	6.33–5.99
C=N	1 560–1 690	6.41–5.92
C — H*	1 350–1 450	7.40–6.90
O — H*	1 260–1 410	7.94–7.09
C — O	1 000–1 310	10.00–7.63
C — C	750–1 300	13.70–7.70
C — N	750–1 300	13.70–7.70
C — Cl	610–800	16.39–12.50

*Denotes a bending or deformation mode.

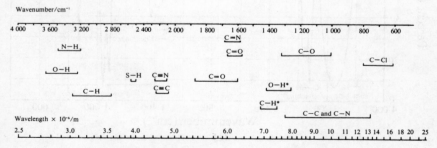

Fig. 5.6 Correlation diagram showing the approximate ranges of characteristic group or bond stretching frequencies.

Identification of compounds

By comparing the infrared spectrum of an unknown compound with those of known compounds, it is often possible to make a rapid identification of the unknown. Of particular use in this method are the large number of peaks occurring in the 1 400–650 cm^{-1} region of most infrared spectra. The peaks arise from complex vibrational and bending motions of the molecule, and each compound has its own characteristic spectral pattern in this region, which is called the 'fingerprint region'. Comparisons of spectra can be made with published libraries of infrared spectra or, nowadays, by using a computer.

Alternatively, the various peaks in an infrared spectrum can be used to identify, tentatively, the different types of bonds and possibly the functional group(s) present in a molecule. Spectral interpretation or correlation in this manner can help enormously in the identification of a compound, even if comparison spectra are unavailable.

It is advisable to examine the 1 500–3 700 cm^{-1} region first. It is here that characteristic absorptions of the bonds associated with some of the common functional groups occur, e.g. O—H, N—H, C=O and C=C. Peaks in this area can usually be assigned fairly easily.

Normally the 1 900–2 750 cm^{-1} region is devoid of peaks. In the 650–1 500 cm^{-1} region many of the peaks correspond to bending vibrations of the molecule (e.g. C—H of CH_2 and CH_3; Fig. 5.5). Some of these bending vibrations are very complex and involve the molecule as a whole rather than individual bonds. A complete assignment of all peaks in terms of individual bonds is therefore impossible. The appearance of the spectrum in this region is very sensitive to molecular structure, which explains the usefulness of the fingerprint region for identification purposes. Compounds with similar molecular structures often have radically different spectra in this region; compare propanone and butanone (Fig. 5.5). One stretching vibration which can usually be unambiguously identified in this region is that of the C—Cl bond.

The existence of peaks in the appropriate places in the spectrum is not proof that a particular bond is present. Sometimes the vibration of one type of bond interferes with that of another and thus affects the appearance of the spectrum. Spectral interpretation therefore needs great care, and confirmation of the identity of a compound is always necessary when this method is used. The fingerprint method, in contrast, is more reliable.

The assignments of the principal peaks in the infrared spectra of some organic compounds is given in Fig. 5.5. Notice that the infrared spectra of two chemically very similar compounds, namely propanone and butanone, are sufficiently different for them to be identified by means of comparison spectra.

Detection of impurities

Impurities in compounds can be detected if the impurity absorbs in a region where the main component does not. For example, pure propanone

78

does not absorb in the region around 3 600 cm^{-1}. The presence of a peak in this region (Fig. 5.5) indicates the presence of an impurity in the propanone. The impurity possibly contains an O—H bond (Table 5.1) and thus could be water.

5.4 Nuclear magnetic resonance spectroscopy

A compass needle is a weak magnet. In pointing to the north pole, it aligns itself with the earth's magnetic field, i.e. it adjusts itself so that its lines of force lie parallel to those of the earth's magnetic field and point in the same direction. This situation represents a stable state of low energy. We could, however, turn the compass needle through 180° so that its magnetic field now opposes that of the earth. (We will neglect the fact that the needle will automatically swing back to its original position if we let it go.) This represents a state of higher energy than the aligned configuration. We can see this from the fact that work must be done (hence energy is required) to turn the needle round from the aligned to the opposed position.

The behaviour of the nuclei of certain atoms can be compared with that of a compass needle. In organic chemistry, the nucleus which particularly concerns us is that of hydrogen-1, i.e. the proton. A proton possesses a magnetic field, and when placed in an external magnetic field it can adopt two positions:

(i)   a low energy configuration, in which the magnetic field of the proton is aligned with the external field;

(ii) a high energy configuration, in which the proton's magnetic field is in opposition to the external field.

So far, we have ignored the fact that a magnetic field is a vector quantity, i.e. it has both direction and magnitude. Strictly speaking we should recognise this, and say, for example, that the earth's 'lines of force' represent its magnetic field vector. Similarly, we should argue that when a proton is in the low energy configuration its magnetic field vector points in the same direction as the vector of the external magnetic field, whereas when the proton is in the high energy state its vector points in the opposite direction to that of the field.

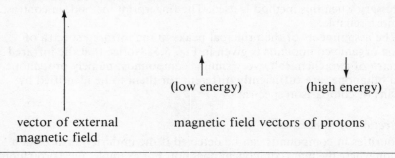

vector of external   magnetic field vectors of protons
magnetic field

When a hydrogen-containing molecule is surrounded by a magnetic field, most of its protons are so arranged that their magnetic field vectors are aligned with the external field, while a smaller proportion have their vectors in opposition to the field. The difference in energy between the low energy (aligned) and high energy (opposition) states depends on the strength of the external field. A compass needle, for example, in the earth's magnetic field is easily turned; hence there is only a small difference in energy between the low and high states. Suppose, however, that we were to place a powerful magnet, whose magnetic field is much stronger than that of the earth, near to the compass. It would now be far more difficult to turn the needle through 180°. Since more work has to be done, more energy is needed, and this shows that the energy difference between the low and high states becomes greater as the strength of the magnetic field increases.

For protons, it can be shown that the energy difference between high and low energy states (ΔE) is directly proportional to the strength of the external magnetic field (H), i.e.

$$\Delta E = E_h - E_l \propto H$$

where E_h and E_l represent the energies of the high and low states respectively.

Suppose that molecules containing protons are subjected to an external magnetic field of strength H_A, and that the difference in energy between high and low energy states is represented by ΔE_A (Fig. 5.7). For any system, consisting of n different parts at equilibrium, the total energy of each of the n parts is the same. In this case, therefore, regardless of the strength of the field, the total energy of nuclei in the low energy

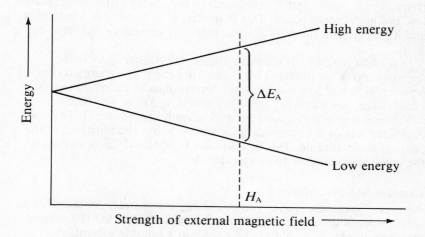

Fig. 5.7 The variation of difference between high and low energy states with strength of the external magnetic field.

state is the same as the total energy of nuclei in the high energy state. Were this not so, a redistribution of nuclei would take place to ensure that the two total energies become equal.

If x and y represent the numbers of nuclei possessing low and high energy respectively,

xE_l = total energy of nuclei in the low energy state,

and yE_h = total energy of nuclei in the high energy state.

Now, $xE_l = yE_h$

but since $E_h > E_l$ then $x > y$

i.e. the number of molecules in the low energy state exceeds that in the high energy state.

If we put some energy into the system, however, we can convert some of the protons from the low to the high energy state. This is most easily achieved by irradiating the molecules with electromagnetic radiation as they lie in the magnetic field.

The energy of electromagnetic radiation is related to its frequency (ν) or its wavelength (λ) by Planck's equation:

$$E = h\nu = \frac{hc}{\lambda}$$

Suppose that radiation of frequency ν_A and energy E_A falls on the molecules. If $E_A = \Delta E_A$ (Fig. 5.7), then radiant energy will be absorbed by the low energy state protons as they become converted to the high energy state.

Thus, protons in a magnetic field will absorb radiant energy whose frequency (or wavelength) corresponds to the energy difference between low and high energy states. This is spoken of as *resonance absorption*, and it gives rise to the term *nuclear magnetic resonance* or n.m.r. for short.

In nuclear magnetic resonance, magnetic field strengths in the approximate range 10 000–15 000 gauss are used. The energy difference between high and low energy states corresponds to radiation in the radio range. For example, at 9 400 gauss (G), $\Delta E \approx 40$ MHz.

Nuclear magnetic resonance is a most useful technique. In organic chemistry it is used to provide information about the positions of the protons in a molecule. This information is often sufficient to enable the molecular structure to be elucidated.

Experimental procedure

An n.m.r. spectrum is obtained by placing the sample, contained in a tube, between the poles of a powerful magnet (Fig. 5.8). The sample may be a liquid, or a solution of a solid in a suitable solvent.

The sample is irradiated with radiation of *fixed* frequency (ν) from the radio range (see above). The magnetic field is then increased over a

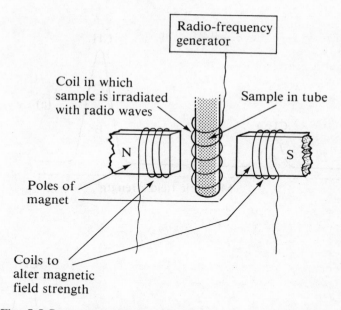

Fig. 5.8 Some essential features of an n.m.r. spectrometer.

small range, up to 100 mG. During the increase, referred to as the *magnetic field sweep*, the energy differences (ΔE, i.e. $E_h - E_l$) for the various protons increase. When, for a particular proton or set of protons, the condition is reached where $\Delta E = h\nu$, absorption of radiation will occur. This absorption is detected by the n.m.r. spectrometer, which feeds a signal to a chart recorder. The strength of the magnetic field at which absorption occurs is characteristic of the type of hydrogen atom involved. It is possible, for instance, to distinguish between hydrogen atoms in CH, CH_2 and CH_3 groups.

An *n.m.r. spectrum* is essentially a graph of the amount of radiation absorbed against the magnetic field strength. Low resolution n.m.r. spectra of ethanal and ethanol are shown in Fig. 5.9.

Peaks A and B correspond to protons in CHO (aldehyde) and CH_3 groups respectively. Intensity of absorption, given by the areas under the peaks, is in the ratio 1 : 3; thus, the relative intensities reflect the numbers of protons of each type present in the molecule. In ethanol, the peaks C, D and E correspond to protons in CH_3, CH_2 and HO groups respectively. The relative intensities of absorption (i.e. areas under the peaks) are in the ratio 3 : 2 : 1, which is in accordance with the formula CH_3CH_2OH.

These spectra show that protons at different positions in a molecule absorb radiation at different magnetic field strengths. This effect arises because electrons in the immediate vicinity of a proton shield it from

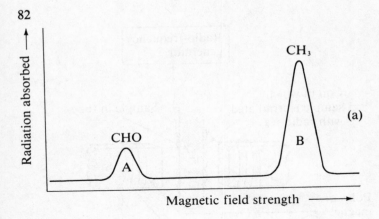

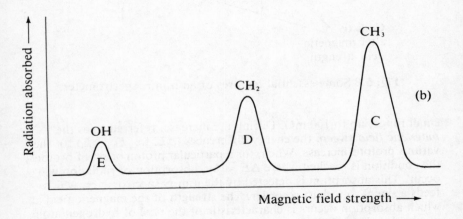

Fig. 5.9 Low resolution n.m.r. spectra of (a) ethanal and (b) ethanol.

the external magnetic field. The greater the electron density, the more effective is this *diamagnetic* shielding effect. Consider two protons in different positions, one of which is poorly shielded and the other well shielded. The former will undergo resonance absorption at a lower field strength than the latter. The HO proton in ethanol, being attached to a highly electronegative oxygen atom, has a low electron density and hence absorbs at a lower field strength than the CH_2 or CH_3 protons.

High resolution n.m.r.
High resolution n.m.r. spectra are obtained by spinning the sample in its tube and using a magnetic field of high strength. Under these conditions, the broad single peaks obtained under low resolution split into a

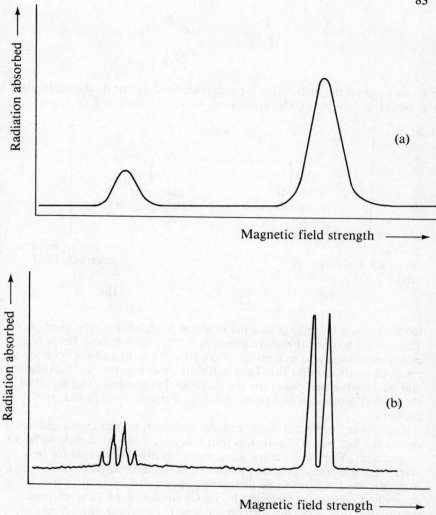

Fig. 5.10 (a) Low resolution n.m.r. spectrum of ethanal. (b) High resolution n.m.r. spectrum of ethanal.

number of sharp peaks. As a simple example, let us consider the low and high resolution n.m.r. spectra of ethanal (Fig. 5.10).

This splitting arises because of a phenomenon known as *spin–spin coupling*, whereby the proton(s) bonded to one carbon atom affect the proton(s) bonded to an adjacent carbon atom. Consider the ethanal molecule:

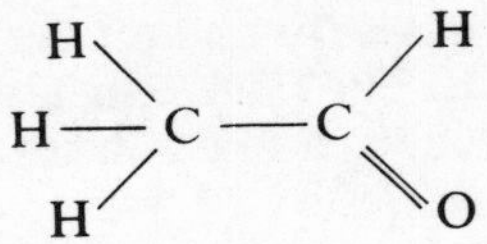

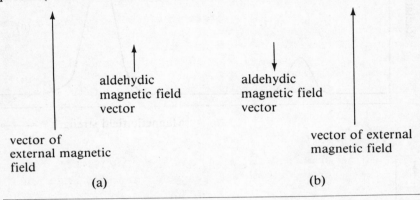

In an external magnetic field the magnetic field vector of the aldehydic proton (i.e. of the —CHO group) can be represented as ↑ or ↓, i.e.

aldehydic
magnetic field
vector

aldehydic
magnetic field
vector

vector of
external magnetic
field

vector of external
magnetic field

(a)

(b)

Configuration (a) strengthens the external field, while (b) weakens it. The *methyl* protons therefore experience external magnetic fields of two magnitudes, one as a result of configuration (a) and the other as a result of configuration (b). They will thus absorb at two field strengths, and we observe two peaks for the three methyl protons. The signal of the methyl protons is said to be split into a *doublet* by the aldehydic proton.

The signal for the aldehydic proton, however, is split into a *quartet*, due to the fact that it experiences four magnetic fields of slightly different magnitudes. The three methyl group protons are responsible for this effect. Each of these protons, in an external magnetic field, has a magnetic field vector in one direction (represented by ↑) or in the opposite direction (represented by ↓). Combination of these various possibilities produces four different magnetic fields, (a), (b), (c) and (d), as follows:

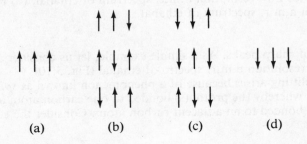

(a)   (b)   (c)   (d)

Notice that in each of the cases (b) and (c) the three different combinations produce the same magnetic field. The aldehydic proton thus experiences four magnetic fields and produces a quartet in the n.m.r. spectrum. The relative intensities of the peaks is in the ratio 1 : 3 : 3 : 1, for reasons which must be discussed at a higher level.

By studying the pattern of splitting due to spin–spin coupling, and the magnetic field strength at which the various protons absorb, it is often possible to determine the structure of a molecule. It is general practice to make a comparison with a *standard substance*. The differences between the magnetic field strengths at which the protons of the standard absorb, and those in the molecule under investigation, are called *chemical shifts*. Tetramethylsilane, $Si(CH_3)_4$, is the standard normally chosen. It gives a single sharp absorption peak, usually well away from the region in which the protons in organic molecules absorb. A small amount of tetramethylsilane is added to the liquid or solution under investigation, and the separation of each peak from the standard is measured.

Applications of nuclear magnetic resonance spectroscopy

Nuclear magnetic resonance is very widely used in organic and organo-metallic chemistry. Only in relatively simple cases, however, can this technique provide sufficient evidence for the structure of a molecule to be elucidated. For complex molecules other physical data are required, e.g. mass spectra, infrared spectra and ultraviolet spectra.

In this section we have concentrated solely on proton n.m.r. Less frequently used, but still of importance, are the n.m.r. spectra derived from the nuclei of $^{13}_{6}C$, $^{19}_{9}F$ and $^{31}_{15}P$.

Chapter 6

The C — C bond in alkanes and other organic compounds

Data on the C—C bond

covalent bond length/nm	0.154
bond dissociation enthalpy/kJ mol^{-1} at 298 K	348

The very high number of organic compounds which exists is due largely to the strength of the C—C covalent bond. Formed by an extensive overlap of s and p orbitals, the bond is stable below about 450 °C (723 K) and enables carbon atoms to join together in chains of virtually any length. The bond is most simply studied in the *alkanes*, saturated aliphatic hydrocarbons of general formula C_nH_{2n+2}, where there are no other bonds (apart from C—H) to influence the chemistry.

6.1 Nomenclature of alkanes

Unbranched chain alkanes

The first four alkanes, CH_4, CH_3CH_3, $CH_3CH_2CH_3$ and $CH_3CH_2CH_2CH_3$, are called methane, ethane, propane and butane respectively. For higher members, the name consists of a term representative of the number of carbon atoms, followed by '-ane', thus:

C_5H_{12} pentane	C_9H_{20} nonane
C_6H_{14} hexane	$C_{10}H_{22}$ decane
C_7H_{16} heptane	$C_{11}H_{24}$ undecane
C_8H_{18} octane	$C_{12}H_{26}$ dodecane

Alkyl radicals

Every alkane can form one or more *alkyl radicals* by the loss of a hydrogen atom. From methane, for example, we get the methyl radical, CH_3—, and from ethane the ethyl radical, CH_3CH_2—, often written as C_2H_5—. From propane can be obtained two radicals, namely propyl,

$CH_3CH_2CH_2$—, and isopropyl, $\begin{matrix} CH_3 \\ \diagdown \\ CH_3 \diagup \end{matrix} CH$—. There are four radicals

derived from butane, with structures and names as follows:

$CH_3CH_2CH_2CH_2$— butyl

$\begin{matrix} CH_3 \\ \diagdown \\ CH_3 \diagup \end{matrix} CHCH_2$— isobutyl

$\begin{matrix} CH_3CH_2CH— \\ | \\ CH_3 \end{matrix}$ *sec*-butyl

$\begin{matrix} CH_3 \\ | \\ CH_3—C— \\ | \\ CH_3 \end{matrix}$ *tert*-butyl

It is not suggested that radicals are formed directly from alkanes by the loss of hydrogen, or that in normal circumstances they are capable of a separate existence, but such groups form part of a great many molecules. In general discussions an alkyl radical is represented by R.

Branched chain alkanes

There is only one way in which one carbon atom and four hydrogen atoms can be arranged, and that is to give a methane molecule. Likewise, there is only one possible form of ethane and one of propane. But for butane there are two ways, I and II, in which the atoms can be arranged without violating bonding requirements.

$CH_3CH_2CH_2CH_3$ I

$\begin{matrix} CH_3 \\ \diagdown \\ CH_3 \diagup \end{matrix} CHCH_3$ II

This is an example of isomerism. In order to distinguish between them, the unbranched form, I, is called butane, while the branched chain version, II, is referred to either by the *trivial name* of isobutane or the *systematic name* of 2-methylpropane. In trival names, the prefix 'iso' is often used to denote a branched chain ending in two methyl groups,

thus: $\begin{matrix} CH_3 \\ \diagdown \\ CH_3 \diagup \end{matrix} CH$—.

There are three isomers of molecular formula C_5H_{12}, with structures and names as follows:

$$CH_3CH_2CH_2CH_2CH_3 \qquad \begin{matrix} CH_3 \\ \diagdown \\ CH_3 \diagup \end{matrix} CHCH_2CH_3 \qquad \begin{matrix} CH_3 \\ | \\ CH_3-C-CH_3 \\ | \\ CH_3 \end{matrix}$$

trivial name	pentane	isopentane	neopentane
systematic name	pentane	2-methylbutane	2,2-dimethylpropane

As the relative molecular mass increases, so the number of possible isomers increases greatly. There are, for example, 5 hexanes, 9 heptanes and 18 octanes. It is difficult to find trivial names for them all, especially names which reflect their structures. For this reason, the higher members of the homologous series must be named systematically in accordance with the following IUPAC rules.

(i) The compound is named as an alkyl substituted alkane, by prefixing the designations of the side chains to the name of the *longest chain* present in the formula.

$$CH_3CHCH_2CH_2CH_2CH_3 \qquad III$$
$$| $$
$$CH_3$$

For example, compound III is a methylhexane. Notice that the name is written as one word, and without a hyphen.

(ii) Rule (i) by itself is insufficient, as will be seen by considering III and IV.

$$\overset{1}{C}H_3\overset{2}{C}H\overset{3}{C}H_2\overset{4}{C}H_2\overset{5}{C}H_2\overset{6}{C}H_3 \text{ III} \qquad \overset{1}{C}H_3\overset{2}{C}H_2\overset{3}{C}H\overset{4}{C}H_2\overset{5}{C}H_2\overset{6}{C}H_3 \text{ IV}$$
$$| \qquad\qquad\qquad\qquad\qquad | $$
$$CH_3 \qquad\qquad\qquad\qquad\qquad CH_3$$

They are clearly both methylhexanes; but we cannot have two compounds with the same name. The carbon atoms of the longest chain are therefore numbered so that we can refer to the positions of the side chains. The direction of numbering should be such that the side chains have the lowest numbers. Form III thus becomes 2-methylhexane (observe the hyphen), and form IV 3-methylhexane. The names 5-methylhexane and 4-methylhexane, obtained by numbering the chain in the opposite direction, are wrong.

$$\overset{6}{C}H_3\overset{5}{C}H\overset{4}{C}H_2\overset{3}{C}H-\overset{2}{C}H\overset{1}{C}H_3 \qquad V$$
$$| \qquad\quad | \quad\ | $$
$$CH_3 \quad\ CH_3\ CH_3$$

With more complex structures confusion may arise over the meaning of the phrase 'lowest numbers'. The correct name is always that which contains the lowest number on the occasion of the first difference. Hence, compound V is 2,3,5-trimethylhexane; not 2,4,5-trimethylhexane. (The sums of the numbers are entirely irrelevant.)

(iii) The presence of identical radicals is indicated by an appropriate multiplying prefix: di-, tri-, tetra-, penta-, hexa-, etc.

$$CH_3CH_2\underset{\overset{|}{CH_3}}{\overset{\overset{CH_3}{|}}{C}}CH_2CH_3 \quad VI$$

For example, compound VI is called 3,3-dimethylpentane: *not* 3,3-methylpentane, and *not* 3-dimethylpentane.

(iv) If two or more side chains of different nature are present, they are cited in alphabetical order. Consider compound VII.

$$\overset{7}{C}H_3\overset{6}{C}H_2\overset{5}{C}H_2\overset{4}{C}H-\underset{\overset{|}{CH_3}}{\overset{\overset{C_2H_5\ \ CH_3}{|\ \ \ \ \ |}}{\overset{3}{C}}}-\overset{2}{C}H_2\overset{1}{C}H_3 \quad VII$$

Since 'e' comes before 'm', the name is 4-ethyl-3,3-dimethyl-heptane, even though the presence of two methyl groups causes us to write 'dimethyl' in accordance with rule (iii).

(v) If two or more side chains are in equivalent positions, the one which is cited first in the name is assigned the lower number.

$$\overset{8}{C}H_3\overset{7}{C}H_2\overset{6}{C}H_2\overset{5}{C}H-\underset{\overset{|}{CH_3}}{\overset{4}{C}}\underset{\overset{|}{C_2H_5}}{\overset{}{H}}\overset{3}{C}H_2\overset{2}{C}H_2\overset{1}{C}H_3 \quad VIII$$

For example, compound VIII is 4-ethyl-5-methyloctane; not 5-ethyl-4-methyloctane.

6.2 Formation of the C — C bond

Wurtz reaction

When an alkyl halide, RX, is treated with sodium (or certain other metals, e.g. copper) in the presence of an aprotic solvent, an alkane, RR, is formed together with a sodium halide. An 'aprotic solvent' is one which does not liberate protons by ionisation; dry ether (ethoxyethane) or cyclohexane is commonly used for the purpose.

$$R\overline{X + 2Na + X}R \xrightarrow[\text{dry ether}]{\text{reflux in}} RR + 2NaX$$

The formation of a sodium halide, with a high lattice enthalpy, drives the reaction to completion, e.g.

$$CH_3CH_2Br + 2Na + BrCH_2CH_3 = CH_3CH_2CH_2CH_3 + 2NaBr$$

The solvent must be aprotic, for in the presence of water or other protic solvent (i.e. one which liberates protons by ionisation) an alkyl halide, RX, is reduced to an alkane, RH. In this example, bromoethane would give ethane rather than butane.

A purification problem arises if a mixture of alkyl halides is used in a *mixed Wurtz reaction* to prepare an alkane with an odd number of carbon atoms, e.g.

$$CH_3I + 2Na + ICH_2CH_3 = CH_3CH_2CH_3 + 2NaI$$

Apart from this reaction, simple Wurtz reactions will occur with each of the alkyl halides:

$$2CH_3I + 2Na = CH_3CH_3 + 2NaI$$

$$2CH_3CH_2I + 2Na = CH_3CH_2CH_2CH_3 + 2NaI$$

Such reactions are not usually attempted, although a *Wurtz–Fittig reaction*, which involves the reaction of an alkyl halide and an aryl halide with sodium, may be quite successful, e.g.

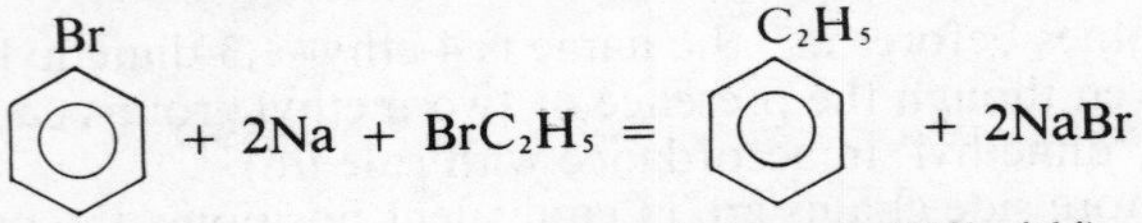

bromobenzene   ethylbenzene (60% yield)

There is no difficulty in purifying the product, for ethylbenzene, which is a liquid, is easily separated from the by-products butane (a gas) and biphenyl, $C_6H_5 \cdot C_6H_5$, (a solid).

Kolbe electrolytic synthesis

Kolbe's method involves the electrolysis of a concentrated aqueous solution of the sodium or potassium salt of a carboxylic acid. The conditions are critical; in particular a smooth platinum anode is needed, and the current density must be high. The overall reaction is as follows:

$$2RCOONa + 2H_2O = RR + 2CO_2 + H_2 + 2NaOH$$

Sodium ethanoate, for example, yields ethane.

Sodium salts of carboxylic acids are ionic, and dissociate in aqueous solution to give hydrated carboxylate ions and sodium ions:

$$RCOONa(s) \rightleftharpoons RCOO^-(aq) + Na^+(aq)$$

Water ionises to a small extent:

$$H_2O(l) \rightleftharpoons H^+(aq) + HO^-(aq)$$

Sodium ions and hydrogen ions migrate to the cathode, where the latter are preferentially discharged:

$$2H^+(aq) + 2e^- = H_2(g)$$

At the anode, carboxylate ions are discharged rather than hydroxide ions, provided that the solution is sufficiently concentrated:

$$RCOO^- = RCOO\cdot + e^-$$

carboxylate free radical

Then: $$RCOO\cdot = R\cdot + CO_2$$

alkyl free radical

Finally: $$2R\cdot = RR$$

Although Kolbe's method is of little importance for the synthesis of hydrocarbons, it is interesting as an example of the use of electrolysis in the preparation of organic compounds. It is also of mechanistic interest, as a means of forming free radicals by anodic oxidation.

Alkyl halides with potassium cyanide or sodium cyanide (§11.5)
A C—C bond is established whenever an alkyl halide undergoes nucleophilic substitution with the cyanide ion:

$$RX + CN^- \xrightarrow[\text{aqueous ethanol}]{\text{reflux in}} RCN + X^-$$

nitrile

Friedel—Crafts reaction (§10.2)
This is the name given to the electrophilic substitution of arenes with alkyl halides or acyl halides. A carbon atom becomes linked to another which forms part of a benzene ring:

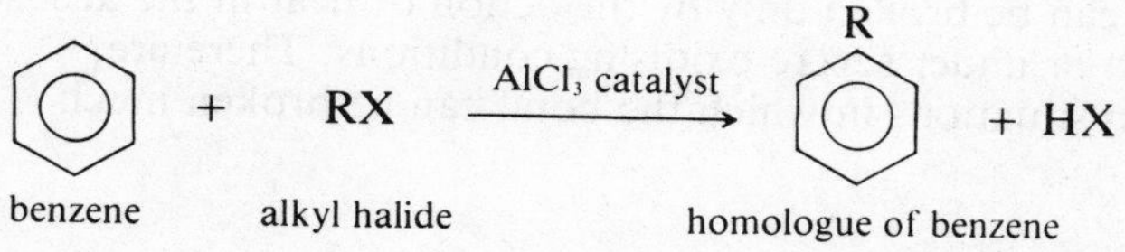

benzene   alkyl halide   homologue of benzene

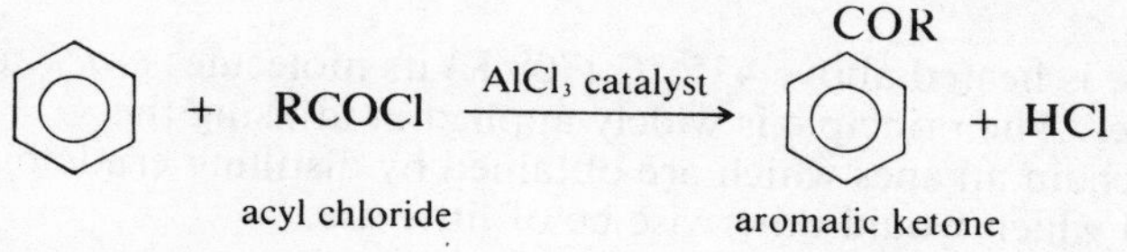

acyl chloride   aromatic ketone

Addition reactions of alkenes (§7.4)

The addition of any reagent to an alkene converts the C=C bond into a C—C bond, e.g.

$$CH_2{=}CH_2 + H_2 \xrightarrow[\text{finely divided Ni catalyst}]{140\ °C\ (413\ K)} CH_3CH_3$$

$C_2H_4 \qquad\qquad\qquad\qquad\qquad\qquad\qquad\quad C_2H_6$

Syntheses with special reagents

In the synthesis of most organic compounds, apart from the very simplest, C—C bonds are usually introduced by means of one of a handful of particularly versatile reagents. The most important are Grignard reagents (Chapter 11), diethyl malonate (Chapter 13) and ethyl acetoacetate (Chapter 13).

6.3 General properties of the alkanes

Methane, ethane, propane and the butanes are gaseous under normal conditions. Then come liquids of gradually increasing boiling temperature, and finally, above $C_{16}H_{34}$, solids of gradually increasing melting temperature. (Both melting temperature and boiling temperature depend on relative molecular mass.)

The alkanes have mild, pleasant smells, and will burn in air or oxygen to give carbon dioxide and water. Being non-polar, they are almost completely insoluble in water but dissolve in less polar solvents.

6.4 Chemical properties of the C—C bond

Because of its high bond dissociation enthalpy, the C—C bond is a difficult one to break. In the majority of organic reactions, particularly where simple molecules are concerned, there are the same number of carbon atoms in the product as in the reactant, and furthermore they are arranged in the same way: rearrangement is uncommon. In general, the C—C bond can be broken only by the action of heat in the absence of air (*pyrolysis*) or under severe oxidising conditions. There are, however, certain situations in which the bond can be broken much more easily.

Thermal cleavage

Cracking

When an alkane is heated above 435 °C (708 K) its molecules *crack* to give smaller ones. The principle is widely applied in utilising the relatively long chain alkanes which are obtained by distilling crude oil (see below) and which would otherwise be of little use.

The principal reaction leads to the formation of a shorter chain alkane and an alkene, e.g.

$$CH_3CH_2CH_2CH_3 \xrightarrow[\text{(708 K)}]{435\ ^\circ C} \begin{array}{l} CH_3CH=CH_2\ +\ CH_4\quad (50\%) \\ \text{propene}\qquad\qquad \text{methane} \\ \\ CH_2=CH_2\ +\ CH_3CH_3\quad (38\%) \\ \text{ethene}\qquad\ \ \text{ethane} \end{array}$$

butane

Important side reactions are isomerisation (see below) and *dehydrogenation*, i.e. the elimination of hydrogen. The latter leads to the formation of alkenes of the original chain length:

$$CH_3CH_2CH_2CH_3 \xrightarrow[\text{(708 K)}]{435\ ^\circ C} CH_3CH_2CH=CH_2\ +\ CH_3CH=CHCH_3$$

$$\text{1-butene}\qquad\qquad \text{2-butene}$$

$$+\ H_2\ (12\%)$$

Dehydrogenation occurs to a lesser extent with the higher alkanes.

In oil refining, mixtures of fairly long chain alkanes are subjected to cracking. Mixtures of shorter chain alkanes are formed, suitable as vehicle fuels, together with ethene, propene and the butenes. These gaseous alkenes can be separated by liquefaction followed by fractional distillation. They are extremely useful, and form the basis of the whole petrochemical industry.

Two main processes are in use, namely *thermal cracking* and *catalytic cracking*. Thermal cracking requires relatively high temperatures (475–600 °C, i.e. 748–873 K) and, like most reactions conducted above 500 °C (773 K), has a mechanism which involves free radicals. Catalytic cracking, in contrast, has an ionic mechanism. An acidic oxide, usually a mixture of silica and alumina, is used to promote the formation of carbonium ions (see below). The process can be carried out at a relatively low temperature (400–500 °C, i.e. 673–773 K), and is particularly well suited to the cracking of *naphthenes*, i.e. cyclopentane and cyclohexane derivatives.

A recent development involves cracking in the presence of steam. The steam cracking of naphtha (see below) to yield ethene and propene is of great importance in view of the demand for plastics. Naphtha is mixed with steam and heated for about a second at 800 °C (1 073 K). The emergent gas mixture is then chilled and fractionated. As well as lower alkanes, lower alkenes and 1,3-butadiene are also obtained.

Thermal cracking proceeds by a *chain reaction*, so called because the reaction of one molecule triggers off the reaction of another, and so on, until all the starting material has been consumed. In every chain reaction there are three stages: (i) initiation; (ii) propagation; and (iii) termination. If, as here, the reaction is propagated by free radicals, it is known as a *free radical chain reaction*.

We shall consider the cracking of butane.

Initiation Free radicals are formed — fairly rapidly above 450 °C (723 K) — by homolytic fission of the C—C bond, which is weaker than the C—H bond:

$$CH_3CH_2CH_2CH_3 \rightarrow 2CH_3CH_2 \cdot$$
ethyl free radicals

or

$$CH_3CH_2CH_2CH_3 \rightarrow CH_3CH_2CH_2 \cdot + CH_3 \cdot$$
propyl and methyl free radicals

Propagation   There are two propagation reactions, in each of which one free radical is consumed and another is formed. Both reactions give, in addition to a free radical, a molecule of one of the final products.

In the first reaction an alkyl free radical from the initiation stage, represented by R·, abstracts an atom of hydrogen from a butane molecule:

$$R \cdot + CH_3CH_2CH_2CH_3 \rightarrow \quad RH \quad + CH_3CH_2\overset{\cdot}{C}HCH_3$$
methane, ethane   *sec*-butyl free radical
or propane

or

$$R \cdot + CH_3CH_2CH_2CH_3 \rightarrow \quad RH \quad + CH_3CH_2CH_2\overset{\cdot}{C}H_2$$
butyl free radical

Observe the formation of alkanes of chain length below that of the original alkane.

In the second propagation reaction, the butyl or *sec*-butyl free radical undergoes homolytic fission at the β-position, i.e. at the bond next but one to C·, to give an alkene and a further methyl or ethyl free radical which propagates the chain:

$$CH_3 \mathbin{\vdots} CH_2 \mathbin{|} \overset{\cdot}{C}HCH_3 \rightarrow CH_3 \cdot \quad + CH_2{=}CHCH_3$$

$$CH_3CH_2 \mathbin{\vdots} CH_2 \mathbin{|} \overset{\cdot}{C}H_2 \rightarrow CH_3\overset{\cdot}{C}H_2 + CH_2{=}CH_2$$

Termination   When the concentration of the original alkane has dropped to a low level, two free radicals may join together to form a molecule, e.g.

$$CH_3 \cdot + C_2H_5 \cdot \rightarrow CH_3CH_2CH_3$$

Decarboxylation of carboxylic acids

The decarboxylation of (i.e. removal of CO_2 from) carboxylic acids to give hydrocarbons involves the rupture of a C—C bond. The reaction is used in both aliphatic and aromatic syntheses:

$$RCOOH \rightarrow RH + CO_2$$

$$ArCOOH \rightarrow ArH + CO_2$$

Heat alone is inadequate, but since carbon dioxide is an acidic oxide its elimination can be achieved by heating with a base. If dry sodium ethanoate is heated with soda lime (a mixture of sodium hydroxide and calcium oxide), methane is obtained in nearly theoretical yield:

$$CH_3COONa + NaOH \xrightarrow{\text{heat dry}} CH_4 + Na_2CO_3$$

The methane may be collected over water.

However, the yields are not nearly so good from the higher carboxylic acids because dehydrogenation occurs to give unsaturated hydrocarbons and free hydrogen. This is a common fault of preparative methods which involve heating organic compounds in the dry state.

Oxidative cleavage

Combustion

Complete oxidation of alkanes occurs on combustion, and gives carbon dioxide, water and a substantial amount of heat energy, e.g.

$$C_2H_6(g) + 3\tfrac{1}{2}O_2(g) = 2CO_2(g) + 3H_2O(l) \quad \Delta H^{\ominus} = -1\ 542 \text{ kJ mol}^{-1}$$

This is the basis of the use of alkanes as fuels both for heating purposes and in the internal combustion engine.

Incomplete combustion in a limited supply of air leads to the formation of water and carbon monoxide or, with less air, to *carbon black*, a pure form of soot which is used as a pigment for paints and inks and as a filler for tyre rubber.

Oxidation by powerful oxidants

Powerful oxidising agents, especially concentrated nitric acid and acidified potassium permanganate solution, are able to break C—C bonds at moderate temperatures. One notable example concerns the oxidative cleavage of ketones. The molecule may break on either side of the carbonyl group to give a mixture of carboxylic acids:

$$\underset{\substack{1 \qquad 2}}{RCH_2|CO|CH_2R'} \rightarrow RCOOH + R'CH_2COOH \qquad \text{(cleavage 1)}$$

$$\text{or } RCH_2COOH + R'COOH \qquad \text{(cleavage 2)}$$

Cleavage in special situations

In certain situations, often when each of the carbon atoms is joined to an oxygen atom, a C—C bond can be broken with remarkable ease.

Oxidative cleavage of 1,2-diols (glycols)

Any common oxidant will cleave a molecule of a 1,2-diol to give two aldehyde or ketone molecules:

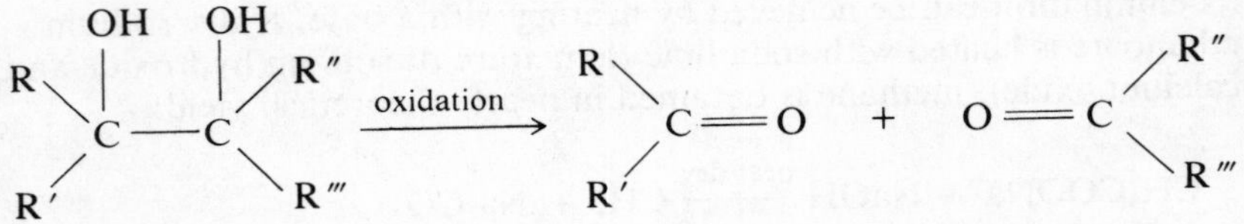

With certain oxidising agents, notably periodic acid, HIO_4, and lead(IV) ethanoate, $(CH_3COO)_4Pb$, no further reactions occur, but with others, such as potassium permanganate or potassium dichromate, aldehydes become oxidised to carboxylic acids.

Oxidative cleavage of 1,2-diketones

In a similar fashion, 1,2-diketones ($RCOCOR'$) are readily oxidised to carboxylic acids, e.g.

$$CH_3COCOCH_3 + H_2O_2 = 2CH_3COOH$$

2,3-butanedione

Oxidation of ethanedioic acid (oxalic acid)

Ethanedioic acid and the ethanedioate ion undergo oxidative cleavage by the permanganate ion in dilute solution to give carbon dioxide:

$$\begin{matrix} COO^- \\ | \\ COO^- \end{matrix} = 2CO_2 + 2e^-$$

The reaction proceeds quantitatively, and can be used for the estimation of ethanedioates or potassium permanganate.

Alkaline cleavage of trihaloaldehydes or trihaloketones

Aqueous alkali readily cleaves the C—C bond in compounds such as trichloroethanal, CCl_3CHO and 1,1,1-trichloropropanone, CCl_3COCH_3:

$$CCl_3CHO + HO^- = CHCl_3 + HCOO^-$$
$$CCl_3COCH_3 + HO^- = CHCl_3 + CH_3COO^-$$

This is the second stage of the *haloform reaction* (§10.2).

Hofmann reaction

The term 'Hofmann reaction' is given to the chemical change whereby an amide is converted into an amine with one less carbon atom, by treatment with bromine and alkali:

$$R-\overset{\displaystyle O}{\underset{\displaystyle NH_2}{C}} \xrightarrow{Br_2\ +\ HO^-} RNH_2$$

e.g. $CH_3CONH_2 \rightarrow CH_3NH_2$

     ethanamide   methylamine

The reaction is thus a degradation, and is commonly called the 'Hofmann degradation of amides'.

Altogether, four separate reactions are involved. Breaking of a C—C bond occurs during the rearrangement of an unstable *nitrene*, i.e. a species possessing six electrons in the outer shell of a nitrogen atom:

$$R-\overset{\displaystyle O}{\underset{\displaystyle NH_2}{C}} \xrightarrow[\text{HO}^-\ \text{catalyst}]{Br_2} R-\overset{\displaystyle O}{\underset{\displaystyle NHBr}{C}} \xrightarrow[-HBr]{\text{heat with HO}^-} R-\overset{\displaystyle O}{\underset{\displaystyle \ddot{N}\!:}{C}}$$

   amide   *N*-bromoamide   acyl nitrene

$$\xrightarrow{\text{rearrangement}} R-N=C=O \xrightarrow[\text{hydrolysis}]{\text{alkaline}} RNH_2 + CO_2$$

                          isocyanate   amine

Isomerisation

In the presence of catalysts which promote the formation of *carbonium ions*, i.e. ions in which a positive charge resides on a carbon atom, alkanes undergo *isomerisation*, i.e. rearrangement to give an isomer of the original alkane. It is a general rule that such rearrangement leads to chain branching. Butane, for example, isomerises to 2-methylpropane at 27 °C (300 K) in the presence of aluminium bromide:

$$CH_3CH_2CH_2CH_3 \underset{\text{catalyst}}{\overset{AlBr_3}{\rightleftharpoons}} (CH_3)_3CH$$

The reaction takes place at this relatively low temperature because a favourable mechanism exists for the conversion.

$$CH_3CH_2CH_2CH_3 + AlBr_3 \rightleftharpoons CH_3CH_2CH_2\overset{\oplus}{C}H_2 + [AlHBr_3]^-$$

$$CH_3CH_2CH_2\overset{\oplus}{C}H_2 \rightleftharpoons \overset{\displaystyle CH_3}{\underset{\displaystyle CH_3}{\diagdown}}\overset{\oplus}{C}-CH_3$$

    primary carbonium ion   tertiary carbonium ion

$$\overset{\displaystyle CH_3}{\underset{\displaystyle CH_3}{\diagdown}}\overset{\oplus}{C}-CH_3 + [AlHBr_3]^- \rightleftharpoons \overset{\displaystyle CH_3}{\underset{\displaystyle CH_3}{\diagdown}}CHCH_3 + AlBr_3$$

The change from a primary carbonium ion to a tertiary carbonium ion is energetically favourable because the latter is stabilised by the delocalisation of its positive charge (§7.4). Carbonium ion rearrangement is fairly common in organic chemistry, and we shall meet it again when discussing the dehydration of alcohols, and the action of nitrous acid on primary amines.

Isomerisation always accompanies catalytic cracking, because carbonium ions are formed when alkanes are heated in the presence of acidic oxides. At elevated temperatures carbonium ions can fragment, which leads to cracking, or they can rearrange, which leads to isomerisation.

In the *reforming* of petroleum, conditions are chosen to encourage rearrangement rather than cracking. MoO_3 and Cr_2O_3 catalysts are used at temperatures of about 500 °C (773 K). The branched chain alkanes so formed have better combustion characteristics than straight chain alkanes in the internal combustion engine.

6.5 Oil refining

Crude oil occurs in many parts of the world, notably the Middle East, Mexico, USA, South America, North Africa, Russia and the North Sea. It consists mainly of hydrocarbons, with small amounts of various compounds of sulphur, oxygen and nitrogen. The hydrocarbons are mainly alkanes, with a very wide range of relative molecular masses. There are also some cycloalkanes or 'naphthenes', and varying proportions of aromatic hydrocarbons.

Primary distillation

The first treatment that crude oil receives at a refinery is fractional distillation. The process can vary, but the following fractions, listed in increasing order of boiling range, are usually collected:

gases
light gasoline
naphtha, or heavy gasoline
kerosene
diesel fuel
(residue)

The technique consists of heating the crude oil in a tubular furnace, so that it is partially vaporised, and then injecting it into the side of a fractionating column which is heated by superheated steam. The various fractions are drawn off at different points of the column (Fig. 6.1).

The fractionating column contains a series of trays and bubble caps, designed to ensure that the rising vapours come into intimate contact with refluxing liquid. The products from the primary distillation are further treated before they are ready for use. Again, the scheme may

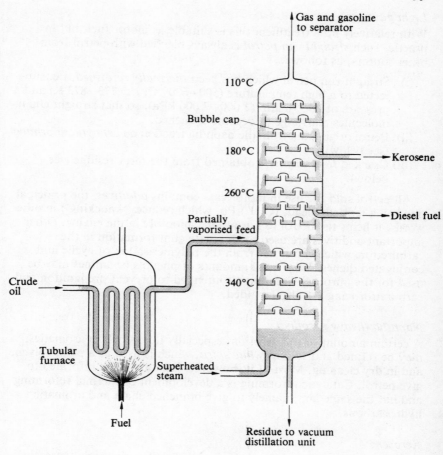

Fig. 6.1 The primary distillation of crude oil.

vary from one refinery to another, and the information which follows
serves only as a guide.

Gases

Originally in solution in the crude oil, gases — mainly propane and
butane — are separated from the gasoline after the primary distillation.
They are readily liquefied by applying pressure at ordinary temperatures,
and in this form they are sold (e.g. 'Calor gas') for industrial and
domestic heating purposes.

Light gasoline

With relatively little treatment this is suitable as motor fuel, but in practice such *straight run petrol* is always blended with petrol from other sources, as follows.

(i) Straight run petrol which has been *thermally reformed*, i.e. subjected to a high temperature (500–600 °C, i.e. 773–873 K) and a pressure of 20–70 atm (2 000–7 000 kPa), so that straight chain molecules rearrange to become branched.

(ii) Petrol originating from the naphtha fraction on *catalytic reforming* (see below).

(iii) Cracked *light distillate* obtained from the tarry residue (see below).

All petrol sold for use in motor cars contains *additives*, the principal one being tetraethyllead, $(C_2H_5)_4Pb$, which reduces 'knocking'; in other words, it helps the petrol to burn more smoothly in the engine. Other important additives are used to eliminate gum formation in the carburettor, which can result from the polymerisation of cyclic and conjugated dienes. Very small amounts of phenols or amines may be used for the purpose. Compounds intended to prevent pre-ignition and carburettor icing are also included.

Naphtha (heavy gasoline)

A certain amount of this fraction, especially if it is rich in aromatics, may be refined and sold as *white spirit*, which is used as a paint solvent and in dry cleaning. Most of it, however, is catalytically reformed to give petrol. Catalytic reforming is a development of thermal reforming and has the same aim, namely to give branched chain and aromatic hydrocarbons.

Kerosene

This fraction is merely refined to give the familiar domestic 'paraffin'. The principal outlet, however, is as a fuel for jet aircraft.

Removal of sulphur compounds constitutes the main treatment, and the modern way of doing this is by *hydrodesulphurisation*. The process utilises cheap hydrogen from the catalytic reforming units to convert thiols, RSH, to hydrogen sulphide:

$$RSH + H_2 \xrightarrow[\substack{15-70 \text{ atm } (1\ 500-7\ 000 \text{ kPa}) \\ Co-Mo-Al_2O_3 \text{ catalyst}}]{300-400\ °C\ (573-673\ K)} RH + H_2S$$

The hydrogen sulphide is removed by distillation or by washing with alkali solution.

Diesel fuel

The diesel fuel fraction is also little changed, the principal treatment again being hydrodesulphurisation.

Residue

Residual hydrocarbons from the base of the primary distillation column flow to a vacuum distillation unit, from which there are three products:

light distillate
lubricating oil
residue

The light distillate is catalytically cracked to give petrol, diesel fuel and gases. The petrol and diesel fuel are blended with the straight run products, while the gases — a mixture of alkanes and alkenes of low relative molecular mass — are separated by liquefaction and distillation.

In practice, three lubricating oil fractions are taken — light, medium and heavy. They are refined in a variety of ways according to the uses to which they are to be put: obviously, oils for medicinal purposes (e.g. 'liquid paraffin') must be treated more drastically than machine oils. The main process is *dewaxing*, in which paraffin wax is precipitated by a suitable solvent, such as 2-butanone. Apart from producing a useful material, this treatment is necessary to prevent the oil from solidifying in the cold.

The tarry residue from the vacuum distillation is not distilled further. Part of it is sold as *bitumen* for road making, and part is employed as fuel oil. For this purpose its viscosity must be reduced, either by blending with diesel fuel from the catalytic cracking units or by cracking in a plant called a *visbreaker*.

Chapter 7

The C═C bond in alkenes

Data on the C═C bond

covalent bond length/nm	0.134
bond dissociation enthalpy/kJ mol^{-1} at 298 K	612

A comparison of this data with that in Chapter 6 shows that the carbon–carbon double bond is considerably shorter and stronger than the carbon–carbon single bond.

Double bonds are found in the *alkenes* (formerly *olefins*), which are unsaturated hydrocarbons of general formula C_nH_{2n}. The word 'unsaturated' in this context means unsaturated with hydrogen; both alkenes and alkynes satisfy this description for they are capable of taking part in addition reactions with hydrogen. Although they do not occur naturally, the lower members of the homologous series are available in very large quantities by the cracking of certain petroleum fractions.

7.1 Nomenclature of alkenes

Alkenes with one double bond, if they are unbranched, are named by replacing the ending '-ane' of the corresponding alkane by '-ene'.

$$CH_3CH═CH_2 \qquad I \qquad \overset{4}{C}H_3\overset{3}{C}H_2\overset{2}{C}H═\overset{1}{C}H_2 \qquad II$$

Thus, compound I is called propene. The chain is numbered in such a direction that the carbon atom on which the double bond starts has the lowest possible locant (i.e. number). Compound II, for example, is called 1-butene; not 3-butene.

$$CH_2{=}CH_2 \quad III \qquad \overset{4}{C}H_2{=}\overset{3}{C}H{-}\overset{2}{C}H{=}\overset{1}{C}H_2 \quad IV$$

The trivial name of ethylene is retained by IUPAC for the first member of the homologous series (III). Compounds with two double bonds are referred to as *alkadienes*; those with three double bonds as *alkatrienes*, etc. For example, compound IV is 1,3-butadiene.

$$\underset{\overset{4}{C}H_3\overset{3}{C}H_2\overset{2}{C}{=}\overset{1}{C}H_2}{\overset{CH_2CH_3}{|}} \quad V \qquad \underset{CH_3C{=}CH_2}{\overset{CH_3}{|}} \quad VI$$

For branched chain alkenes, the chain on which the name is based must contain the double bond, even though it may not necessarily be the longest chain in the molecule. For example, compound V is called 2-ethyl-1-butene, even though the longest carbon chain possesses five atoms. The name of compound VI is 2-methylpropene; the trivial name of isobutene is no longer recognised.

7.2 Formation of the C═C bond

Apart from addition to the C≡C bond, all the basic methods of forming the C═C bond are elimination reactions. Such reactions often compete with substitution reactions and may be favoured by adjustment of the conditions — principally, by raising the temperature and altering the solvent.

Elimination methods

From $\overset{H}{\underset{}{\diagdown}}\overset{X}{\underset{}{\diagup}}$ $-C{-}C-$ *(X = Cl, Br or I) by the elimination of HX (§11.5)*

When an alkyl halide is treated with a base, substitution usually occurs to give an alcohol:

$$RX + HO^-(aq) = ROH + X^-$$

However, at higher temperatures, and with a strong base in concentrated solution in ethanol, dehydrohalogenation takes place to give an alkene:

$$\underset{H}{\overset{H}{\diagdown}}C{-}C\underset{H}{\overset{X}{\diagup}}R' + HO^-(alc) = \underset{H}{\overset{R}{\diagdown}}C{=}C\underset{H}{\overset{R'}{\diagup}} + X^- + H_2O$$

From $\overset{H}{\underset{}{\diagdown}}\overset{OH}{\underset{}{\diagup}}$ $-C{-}C-$ *by the elimination of H_2O (§12.5)*

In the presence of an acid catalyst, e.g. conc. H_2SO_4 or Al_2O_3, alcohols can be dehydrated to give either ethers or alkenes according to the severity of the conditions. Under mild conditions (relatively low temperature and limited acid) a substitution reaction leads to etherification:

$$2ROH \xrightarrow[\text{130 °C (403 K)}]{\text{conc. } H_2SO_4} R-O-R + H_2O$$

Under more drastic conditions, i.e. at a relatively high temperature and with excess acid, an alkene is formed in an elimination reaction:

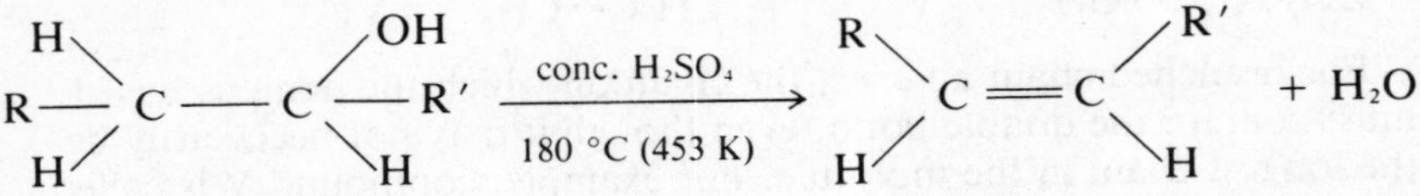

The mechanism in both cases involves protonation of the alcohol, and is often followed by the loss of water to give a carbonium ion. In etherification, the carbonium ion reacts with another alcohol molecule, but in the formation of alkenes the carbonium ion loses a proton.

From $\overset{X}{\underset{/}{-}}C\overset{}{\underset{}{-}}C\overset{X}{\underset{\backslash}{-}}$ *or* $\overset{}{\underset{/}{-}}C\overset{}{\underset{}{-}}C\overset{X}{\underset{\backslash}{-}}X$ *(X = Cl, Br or I) by the elimination of* X_2

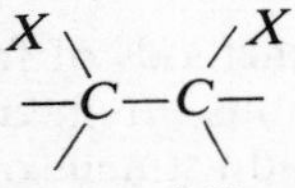

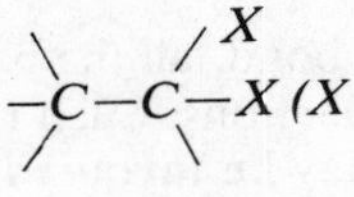

A double bond can be formed by eliminating two halogen atoms from a suitable dihalide by means of a metal — usually powdered zinc, e.g.

$$\left.\begin{array}{l} CH_3CHBrCH_2Br \\ \text{a 1,2-dihalide} \\ \text{or} \\ CH_3CH_2CHBr_2 \\ \text{a 1,1-dihalide} \end{array}\right\} \xrightarrow[\text{in methanol}]{\text{Zn dust}} CH_3CH{=}CH_2 + ZnBr_2$$

From $\overset{H}{\underset{/}{-}}C\overset{}{\underset{}{-}}C\overset{H}{\underset{\backslash}{-}}$ *by the elimination of* H_2 *(§6.4)*

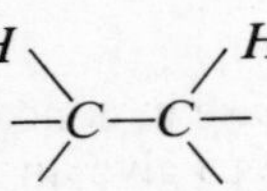

The cracking of alkanes at high temperatures is accompanied by dehydrogenation to alkenes. The method is of importance in the industrial synthesis of butenes and 1,3-butadiene (cf. steam cracking of naphtha), in which butane is passed over heated metal oxide catalysts.

Addition methods

The partial hydrogenation of an alkyne converts the carbon–carbon triple bond to a double bond, e.g.

$$CH{\equiv}CH \xrightarrow{H_2} CH_2{=}CH_2 \left[\xrightarrow{H_2} CH_3-CH_3\right]$$
$$\text{ethyne} \qquad \text{ethene} \qquad \text{ethane}$$

The addition can be stopped at the alkene stage by the use of a *Lindlar catalyst*, i.e. palladium, partly poisoned by lead, on calcium carbonate.

Partial halogenation, likewise, results in the establishment of a double bond. Partial chlorination is barely practicable, but partial bromination and iodination are not difficult to achieve. A variety of other reagents (e.g. HCl, HCN, ROH, RCOOH) also add across a $C\equiv C$ bond (§8.4).

7.3 General properties of the alkenes

In their physical characteristics the alkenes closely resemble the alkanes. Thus, the first few members of the series are gases, then come liquids of gradually increasing boiling temperature, and finally solids of gradually increasing melting temperature.

They have mild, sweet smells, and are non-polar compounds which are insoluble in water. They burn readily to give carbon dioxide and water, although with more yellow, smoky flames than the alkanes. This is a reflection of their higher C : H ratio.

7.4 Chemical properties of the C=C bond

The $C=C$ bond, in contrast to the $C-C$ bond, is characterised by a high reactivity. Alkenes undergo a considerable number of addition reactions across the double bond, as a result of which saturated *adducts*, i.e. addition products, are formed:

$$\underset{\text{alkene}}{\diagdown C = C \diagup} + \underset{\text{reagent}}{AB} = \underset{\text{adduct}}{\overset{A}{\diagdown} \overset{B}{\underset{\diagup}{C - C}} \diagdown}$$

Three principal mechanistic routes are possible, namely electrophilic addition, radical addition and nucleophilic addition. Nucleophilic addition is rare and will not be discussed here. Hydrogenation (i.e. the addition of hydrogen) is a special case in that it takes place by an adsorption mechanism.

Electrophilic addition

The electron-rich region between the carbon atoms of a double bond is bound to attract electrophilic reagents. These are generally polarised molecules, $\overset{\delta^+}{A}-\overset{\delta^-}{B}$, which act as electrophiles because of the partial positive charge on one of their atoms. An example is $\overset{\delta^+}{H}-\overset{\delta^-}{Br}$. Certain non-polar molecules, e.g. Br—Br, can also act as electrophiles because they become polarised under the prevailing conditions.

The π bond of an alkene (§1.2) is readily polarised at the approach of such a reagent, and a weak coordinate bond can be formed by the donation of a pair of electrons from the alkene to the electrophilic

106

reagent. Because π electrons are involved in its formation, the resulting complex is known as a *π-complex* (Fig. 7.1).

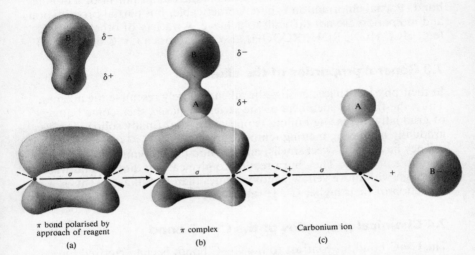

π bond polarised by approach of reagent

(a)

π complex

(b)

Carbonium ion

(c)

Fig. 7.1 (a) The approach of an electrophile to an alkene. (b) Coordination of π electrons of the C=C bond to the electrophile to give a π-complex. (c) Formation of a carbonium ion.

The reaction can be represented by the equation:

$$\underset{}{>}C=C\underset{}{<} \quad + \quad \overset{\delta^+}{A}—\overset{\delta^-}{B} \quad \longrightarrow \quad \underset{\pi\text{-complex}}{\overset{\overset{\textstyle\overset{\delta^+}{A}—\overset{\delta^-}{B}}{\uparrow}}{>C=C<}}$$

Slowly, the π-complex changes into a carbonium ion. This happens by the breaking of one bond, A—B, and the formation of another, A—C. At the same time, an anion B⁻ is released into the solvent:

$$\underset{\pi\text{-complex}}{\overset{\overset{\textstyle\overset{\delta^+}{A}—\overset{\delta^-}{B}}{\uparrow}}{>C=C<}} \quad \xrightarrow{\text{slowly}} \quad \underset{\text{carbonium ion}}{—\overset{\overset{\textstyle A}{|}}{C}—\overset{\oplus}{C}<} \quad + \quad B^-$$

During this reaction, the covalent bond A—B in the reagent is effectively broken into two oppositely charged ions, A^+ and B^-. This is an example of heterolytic cleavage. There is no evidence to suggest that A^+ exists in the free state before bonding to a carbon atom.

Reaction is completed by the rapid addition of B^- to the carbonium ion:

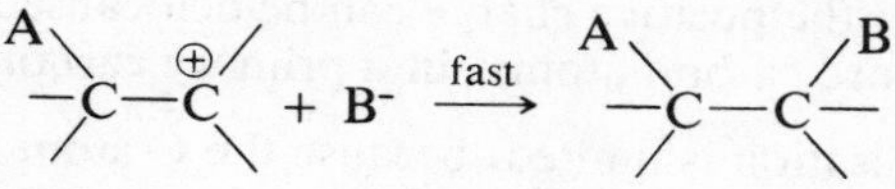

$$-\overset{|}{\underset{|}{C}}-\overset{\oplus}{\underset{|}{C}}\diagup + B^- \xrightarrow{\text{fast}} -\overset{A}{\underset{|}{C}}-\overset{B}{\underset{|}{C}}-$$

In principle, any reagent which satisfies the three conditions for an electrophile (§4.1) can add across a C=C bond. The commonest, which we shall now study in detail, are the halogens (e.g. Br_2), various oxidising agents (e.g. O_3) and strong protic acids (e.g. HBr and H_2SO_4). Weak acids, such as HCN or CH_3COOH, do not react with alkenes because of the difficulty of heterolytic cleavage into protons and anions.

HF, HCl, HBr and HI

The addition of hydrogen halides follows the general pattern shown above:

$$CH_2{=}CH_2 + \overset{\delta^+}{H}{-}\overset{\delta^-}{X} \xrightarrow{\text{fast}} CH_2{=}CH_2 \overset{\overset{\displaystyle\uparrow}{\overset{\delta^+}{H}-\overset{\delta^-}{X}}}{\xrightarrow{\text{slow}}} CH_3{-}\overset{\oplus}{C}H_2 + X^- \xrightarrow{\text{fast}} CH_3CH_2X$$

ethene   π-complex   carbonium ion   ethyl halide

The order of reactivity of the hydrogen halides towards alkenes is HI > HBr > HCl ≫ HF, which corresponds to the order of their strength as acids. Concentrated aqueous solutions of HI and HBr absorb alkenes at room temperature, but concentrated hydrochloric acid attacks only the higher alkenes. Gaseous hydrogen chloride at a temperature of 200 °C (473 K) can be used to convert the lower alkenes to alkyl chlorides, e.g. $CH_2{=}CH_2$ to CH_3CH_2Cl. Hydrogen fluoride gas adds to alkenes only under pressure.

The addition of a hydrogen halide to propene (or any other unsymmetrical alkene) could conceivably give two products, e.g.

$$CH_3CH{=}CH_2 + HBr \diagup^{\displaystyle \rightarrow CH_3CH_2CH_2Br}_{\displaystyle \rightarrow CH_3CHBrCH_3}$$

                                                                  1-bromopropane

                                                                  2-bromopropane

In practice, electrophilic addition leads exclusively to 2-bromopropane. The reason hinges on the difference between the stabilities of primary, secondary and tertiary carbonium ions. In a *primary carbonium ion* the charged carbon atom is attached to one hydrocarbon radical; in a *secondary carbonium ion* it is attached to two, and in a *tertiary carbonium ion* to three.

It is a law of physics that the stability of any charged system is increased by dispersal of the charge. Thus, an electrical charge on a brass sphere does not remain concentrated at one locality, but becomes evenly distributed over the entire sphere. Carbonium ions, likewise, will gain in stability if the positive charge can be delocalised, i.e. shared among two or more carbon atoms. In a primary carbonium ion the scope for delocalisation is limited, because the $\overset{\oplus}{C}$ atom is joined to only one other carbon atom, but in secondary and especially in tertiary carbonium ions dispersal of charge can occur to a greater extent.

| primary carbonium ion | secondary carbonium ion | tertiary carbonium ion |

Increasing delocalisation of positive charge

→

Increasing stability and increasing ease of formation

Charge sharing is possible with neighbouring carbon atoms but not with hydrogen atoms because alkyl groups have a positive inductive effect, i.e. an electron releasing effect (§1.4).

Tertiary carbonium ions are thus formed most easily, and are relatively common intermediates in reaction mechanisms. Secondary carbonium ions are formed less readily, and primary carbonium ions are formed only if there is no alternative mechanism.

Let us now reconsider the addition of hydrogen bromide to propene.

$$CH_3CH{=}CH_2 + H^+ \rightarrow CH_3CH_2\overset{\oplus}{C}H_2 \xrightarrow{Br^-} CH_3CH_2CH_2Br$$

primary carbonium ion   1-bromopropane

$$CH_3CH{=}CH_2 + H^+ \rightarrow CH_3\overset{\oplus}{C}HCH_3 \xrightarrow{Br^-} CH_3CHBrCH_3$$

secondary carbonium ion   2-bromopropane

2-Bromopropane is obtained because its formation involves a secondary carbonium ion, whereas the formation of 1-bromopropane would proceed via a relatively unstable primary carbonium ion.

Other unsymmetrical reagents, e.g. H_2SO_4, act in the same way, a fact which is summarised by *Markownikoff's rule*: 'In the electrophilic addition of protic acids to unsymmetrically substituted alkenes, the hydrogen atom adds to the carbon atom of the double bond bearing the most hydrogen'.

Markownikoff's rule applies only when electrophilic addition occurs. In cases of free radical addition, e.g. the addition of hydrogen bromide in the presence of a peroxide catalyst, the rule is reversed (see below).

H_2SO_4

Chemical reactions occur when alkenes are absorbed by sulphuric acid. The products depend on whether the sulphuric acid is anhydrous or mixed with water. In both cases a carbonium ion is formed at first, but various nucleophiles compete for the carbonium ion.

Anhydrous sulphuric acid

Alkenes are absorbed at an appreciable rate by anhydrous or highly concentrated sulphuric acid. *Monoalkyl sulphates*, which are half-esters of sulphuric acid (§12.5), are formed by the electrophilic addition of H_2SO_4 across the C=C bond, e.g.

$$\overset{\delta^+}{H}-\overset{\delta^-}{O}-SO_2-OH$$

$$CH_2=CH_2 + \overset{\delta^+}{H}-\overset{\delta^-}{O}-SO_2-OH \rightarrow CH_2\overset{\uparrow}{=}CH_2$$

i.e. H_2SO_4 π-complex

$$\rightarrow CH_3-\overset{\oplus}{C}H_2 + \overset{\ominus}{O}-SO_2-OH \rightarrow CH_3CH_2O-SO_2-OH$$

i.e. HSO_4^- ethyl hydrogen sulphate

With unsymmetrical alkenes Markownikoff's rule is obeyed, e.g.

$$CH_3CH=CH_2 + H_2SO_4 = (CH_3)_2CHO-SO_2-OH$$

isopropyl hydrogen sulphate

Monoalkyl sulphates can be hydrolysed by hot water to give alcohols, e.g.

$$CH_3CH_2O-SO_2-OH + H_2O = CH_3CH_2OH + H_2SO_4$$

Dialkyl sulphates, which are full esters of sulphuric acid, are formed as by-products whenever alkenes react with anhydrous sulphuric acid. The reason is that monoalkyl sulphates are electrophilic reagents and attack alkenes in exactly the same way as sulphuric acid, e.g.

$$\overset{\delta^+}{H}-\overset{\delta^-}{O}-SO_2-OCH_2CH_3$$

$$CH_2=CH_2 + \overset{\delta^+}{H}-\overset{\delta^-}{O}-SO_2-OCH_2CH_3 \rightarrow CH_2\overset{\uparrow}{=}CH_2$$

$$\rightarrow CH_3-\overset{\oplus}{C}H_2 + \overset{\ominus}{O}-SO_2-OCH_2CH_3 \rightarrow CH_3CH_2-O-SO_2-O-CH_2CH_3$$

diethyl sulphate

Aqueous sulphuric acid

In the presence of water, and at a high temperature, sulphuric acid converts alkenes into alcohols directly, i.e. not via monoalkyl sulphates. This is because a carbonium ion is attacked by water in preference to HSO_4^- ions, e.g.

$$CH_3\overset{\oplus}{C}H_2 + H_2O \rightarrow CH_3CH_2\overset{\overset{\displaystyle \overset{\oplus}{O}H_2}{|}}{} \rightarrow CH_3CH_2OH + H^+$$

protonated   ethanol
ethanol

This forms the basis of the manufacture of alcohols by the acid catalysed hydration of alkenes (§12.3). Water alone will not bring about the conversion because its acidity is far too low. The catalyst may be either a protic acid, e.g. H_2SO_4 or H_3PO_4, or a Lewis acid, e.g. WO_3.

When alcohols are prepared in this way ethers always appear as by-products, because alcohols (like water) can react with carbonium ions, e.g.

$$CH_3\overset{\oplus}{C}H_2 + CH_3CH_2OH \rightarrow CH_3CH_2\overset{\overset{\displaystyle \overset{\oplus}{H}OCH_2CH_3}{|}}{} \rightarrow CH_3CH_2-O-CH_2CH_3 + H^+$$

protonated   ethoxyethane
ethoxyethane   (diethyl ether)

Cl_2 and Br_2

Chlorine or bromine will react readily with alkenes at room temperature. For example, when ethene is bubbled through a solution of bromine in a suitable solvent, such as tetrachloromethane, the red colour of the bromine is soon discharged.

$$CH_2=CH_2 + Br_2 = CH_2BrCH_2Br$$

1,2-dibromoethane

Similarly, if cyclohexene (a liquid) is used the red colour again disappears, and does so *without the evolution of hydrogen bromide*. This indicates that the reaction with bromine must be addition; not substitution.

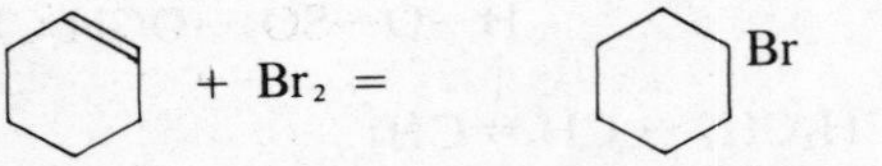

1,2-dibromocyclohexane

The reason that molecular bromine, a non-polar substance, behaves as an electrophilic reagent is not immediately obvious. Indeed, pure halogens do not react with pure alkenes except in the presence of light, when free radical addition can take place. However, the bromine molecule can become temporarily polarised by means of:

(i)   a polar solvent, such as ethanol (§1.4);
(ii) traces of moisture;
(iii) glass surfaces;
(iv) catalysts such as anhydrous aluminium bromide.

Once the bromine molecule has become polarised, reaction with the alkene can occur in the normal way, e.g.

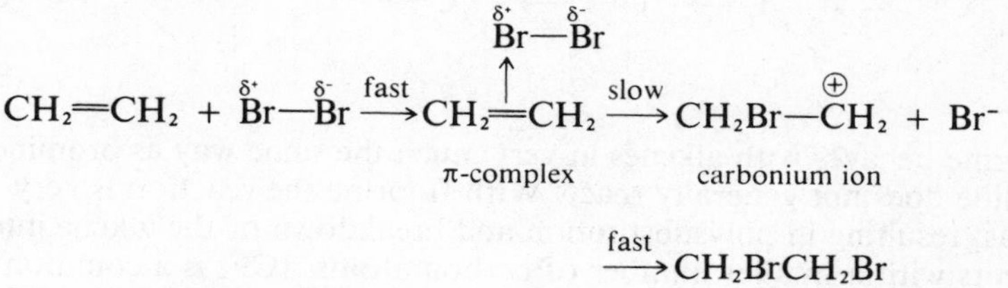

$$CH_2{=}CH_2 + \overset{\delta^+}{Br}{-}\overset{\delta^-}{Br} \xrightarrow{\text{fast}} \underset{\pi\text{-complex}}{CH_2{=}CH_2 \cdot \overset{\overset{\overset{\delta^+}{Br}{-}\overset{\delta^-}{Br}}{\uparrow}}{}} \xrightarrow{\text{slow}} \underset{\text{carbonium ion}}{CH_2Br{-}\overset{\oplus}{C}H_2} + Br^-$$

$$\xrightarrow{\text{fast}} CH_2BrCH_2Br$$

The positive bromine ion, Br^+, is not involved in this mechanism. Not only is there no experimental evidence for its existence, but the first ionisation enthalpy of bromine is very high (1 140 kJ mol^{-1} at 298 K) and comparable with the values for noble gases. The Br^- ion, however, *is* involved, as may be proved by conducting the reaction in the presence of chloride ions. Br^- and Cl^- ions compete for the carbonium ion, and the product is a mixture of CH_2BrCH_2Br and CH_2BrCH_2Cl.

This mechanism, although adequate for most purposes, does not explain the fact that the reaction between bromine and an alkene usually takes place exclusively by a *trans* addition, i.e.

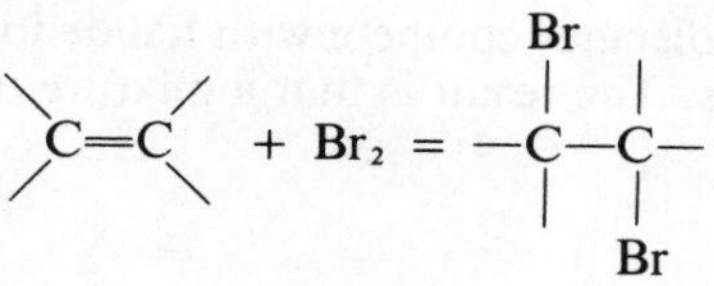

$$\underset{/}{\overset{\backslash}{}}C{=}C\underset{\backslash}{\overset{/}{}} + Br_2 = \underset{\overset{|}{Br}}{\overset{\overset{Br}{|}}{-C}}{-}\underset{|}{\overset{|}{C}}{-}$$

This principle has been established by studying reactions in which the product either possesses two asymmetric carbon atoms, or cannot undergo free rotation because of a ring structure. In simple cases, such as the reaction between bromine and ethene, the stereochemistry of addition is not apparent. (There is only one form of 1,2-dibromoethane.)

The observation is explained by the theory that the carbonium ion undergoes rearrangement to a cyclic form known as a *bromonium ion*:

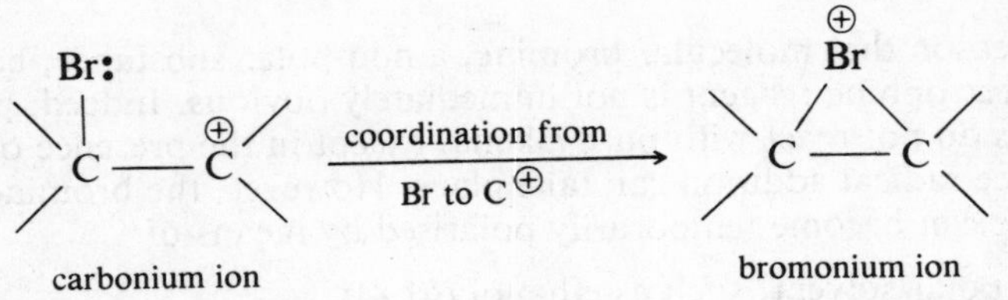

carbonium ion → bromonium ion

In the final stage of the addition the bulky bromine atom hinders the approach of the Br^- ion to one side of the bromonium ion, and leads to attack 'from the rear'. The complete mechanism is thus as follows:

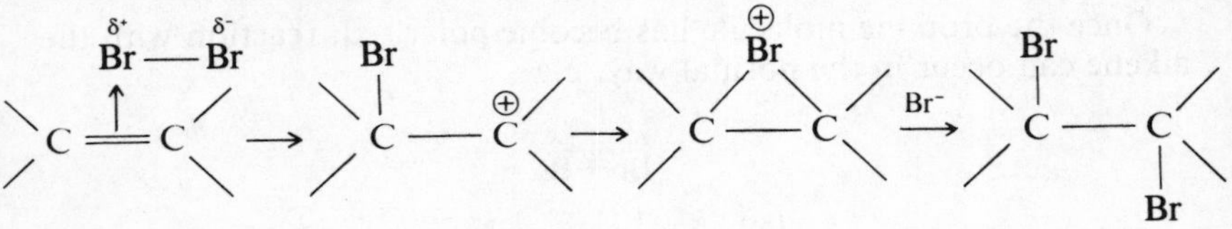

Chlorine behaves with alkenes in very much the same way as bromine, but iodine does not generally react. With fluorine the reaction is very vigorous, resulting in polysubstitution and breakdown of the alkene into fragments with a smaller number of carbon atoms. (CF_4 is a common product.) A consideration of the enthalpy changes involved helps to explain these differences between the halogens (Table 7.1).

Although Table 7.1 is instructive, we must remember that a reliable guide to the likelihood of a reaction taking place is provided only by the free energy change, ΔG; not by the enthalpy change, ΔH.

Cl_2 and Br_2 in aqueous or ethanolic solution

The preceding discussion has rested on the assumption that the alkene is reacting with a solution of chlorine or bromine in an inert solvent, such as tetrachloromethane. However, if the halogen is dissolved in a solvent which can act as an electrophilic reagent, such as water or ethanol, the situation is rather different, because in the final stage of the mechanism the water or ethanol molecules compete with halide ions for coordination to the carbonium ions. The result is that a mixture of adducts is obtained, e.g.

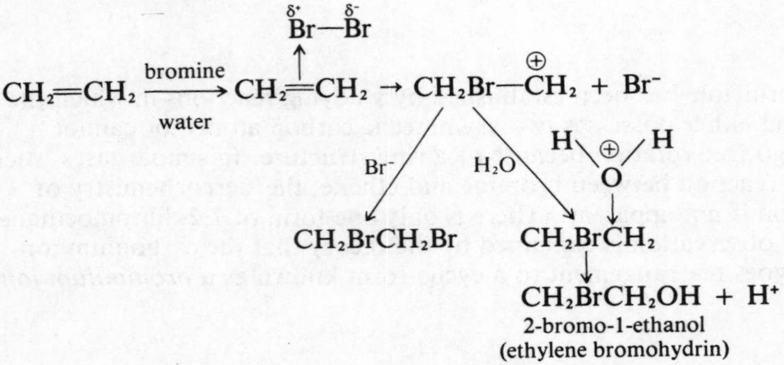

2-bromo-1-ethanol
(ethylene bromohydrin)

Table 7.1 Standard enthalpy changes on the addition of halogens to a $C = C$ bond

Reagent	Bond dissociation enthalpy/kJ mol⁻¹ ΔH_1	Enthalpy change associated with $C = C \rightarrow C - C$ / kJ mol⁻¹ ΔH_2†	Enthalpy change on forming new bonds/ kJ mol⁻¹ ΔH_3	Overall enthalpy change/kJ mol⁻¹ $\Delta H_1 + \Delta H_2 + \Delta H_3$	Observation
F_2	$+158$	$+264$	$2 \times (-484)$ (2 C—F) $= -968$	-546	polyfluorination
Cl_2	$+242$	$+264$	$2 \times (-338)$ (2 C—Cl) $= -676$	-170	chlorine addition
Br_2	$+193$	$+264$	$2 \times (-276)$ (2 C—Br) $= -552$	-95	bromine addition
I_2	$+151$	$+264$	$2 \times (-238)$ (2 C—I) $= -476$	-61	iodine addition is very difficult

†The figure of $+264$ kJ mol⁻¹ represents the difference between the bond dissociation enthalpies of $C = C$ and $C - C$.

Note: These calculations strictly apply only to free radical addition in the gaseous state, as they are based on the assumption that the reagent undergoes homolytic fission.

If the bromide ion concentration is low, the main product is 2-bromo-1-ethanol. The effect is as though the ethene were attacked by hypobromous acid,

$$CH_2{=}CH_2 + H\overset{\delta^-}{O}{-}\overset{\delta^+}{Br} = CH_2BrCH_2OH$$

but because of the low concentration of hypobromous acid in bromine water it is unlikely that HBrO molecules are involved in the mechanism.

With chlorine water the principal product is 2-chloro-1-ethanol (ethylene chlorohydrin), CH_2ClCH_2OH.

When reaction occurs with an unsymmetrical alkene, the *halogen atom* becomes attached to the carbon atom bearing the most hydrogen. This is consistent with the theoretical basis of Markownikoff's rule, namely that addition proceeds via the most stable carbonium ion, e.g.

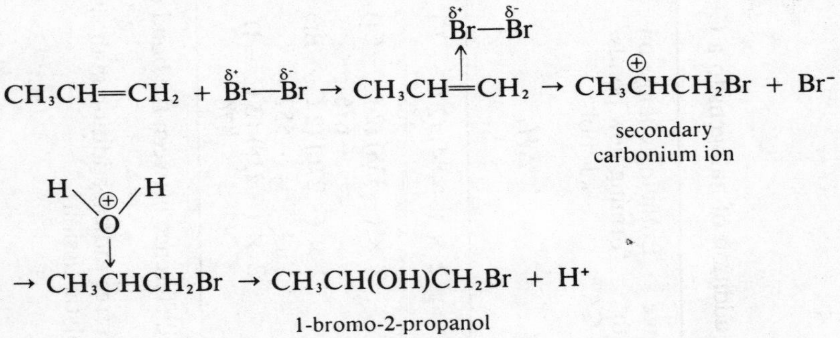

$$CH_3CH{=}CH_2 + \overset{\delta^+}{Br}{-}\overset{\delta^-}{Br} \rightarrow CH_3CH{=}CH_2 \rightarrow CH_3\overset{\oplus}{C}HCH_2Br + Br^-$$

secondary carbonium ion

$$\rightarrow CH_3\overset{}{C}HCH_2Br \rightarrow CH_3CH(OH)CH_2Br + H^+$$

1-bromo-2-propanol

When alkenes react with halogens in alcoholic solution the same principles apply and the product is a mixture of a dihalide and a haloether, e.g.

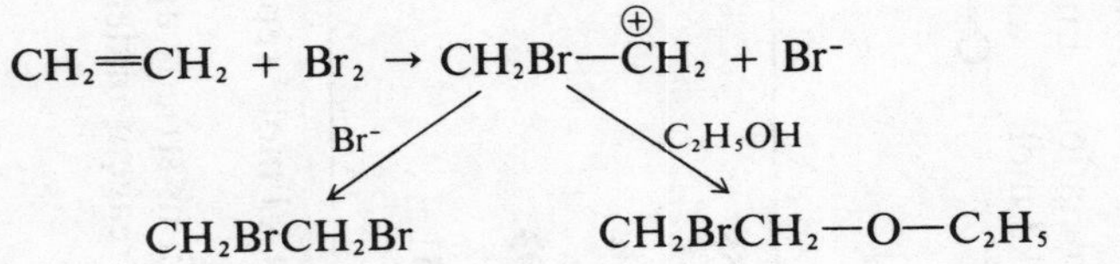

$$CH_2{=}CH_2 + Br_2 \rightarrow CH_2Br{-}\overset{\oplus}{C}H_2 + Br^-$$

Br^- C_2H_5OH

$$CH_2BrCH_2Br \qquad CH_2BrCH_2{-}O{-}C_2H_5$$

Oxidation

Alkenes are very susceptible to oxidation, by the electrophilic addition of an oxidising agent to the C=C bond. The nature of the product depends very much on the reagent, thus:

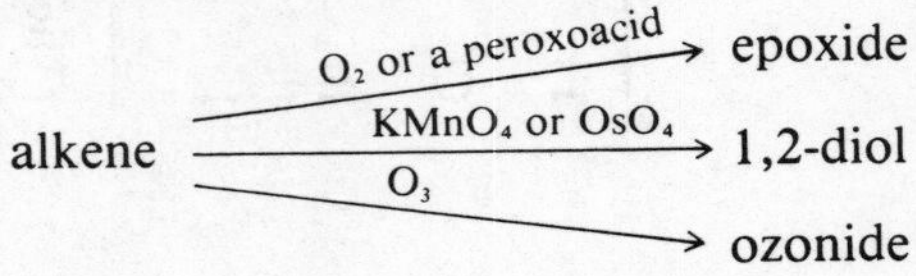

alkene
- $\xrightarrow{\text{O}_2 \text{ or a peroxoacid}}$ epoxide
- $\xrightarrow{\text{KMnO}_4 \text{ or OsO}_4}$ 1,2-diol
- $\xrightarrow{\text{O}_3}$ ozonide

Secondary changes may occur, so that a variety of final products is possible.

O₂ or peroxoacids

Reaction with oxygen occurs at high temperatures and pressures in the presence of a silver catalyst. The products, known as *epoxides*, are cyclic ethers, e.g.

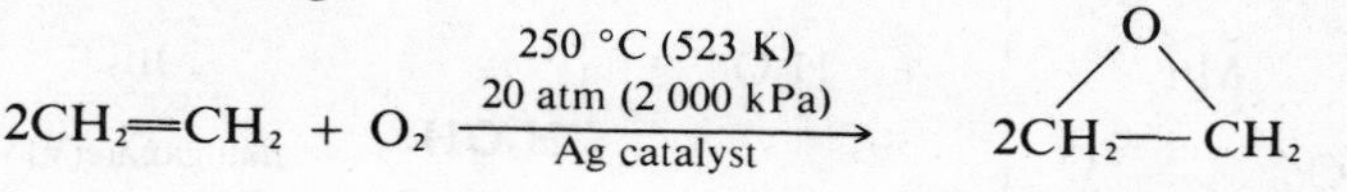

$$2CH_2{=}CH_2 + O_2 \xrightarrow[\text{Ag catalyst}]{\substack{\text{250 °C (523 K)} \\ \text{20 atm (2 000 kPa)}}} 2CH_2{-}CH_2 \text{ (with O bridging)}$$

oxirane (ethylene oxide)

This reaction is of commercial importance. Oxirane is used partly as a fumigant for preserving food and tobacco and partly as a chemical intermediate, especially in the manufacture of 1,2-ethanediol.

Reactions with peroxoacids (e.g. perbenzoic acid, $C_6H_5C\overset{O}{\underset{O-O-H}{}}$) also give epoxides when conducted in inert solvents such as ethoxyethane (diethyl ether) or trichloromethane (chloroform). If the reaction is carried out in aqueous solution the epoxide undergoes acidic hydrolysis to give a 1,2-diol; the latter may subsequently become esterified.

KMnO₄ or OsO₄

Both potassium permanganate and osmium(VIII) oxide bring about the *hydroxylation* of alkenes, i.e. they result in the addition of two hydroxyl groups across the C=C bond. The products are thus 1,2-diols, commonly called 'glycols', e.g.

$$\begin{array}{c} CH_2 \\ \| \\ CH_2 \end{array} \xrightarrow[\text{KMnO}_4 \text{ solution}]{\text{cold dilute alkaline}} \begin{array}{c} CH_2OH \\ | \\ CH_2OH \end{array}$$

1,2-ethanediol (ethylene glycol)

It is considered likely that two reactions occur. In the first, the permanganate ion adds to the alkene to give an anion which is, in essence, a cyclic ester:

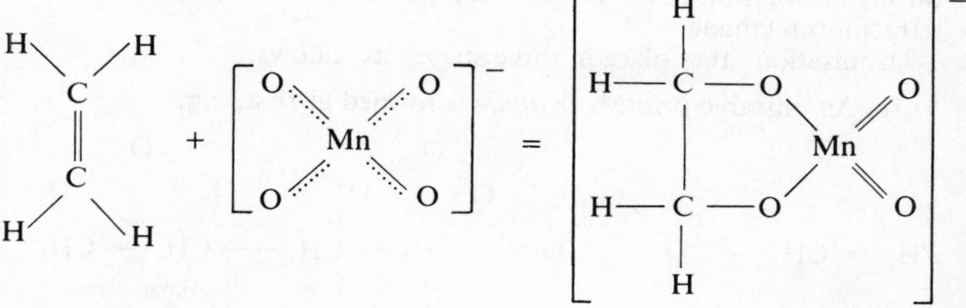

In this reaction the permanganate ion acts as an electrophile, despite its negative charge, because the high oxidation number of manganese ($+7$) causes the electrons of the Mn—O bonds to be drawn towards the manganese atom.

In the second reaction the cyclic ester is hydrolysed:

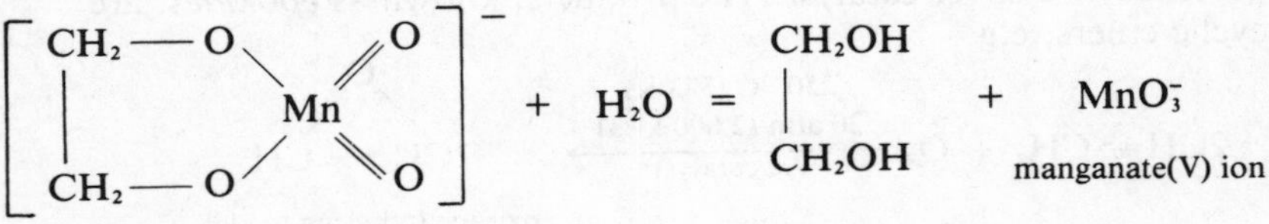

$$
\left[
\begin{array}{c}
\text{CH}_2\text{—O} \\
| \\
\text{CH}_2\text{—O}
\end{array}
\underset{\text{Mn}}{\diagup\diagdown}
\begin{array}{c}
\diagdown\text{O} \\
\diagup\text{O}
\end{array}
\right]^{-}
+ \text{H}_2\text{O} =
\begin{array}{c}
\text{CH}_2\text{OH} \\
| \\
\text{CH}_2\text{OH}
\end{array}
+ \underset{\text{manganate(V) ion}}{\text{MnO}_3^{-}}
$$

Disproportionation of manganese(V) leads to the formation of a green solution of a manganate(VI) and a brown precipitate of manganese(IV) oxide.

Yields of 1,2-diols are good when osmium(VIII) oxide is used as the oxidant, but are poor for potassium permanganate because 1,2-diols undergo oxidative cleavage to give aldehydes and ketones (§6.4). Ketones formed in this way remain as such, for they are not easily oxidised, but aldehydes undergo further oxidation by the permanganate to give carboxylic acids.

O_3

In the presence of traces of water, or on contact with glass apparatus, ozone (trioxygen) becomes polarised (cf. bromine in similar circumstances) and acts as an electrophilic reagent. It will attack the C=C bond of an alkene to give an *ozonide*, e.g.

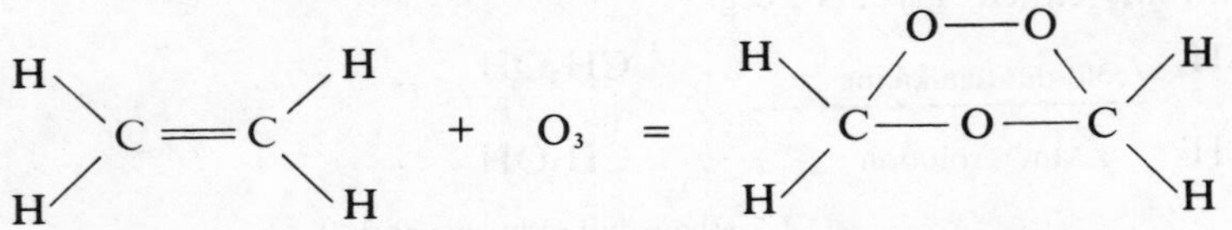

ethene ozonide

The addition of ozone, known as *ozonisation*, is brought about at room temperature by bubbling ozonised oxygen, from an ozoniser, through a solution of an alkene in an inert solvent such as hexane or tetrachloromethane.

Ozonisation takes place in three stages, as follows.

(i) An unstable *primary ozonide* is formed at first, e.g.

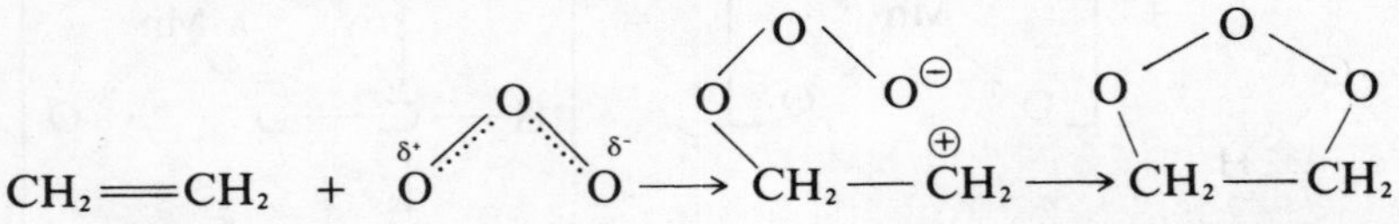

primary ozonide

(ii) Rapid decomposition of the primary ozonide gives a molecule of an aldehyde (or a ketone) and a *zwitterion*, i.e. an ion possessing two charges, one positive and one negative:

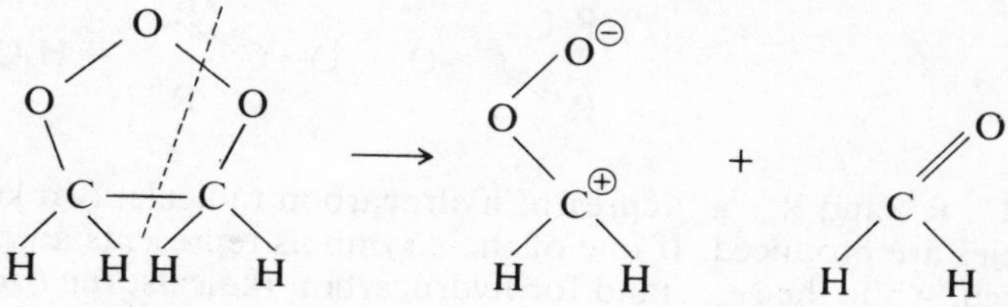

zwitterion   methanal

(iii) The zwitterion and the aldehyde then react together to give the ozonide:

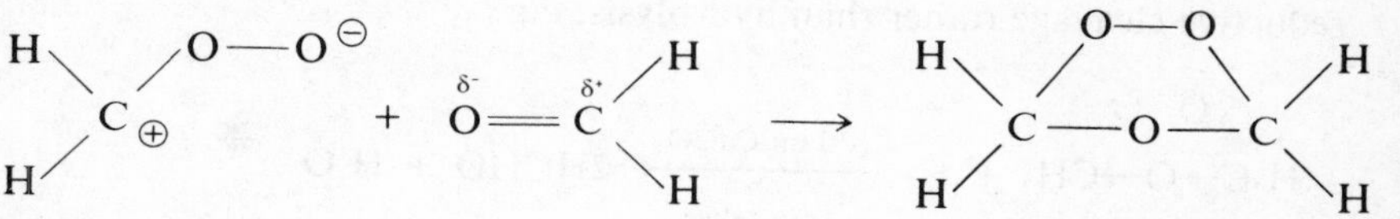

Ozones are likely to be contaminated by *polymeric ozonides*, formed by the head-to-tail addition of zwitterions:

$$\overset{\oplus}{C}H_2{-}O{-}\overset{\ominus}{O} + \overset{\oplus}{C}H_2{-}O{-}\overset{\ominus}{O} \rightarrow \overset{\oplus}{C}H_2{-}O{-}O{-}CH_2{-}O{-}\overset{\ominus}{O} \text{ etc}$$

polymeric ozonide

Polymeric ozonides are undesirable, partly because they are highly explosive and partly because they are resistant to hydrolysis (see below). Their formation can be minimised by conducting ozonisation in a strongly polar solvent such as methanol or ethanoic acid.

To avoid the risk of explosion, ozonides are seldom isolated. They are usually decomposed by the addition of water, e.g.

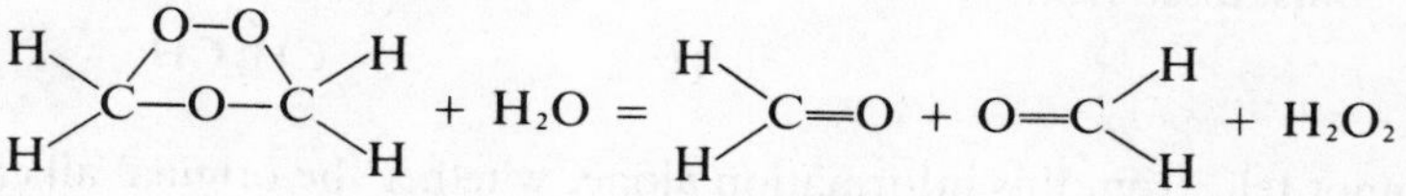

two molecules of methanal

Ozonisation, followed by hydrolysis, is known as *ozonolysis*. Ozonolysis can give aldehydes or ketones or a mixture of the two, as shown by the following general scheme:

$$\underset{R'}{\overset{R}{>}}C=C\underset{R'''}{\overset{R''}{<}} \xrightarrow{O_3} \underset{R'}{\overset{R}{>}}C\underset{O}{\overset{O-O}{\underset{|}{-}}}C\underset{R'''}{\overset{R''}{<}} \xrightarrow{H_2O}$$

$$\underset{R'}{\overset{R}{>}}C=O + O=C\underset{R'''}{\overset{R''}{<}} + H_2O_2$$

If R, R′, R″ and R‴ all represent hydrocarbon radicals, two ketone molecules are produced. If one of these symbols represents an atom of hydrogen, while the rest stand for hydrocarbon radicals, the ozonolysis products are one molecule of aldehyde and one of ketone. If R or R′ is H, and R″ or R‴ is also H, the products are two aldehyde molecules.

The hydrogen peroxide formed during the hydrolysis of an ozonide is an oxidising agent and is liable to convert aldehydes into carboxylic acids. This can be minimised by hydrolysing the ozonide in the presence of a reducing agent, e.g sodium hydrogensulphite or zinc dust, which destroys the peroxide. Alternatively, the ozonide can be subjected to reductive cleavage rather than hydrolysis, e.g.

$$H_2C\underset{O}{\overset{O-O}{\underset{}{-}}}CH_2 + H_2 \xrightarrow[\text{as catalyst}]{\text{Pd on CaCO}_3} 2HCHO + H_2O$$

The ozonolysis of alkenes provides a means of establishing their structures. The products have merely to be separated and identified, and their formulae written down in such a way that the carbonyl groups face each other. Drawing in the double bond and deleting the oxygen atoms then gives the alkene. Consider, for example, an unknown alkene which, on ozonolysis, gives propanal and 2-butanone:

$$\underset{H}{\overset{CH_3CH_2}{>}}C=O \quad \text{and} \quad O=C\underset{CH_2CH_3}{\overset{CH_3}{<}}$$

must arise from

$$\underset{H}{\overset{CH_3CH_2}{>}}C=C\underset{CH_2CH_3}{\overset{CH_3}{<}}$$

We cannot tell, from this information alone, whether the original alkene is a *cis* or *trans* isomer.

Radical addition

Under conditions which favour the formation of free radicals, certain reagents add to the C=C bond by a free radical chain reaction. As with radical substitution (§10.2), the mechanism involves three stages: initiation, propagation and termination. The commonest reagents to act in this way are chlorine, bromine and hydrogen bromide.

Cl_2 and Br_2

In dry non-polar solvents (e.g. tetrachloromethane), and in the absence of Lewis acid catalysts such as aluminium chloride or bromide, chlorine and bromine molecules are not polarised and are unlikely to add to the C=C bond by an electrophilic mechanism. Because the reactions that occur under such conditions are accelerated by light and catalysts (e.g. dibenzoyl peroxide) which favour the formation of free radicals, it is believed that the reactions take place mainly by a free radical mechanism.

Initiation   Ultraviolet light (or a peroxide catalyst) promotes the homolytic fission of the halogen molecule into two halogen free radicals, e.g.

$$Br_2 \rightarrow 2Br\cdot$$

Propagation   As for radical substitution there are two propagation reactions, in each of which a free radical is consumed while another is formed. In only *one* of these reactions does a molecule of product appear. (Addition reactions, in contrast to substitution, give only one product.)

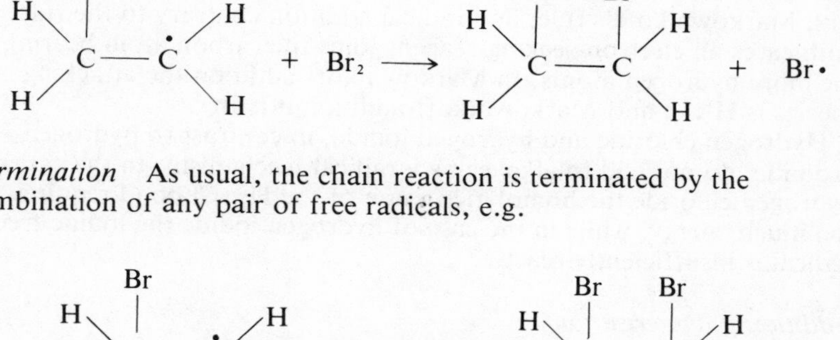

Termination   As usual, the chain reaction is terminated by the combination of any pair of free radicals, e.g.

HBr

The radical addition of hydrogen bromide to alkenes in the presence of a peroxide catalyst is noteworthy in that it is *anti-Markownikoff*, i.e. it takes place contrary to Markownikoff's rule. This is known as the *Kharasch effect* or the *peroxide effect*.

Initiation   A radical R• from the peroxide reacts with hydrogen bromide to give a bromine free radical:

R• + HBr → RH + Br•

Propagation

Reaction 1   Br• + CH_3CH=CH_2 → $CH_3\overset{\bullet}{C}HCH_2Br$

a *secondary* free radical

Reaction 2 $CH_3\overset{\bullet}{C}HCH_2Br$ + HBr → $CH_3CH_2CH_2Br$ + Br•

1-bromopropane

Termination   By the combination of any pair of free radicals.

In reaction 1 of the propagation stage, the formation of a primary free radical, $CH_3CHBr\overset{\bullet}{C}H_2$, which would lead to 2-bromopropane in accordance with Markownikoff's rule, is discounted on grounds of instability. The order of stability of free radicals is the same as that of carbonium ions, i.e. tertiary > secondary > primary, although the actual differences in stability are less marked for free radicals than for carbonium ions.

It can be argued that there is basically little difference between radical addition and electrophilic addition to alkenes. Both free radicals and electrophiles are attracted to the π electrons of a double bond because they are deficient in electrons. If we accept this similarity, we can see that there is no real conflict between electrophilic addition in accordance with Markownikoff's rule, and radical addition contrary to the rule. In both cases an electron-seeking reagent joins the carbon atom bearing the more hydrogen atoms. In Markownikoff addition the attacking species is H^+; in anti-Markownikoff addition it is Br•.

Hydrogen chloride and hydrogen iodide, in contrast to hydrogen bromide, do not add to alkenes by a radical mechanism. In the case of hydrogen chloride the homolytic fission of the H—Cl bond requires too much energy, while in the case of hydrogen iodide the iodine free radical is insufficiently reactive.

Addition polymerisation

Addition polymerisation is the principal way of making plastics. It is not the only way; polyesters and polyamides, for instance, are prepared by *condensation polymerisation*.

In the presence of a catalyst, molecules of simple alkenes or substituted alkenes can join together to give larger molecules. The alkenic starting material is known as a *monomer*, the product is a *polymer*, and the addition reaction by which the polymer is formed is referred to as *polymerisation*. Some of the commonest monomers are ethene, propene, chloroethene (vinyl chloride), phenylethene (styrene) and propenenitrile (acrylonitrile). In the discussion which follows we shall use the general formula $CH_2\text{=}CHX$.

If only two molecules of monomer become joined together the product is called a *dimer*:

$$CH_2\text{=}CHX + H/CH\text{=}CHX = CH_3CHXCH\text{=}CHX$$

Three molecules of the monomer give one of the *trimer*:

$$CH_3CHXCH\text{=}CHX + CH_2\text{=}CHX = CH_3CHXCH_2CHXCH\text{=}CHX$$

Four molecules of the monomer give one of the *tetramer*, and so on.

These 'low polymers' are gases, liquids or solids with low melting temperatures and no plastic properties. However, if a large number of molecules (500 or more) join together, the product is a solid with the familiar properties of a plastic and is referred to as a 'high polymer'. Its formation can be represented by the equation:

$$nCH_2\text{=}CHX = (-CH_2-CHX-)_n$$

The relative molecular mass of the product depends on the conditions of polymerisation, namely temperature, pressure, catalyst, and the presence or absence of light. Perhaps the most important factor is the nature of the catalyst. Ethene, for instance, when polymerised in the presence of oxygen, will give poly(ethene) only at pressures above 1 000 atm (100 000 kPa), but with an organometallic *Ziegler catalyst*, e.g. a pentyllithium−titanium(IV) chloride complex, the reaction may be conducted at atmospheric pressure.

Addition polymerisation proceeds by a chain reaction which may be propagated by either free radicals or ions depending on the type of catalyst that is employed. In the presence of a peroxide catalyst it occurs by the former route. To illustrate the principle we shall now consider the polymerisation of phenylethene (styrene) in the presence of dibenzoyl peroxide.

Initiation   Dibenzoyl peroxide decomposes at temperatures above 70 °C (343 K) to produce benzoate and phenyl free radicals, represented by Ar· in subsequent discussion:

$$C_6H_5CO-O-O-COC_6H_5 \rightarrow 2C_6H_5COO\cdot$$

dibenzoyl peroxide   benzoate free radical

$$C_6H_5COO\cdot \rightarrow C_6H_5\cdot + CO_2$$

phenyl free radical

Strictly speaking, the dibenzoyl peroxide is not a catalyst, because it enters permanently into the polymer chain and is not regenerated at the end of the reaction. It is better referred to as an *initiator*, because it starts the chain reaction.

Propagation The propagation stage of a radical polymerisation, in contrast to that of a radical addition, does not involve two alternating reactions. Instead, a whole series of reactions occurs, in each of which a molecule of the monomer joins a radical to give a longer chain radical. Product molecules are not formed at the propagation stage, but appear only at termination.

First: $Ar\cdot + CH_2{=}CHPh \rightarrow ArCH_2\overset{\cdot}{C}HPh$ $\qquad(Ph = C_6H_5)$

Then: $ArCH_2\overset{\cdot}{C}HPh + CH_2{=}CHPh \rightarrow ArCH_2CHPhCH_2\overset{\cdot}{C}HPh$

Eventually: $Ar(CH_2CHPh)_nCH_2\overset{\cdot}{C}HPh + CH_2{=}CHPh$

$$\rightarrow Ar(CH_2CHPh)_{n+1}CH_2\overset{\cdot}{C}HPh$$

The propagation stage is so rapid that polymers of high relative molecular mass are produced early in the reaction, at a time when much monomeric phenylethene is still present.

Termination Several types of reaction are possible, e.g.

$$2Ar(CH_2CHPh)_nCH_2\overset{\cdot}{C}HPh \rightarrow Ar(CH_2CHPh)_{2n+2}Ar$$

or

$$Ar(CH_2CHPh)_nCH_2\overset{\cdot}{C}HPh + Ar\cdot \rightarrow Ar(CH_2CHPh)_{n+1}Ar$$

A growing chain can also be terminated by reaction with a compound known as an *inhibitor*. Examples are hydroquinone and 1,4-benzo-quinone. They are, in the main, hydrogen-containing compounds which react with chains in the following way:

$$Ar(CH_2CHPh)_nCH_2\overset{\cdot}{C}HPh + InH \rightarrow Ar(CH_2CHPh)_nCH_2CH_2Ph + In\cdot$$

inhibitor

followed by

$$2In\cdot \rightarrow InIn$$

For a compound to be effective as a chain terminator, the radical $In\cdot$ should be insufficiently reactive to serve as an initiator.

In the presence of catalysts other than peroxides, e.g. boron trifluoride or aluminium chloride, the mechanism of polymerisation is somewhat different, in that the chain reaction is propagated by ions rather than free radicals.

All addition polymers are *thermoplastic*, i.e. they soften and become more plastic on heating. Eventually they melt, but return to their original form on cooling. Their plasticity is due to the ability of the long chain molecules to slide over one another; there is little or no *cross-linking* from one chain to another.

Coloured plastics are produced by dispersing pigments such as titanium(IV) oxide (white) or carbon black into the material.

Certain plastics, notably poly(chloroethene), i.e. PVC, are hard and brittle in the pure state, but may be *plasticised* by mixing them with oily liquids such as dibutyl 1,2-benzenedicarboxylate (dibutyl phthalate). As much as 50 per cent may be required. Plasticisers are not without their disadvantages, for they tend to give the plastic a slightly sticky feel, and they may evaporate over the years so that the plastic gradually hardens.

The commonest plastics made from single monomers are listed in Table 7.2. Many modern plastics, however, are 'tailor made' for particular purposes by *copolymerising* a mixture of two or more monomers. For instance, a mixture of chloroethene and ethenyl ethanoate copolymerises to give a plastic which has a lower softening temperature and better flow than pure poly(chloroethene). (The toughness and non-flammability of PVC are retained.) The properties of the copolymer are better than those of a mixture or 'alloy' of the two plastics. This is because the chains of the copolymer contain both types of monomer unit linked together in a random fashion. Only very rarely do monomer molecules in a mixture polymerise independently of each other to give a mixture of two simple plastics.

Another copolymer of commercial importance is made from a phenylethene — 1,3-butadiene mixture. Because poly(1,3-butadiene) is rubbery, the addition of 1,3-butadiene to phenylethene in the ratio 1 : 3 yields a plastic, 'toughened polystyrene', which is less brittle than pure poly(phenylethene). Reversing the ratio of monomers gives a useful synthetic rubber, called SBR rubber.

Propenenitrile is used in making numerous copolymers. Added to phenylethene, it gives a copolymer with a melting temperature higher than that of poly(phenylethene). With 1,3-butadiene, it forms a synthetic rubber with some outstandingly good properties, and when copolymerised with both 1,3-butadiene and phenylethene it yields the so-called ABS polymer, which is very hard and tough and which is particularly useful because it can be electroplated.

Hydrogenation

Compounds containing C=C bonds react with hydrogen in the presence of certain metal catalysts, notably nickel, platinum or palladium,

Table 7.2 Addition polymers

MONOMER			
Formula	Name	Source	Polymerisation conditions
$CH_2=CH_2$	ethene or ethylene	cracking of petroleum fractions	1 000–2 000 atm (100 000–200 000 kPa); 100–300°C (373–573 K)
$CH_2=CH_2$	ethene or ethylene	cracking of petroleum fractions	in solution at 5–30 atm (500–3 000 kPa) with a Ziegler catalyst, e.g. $(C_2H_5)_3Al + TiCl_4$
$CH_3CH=CH_2$	propene or propylene	cracking of petroleum fractions	in solution with a Zeigler catalyst
$CH_2=CHCl$	chloroethene or vinyl chloride	$CH\equiv CH + HCl \xrightarrow[\text{(423 K)}]{HgCl_2 \atop 150°C} CH_2=CHCl$ $CH_2=CH_2 + Cl_2 = CH_2ClCH_2Cl$ $\downarrow \begin{array}{c} 500°C \\ (773\ K)\end{array}$ $CH_2=CHCl$	emulsion polymerisation with a peroxide catalyst
$C_6H_5CH=CH_2$	phenylethene or styrene	$C_6H_6 + CH_2=CH_2 \xrightarrow[\text{(363–373 K)}]{AlCl_3 \atop 90–100°C} C_6H_5CH_2CH_3$ $\Big\downarrow \begin{array}{c} ZnO\ or\ Fe_2O_3 \\ 600–630°C \\ (873–903\ K)\end{array}$ $C_6H_5CH=CH_2 + H_2$	mass polymerisation with a peroxide catalyst
$CH_2=CHCN$	propenenitrile or acrylonitrile	$2CH_3CH=CH_2 + 2NH_3 + 3O_2$ $\downarrow \begin{array}{c}\text{bismuth molybdate}\\ 450°C\ (723\ K)\end{array}$ $2CH_2=CHCN + 6H_2O$	in water, with a peroxide catalyst
$CF_2=CF_2$	tetrafluoroethene	$CHCl_3 + 2HF = CHClF_2 + 2HCl$ $2CHClF_2 \xrightarrow[\text{(973 K)}]{700°C} CF_2=CF_2 + 2HCl$	in water, under pressure, with $(NH_4)_2S_2O_8$ catalyst
$CH_2=\overset{\displaystyle CH_3}{\underset{\displaystyle \vert}{C}}-COOCH_3$	methyl 2-methyl-propenoate or methyl methacrylate	$\overset{CH_3}{\underset{CH_3}{>}}C=O + HCN \xrightarrow[\text{(293 K)}]{20°C} \overset{CH_3}{\underset{CH_3}{>}}\overset{\displaystyle OH}{\underset{\displaystyle CN}{C}}$ $CH_2=\overset{\displaystyle CH_3}{\underset{\displaystyle \vert}{C}}-C\overset{\displaystyle O}{\underset{\displaystyle OCH_3}{<}} \xleftarrow[H_2SO_4]{CH_3OH}$	mass polymerisation with a peroxide catalyst
$CH_3COOCH=CH_2$	ethenyl ethanoate or vinyl acetate	$CH\equiv CH + CH_3COOH \xrightarrow[\text{(433 K)}]{(CH_3COO)_2Zn \atop 160°C}$ $CH_3COOCH=CH_2 \leftarrow$	emulsion polymerisation with a peroxide catalyst

MONOMER

POLYMER				
Name	Trade names	Outstanding advantages	Outstanding disadvantages	Main uses
low density poly(ethene), polyethylene or polythene		cheap and tough, electrically insulating	low melting temperature, partially opaque	squeezy bottles, plastic bags, hose pipes
high density poly(ethene), polyethylene or polythene	Alkathene	higher melting and more transparent than low density poly(ethene)		buckets, washing-up bowls, food boxes
poly(propene) or polypropylene	Propathene Ulstron (fibre)	higher melting and tougher than poly(ethene); no odour		crates, food containers; as fibre, for ropes and blankets
poly(chloroethene) or polyvinyl chloride; PVC		tough and weather resistant; non-flammable		unplasticised, for pipes and gutters: plasticised, for raincoats, upholstery, records, floor tiles
poly(phenylethene) or polystyrene		ideal for injection mouldings	brittle, flammable	toys, pens and other small articles: expanded polystyrene for packaging and ceiling tiles
poly(propenenitrile) or polyacrylonitrile	Acrilan Orlon Courtelle	as fibre, resembles wool		carpets, blankets, sweaters
poly(tetrafluoroethene); PTFE	Fluon Teflon	chemically inert, low coefficient of friction	expensive	non-stick utensils, sleeves for ground glass joints
poly(methyl 2-methylpropenoate) or polymethyl methacrylate	Perspex Plexiglas	glass-like and tough		baths, aircraft windows, car rear lights
poly(ethenyl ethanoate) or polyvinyl acetate; PVA			low melting temperature	as an emulsion for emulsion paints and adhesives

according to the general equation:

$$\underset{R'}{\overset{R}{}}C = C\underset{R'''}{\overset{R''}{}} \quad + \quad H_2 \quad = \quad \underset{R'}{\overset{R}{}}\overset{\overset{H}{|}}{C} - \overset{\overset{H}{|}}{C}\underset{R'''}{\overset{R''}{}}$$

R, R′, etc represent hydrocarbon radicals or atoms of hydrogen

The reaction proceeds by the chemisorption of hydrogen and the alkene on to the surface of the catalyst, followed by the transfer of hydrogen atoms and desorption of the product:

Because hydrogenation takes place on its surface, the catalyst must have the greatest possible surface area. This is usually achieved by using the metal in a finely divided form. Finely divided platinum is made *in situ* by the reduction (by H_2) of powdered platinum(IV) oxide, $PtO_2(aq)$, known as 'Adams' catalyst'. Finely divided nickel, called 'Raney nickel', is made from Raney alloy, which is a 50 : 50 mixture of nickel and aluminium. Treatment with aqueous sodium hydroxide dissolves away the aluminium as sodium aluminate, leaving a fine deposit of nickel. This is filtered off and washed, first with water and then with ethanol. **Warning: Raney nickel is pyrophoric, i.e. spontaneously combustible in air. It should be stored under ethanol and not be allowed to dry out.**

The necessary conditions for hydrogenation depend partly on the structure of the alkenic compound and partly on the nature of the catalyst. With finely divided platinum the reaction can normally be carried out at room temperature and atmospheric pressure in a specially designed 'hydrogenation apparatus'. This consists essentially of one or more graduated glass tubes, which can be repeatedly filled with hydrogen from a cylinder. The tubes are connected to a flask containing the alkene, usually in solution in ethanol, together with the catalyst. The flask is continuously shaken or stirred until the calculated volume of hydrogen has been absorbed or until there is no further reaction.

When nickel is used as the catalyst, the reaction may not take place satisfactorily under these conditions. Hydrogenation at an elevated temperature and pressure can be performed in a stainless steel autoclave.

Warning: Mixtures of hydrogen and air are spontaneously explosive in the presence of a catalyst. A hydrogenation apparatus must only be used under supervision.

The reduction of alkenes to alkanes without using hydrogen can be achieved by means of diborane, B_2H_6. This compound reacts with alkenes to give trialkylboron compounds in a reaction known as *hydroboration*, e.g.

$$6CH_3CH{=}CH_2 + B_2H_6 \xrightarrow[\text{ether solution}]{0\ °C\ (273\ K)\ in} 2(CH_3CH_2CH_2)_3B$$

tripropylboron

Diborane, made by reaction between sodium tetrahydroborate(III) and boron trifluoride, must be handled with care because it is spontaneously flammable in air.

To liberate the alkane, the trialkylboron compound is heated with an anhydrous carboxylic acid:

$$(CH_3CH_2CH_2)_3B \xrightarrow{RCOOH} CH_3CH_2CH_3$$

7.5 Alkadienes

Classification of dienes

There are three possible arrangements of the $C{=}C$ bonds in a diene molecule:

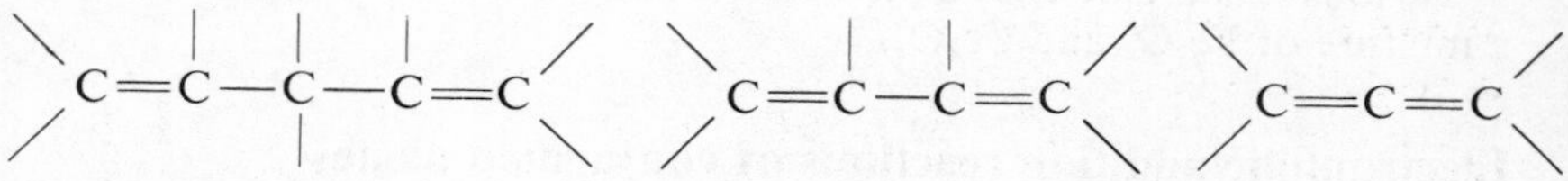

isolated	*conjugated*	*cumulated*
double bonds separated by more than one single bond	double bonds separated by one single bond	both double bonds originating from a single carbon atom

There are therefore three types of dienes.

(i) *Unconjugated dienes*, in which the two $C{=}C$ bonds are isolated. An example is

$$CH_2{=}CH{-}CH_2{-}CH_2{-}CH{=}CH_2$$

1,5-hexadiene

Such compounds are of relatively little interest, for in their addition reactions each double bond behaves independently of the other.

(ii) *Conjugated dienes*, with alternate double and single bonds (§1.3). The most important examples are

$$CH_2{=}CH{-}CH{=}CH_2 \quad \text{and} \quad CH_2{=}\overset{\overset{\displaystyle CH_3}{|}}{C}{-}CH{=}CH_2$$

1,3-butadiene   2-methyl-1,3-butadiene (isoprene)

These compounds are interesting because, in their addition reactions, the double bonds influence each other. They are important in laboratory syntheses and also in industry, where they are used as monomers in the manufacture of rubber.

(iii) *Allenes*, in which the double bonds are cumulated. They are called after allene, which is the trivial name for propadiene, $CH_2{=}C{=}CH_2$, the simplest compound of this type. Allenes are seldom encountered because their high reactivity makes preparation difficult.

Manufacture of 1,3-butadiene

Butadiene is prepared on a very large scale by the catalytic dehydrogenation of 1-butene and 2-butene. (These compounds are obtained by the cracking of petroleum fractions.)

$$\left.\begin{array}{l} CH_2{=}CHCH_2CH_3 \\ CH_3CH{=}CHCH_3 \end{array}\right\rangle \xrightarrow[\text{(868–948 K)}]{595\text{–}675\ ^\circ C} CH_2{=}CH{-}CH{=}CH_2 + H_2$$

Various metal oxides and phosphates can be used as the catalyst, e.g. a mixture of Fe_2O_3 and Cr_2O_3.

Electrophilic addition reactions of conjugated dienes

Complete addition, i.e 2 mol of reagent to 1 mol of diene, occurs with an excess of electrophilic reagents such as Cl_2, Br_2 or HCl, e.g.

$$CH_2{=}CH{-}CH{=}CH_2 + 2Br_2 = CH_2BrCHBrCHBrCH_2Br$$

Partial addition, i.e. 1 mol of reagent to 1 mol of diene, can take place in two ways, as follows.

(i) *1,2-addition*, i.e. addition across the first two carbon atoms of the conjugated system.

Here, the reagent attacks one of the C=C bonds in the expected manner, e.g.

$$CH_2{=}CH{-}CH{=}CH_2 + Br_2 = CH_2BrCHBrCH{=}CH_2$$

(ii) *1,4-addition*, i.e. addition to carbon atoms numbers 1 and 4 of the conjugated system.

This is not predictable from a knowledge of monoenes. Both double bonds are consumed in the addition, but at the same time a new one is formed between carbon atoms 2 and 3 of the conjugated system, e.g.

$$CH_2=CH-CH=CH_2 + Br_2 = CHBrCH=CHCH_2Br$$

Both modes of addition occur simultaneously. The reason lies in the fact that electrophilic addition proceeds via an *allyl carbonium ion*.

This is the term given to an ion of the type $CH_2=CH-\overset{\oplus}{C}HR$, in which the carbon atom bearing the positive charge is joined to a doubly bonded carbon atom. The π electrons of the double bond are drawn towards the $\overset{\oplus}{C}$ atom, thus:

$$CH_2=CH-\overset{\oplus}{C}HR$$

As a result, the π orbital becomes delocalised. Both the carbon–carbon bonds shown here are equivalent to each other, and the formula is best written as

$$\overbrace{CH_2\cdots CH\cdots CHR}^{\oplus}$$

As always, delocalisation leads to stability, and allyl carbonium ions are as stable as tertiary carbonium ions. (This is despite the fact that the formula $CH_2=CH-\overset{\oplus}{C}HR$ represents a secondary carbonium ion.)

Because of the high stability of allyl carbonium ions and their consequent strong tendency to be formed, addition to a conjugated diene always commences to one of the ends of the system, i.e. the electrophile joins the first carbon atom of the system. For example,

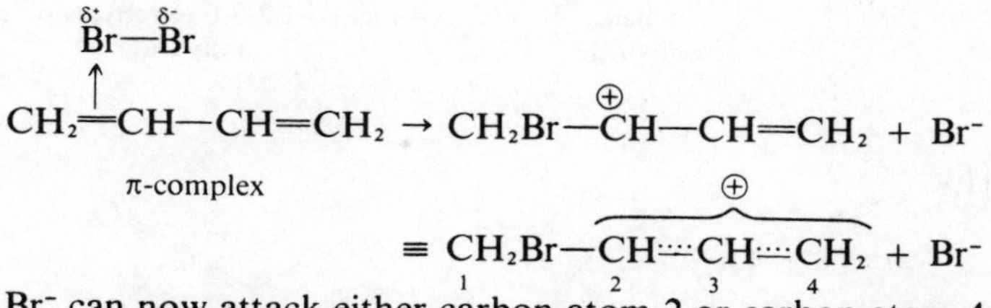

$$\underset{\underset{\pi\text{-complex}}{}}{\overset{\overset{\delta^+}{Br}-\overset{\delta^-}{Br}}{\underset{\uparrow}{}}}\ CH_2=CH-CH=CH_2 \rightarrow CH_2Br-\overset{\oplus}{C}H-CH=CH_2 + Br^-$$

$$\equiv CH_2Br-\overbrace{\underset{1}{CH}\cdots\underset{2}{CH}\cdots\underset{3}{CH}_2}^{\oplus} + Br^-$$

Br^- can now attack either carbon atom 2 or carbon atom 4:

$$CH_2Br-\overbrace{CH\cdots CH\cdots CH_2}^{\oplus} + Br^- \underset{1,4\text{-addition}}{\overset{1,2\text{-addition}}{\longrightarrow}} \begin{array}{l} CH_2BrCHBrCH=CH_2 \\ CH_2BrCH=CHCH_2Br \end{array}$$

The ratio of the 1,2- to the 1,4-adduct is not fixed but depends on the reaction conditions, notably the temperature. Experiments show that at relatively high temperatures 1,4-addition predominates, while at lower temperatures the product is mainly the 1,2-adduct. This is despite the fact that the 1,4-adduct is the more stable of the two, and on thermodynamic grounds would be expected to predominate under all conditions. The activation energy required for the formation of the 1,4-adduct is relatively high, so that at low temperatures its rate of formation is low. Raising the temperature increases the rate of both modes of addition, but 1,4-addition is assisted proportionately more than 1,2-addition.

Syntheses from conjugated dienes

One example of 1,4-addition, namely the Diels−Alder reaction, has already been mentioned (§2.2). The reaction provides a versatile means of synthesising six-membered ring compounds. Any conjugated diene can be used, with any unsaturated reagent in which the C=C bond is activated by an adjoining electron-withdrawing group, e.g. Cl, C=O, C≡N or OR. Such reagents are termed *dienophiles* or *philodienes*, and include maleic anhydride, acrolein, acrylic acid (CH_2=CHCOOH), 1,4-benzoquinone and ethyl vinyl ether.

Further examples of Diels−Alder reactions are as follows.

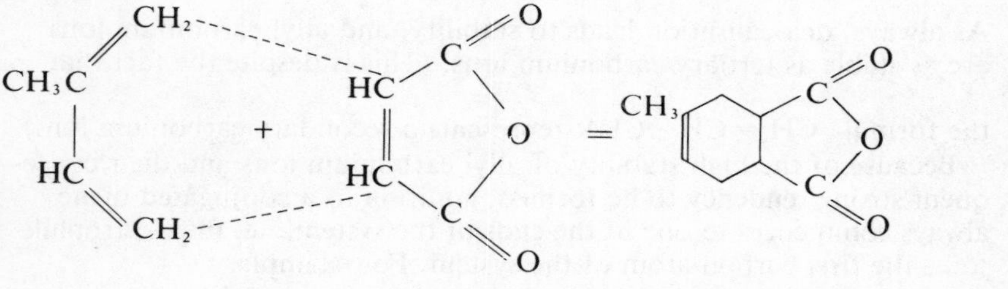

isoprene   maleic anhydride   4-methyl-1,2,3,6-tetrahydrophthalic anhydride

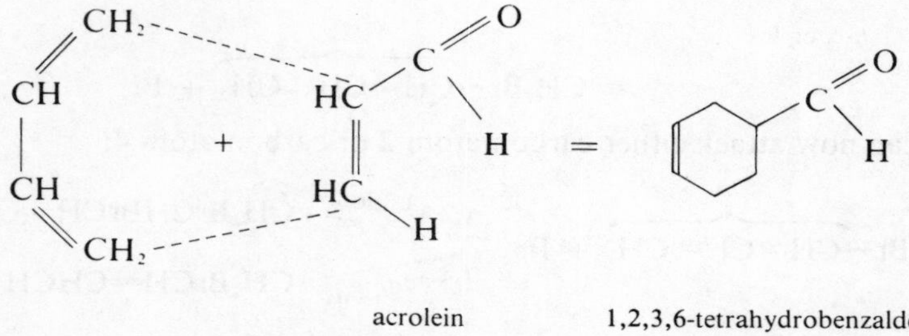

acrolein   1,2,3,6-tetrahydrobenzaldehyde

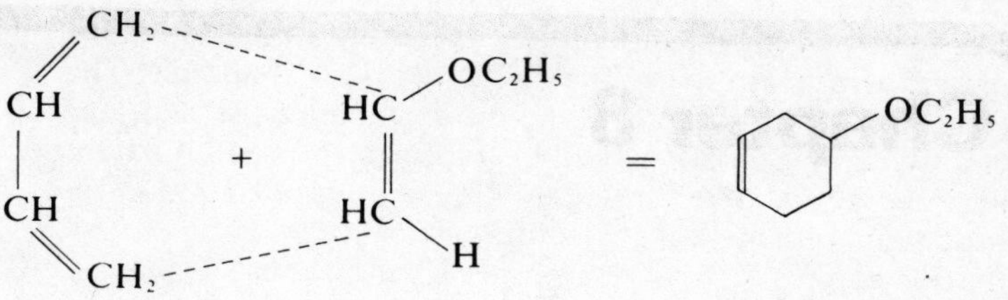

ethyl vinyl ether   1,2,3,6-tetrahydrophenetole

Chapter 8

The C≡C bond in alkynes

Data on the C≡C bond

covalent bond length/nm	0.120
bond dissociation enthalpy/kJ mol⁻¹ at 298 K	837

The carbon—carbon triple bond comprises a σ bond and two interacting π bonds (§1.2) and is characterised by its great strength. Its bond dissociation enthalpy, shown above, is exceeded only by that of C≡N (890 kJ mol⁻¹) and N≡N (944 kJ mol⁻¹).

The only common alkyne is ethyne or 'acetylene', CH≡CH, a gas which burns with a hot flame and which is widely used in the cutting and welding of metals. In the past, ethyne was made exclusively by the hydrolysis of calcium acetylide, CaC_2, but is now manufactured also from methane by pyrolysis at 1 500 °C (1 773 K):

$$2CH_4 = CH≡CH + 3H_2$$

In order to avoid further dehydrogenation to carbon, the reaction time is extremely short (<0.01 s), and the product is rapidly cooled.

8.1 Nomenclature of alkynes

Alkynes are named exactly like alkenes, except that the name ends in

'-yne'. Thus, the compound $\overset{4}{C}H_3\overset{3}{C}H_2\overset{2}{C}≡\overset{1}{C}H$ is named 1-butyne. Compounds with more than one triple bond are called *alkadiynes, alkatriynes,* etc, and named accordingly. The trivial name of acetylene is retained by IUPAC for the first member of the homologous series.

8.2 Formation of the C≡C bond

Elimination methods

The C≡C bond is usually formed by the elimination of two molecules of a hydrogen halide from a dihalide, e.g.

$$[CH_2{=}CH_2 \xrightarrow{Br_2}] \quad CH_2BrCH_2Br \xrightarrow[-HBr]{\text{ethanolic KOH}} CH_2{=}CHBr$$

ethene   1,2-dibromoethane   bromoethene

$$\xrightarrow[-HBr]{\text{ethanolic KOH}} CH{\equiv}CH$$

ethyne

$$[CH_3CHO \xrightarrow{PCl_5}] \quad CH_3CHCl_2 \xrightarrow[-HCl]{\text{ethanolic KOH}} CH_2{=}CHCl$$

ethanal   1,1-dichloroethane   chloroethene

$$\xrightarrow[-HCl]{\text{ethanolic KOH}} CH{\equiv}CH$$

Elimination of the second molecule of hydrogen halide is very slow. The yield is poor, owing to the occurrence of side reactions with the ethanolic alkali, but can be improved by the use of sodium amide, $NaNH_2$, in place of potassium hydroxide.

Other elimination methods, which may seem theoretically possible, do not work well in practice. Thus, the elimination (with concentrated sulphuric acid) of two molecules of water from a diol is more likely to result in oxidative cleavage of the diol (§6.4) than in alkyne formation; and the elimination of hydrogen from an alkene proceeds only at a high temperature and is difficult to control. (The conversion of an alkene to an alkyne is best achieved via a 1,2-dihalide, as shown above.)

Hydrolysis of carbides

The lower alkynes are conveniently prepared by the hydrolysis, with cold water, of ionic metallic carbides:

$$CaC_2 + 2H_2O = CH{\equiv}CH + Ca(OH)_2$$
$$Mg_2C_3 + 4H_2O = CH_3\,C{\equiv}CH + 2Mg(OH)_2$$

Aluminium carbide, however, hydrolyses to give methane, and interstitial carbides (e.g. tungsten carbide) do not hydrolyse at all.

8.3 General properties of the alkynes

In their physical properties, the alkynes closely resemble the alkanes and the alkenes. Thus, the lower members exist as gases and liquids. Their solubility in water is low, as a result of non-polarity, and they burn readily in air with a smoky yellow flame.

Ethyne is a gas; its boiling temperature is −83.6 °C (189.6 K). It can be liquefied by applying pressure at room temperature, but the liquid is unstable and explodes on shock:

$$CH{\equiv}CH(g) \rightleftharpoons 2C(s) + H_2(g) \qquad \Delta H^{\ominus} = -227 \text{ kJ mol}^{-1}$$

It is more stable in solution in propanone, and it is in this form that liquid ethyne is normally stored. One volume of propanone is capable of dissolving 300 volumes of ethyne at a presssure of 12 atm (1 200 kPa), to give a solution that can be kept with reasonable safety in steel cylinders packed with an absorbent material such as kieselguhr.

8.4 Chemical properties of the C≡C bond

The C≡C bond, like the C=C bond, undergoes addition reactions, although often by a different mechanism. Nearly all reagents which add across a C=C bond, and certain others besides, can add to a C≡C bond:

$$-C{\equiv}C- \;\; + \;\; XY \;\; = \;\; \overset{\displaystyle X \quad Y}{\underset{\displaystyle \; }{-\overset{|}{C}=\overset{|}{C}-}}$$

Markownikoff's rule (§7.4) is obeyed when protic acids attack an unsymmetrical alkyne.

If the reagent XY can add to a C=C bond, the addition continues:

$$\overset{\displaystyle X \quad Y}{-\overset{|}{C}=\overset{|}{C}-} \;\; + \;\; XY \;\; = \;\; X-\overset{\displaystyle X}{\overset{|}{C}}-\overset{\displaystyle Y}{\overset{|}{C}}-Y$$

Addition can usually be arrested at the end of the first stage.

Because of the similarity in structure between alkenes and alkynes, we should expect the C≡C bond to react by electrophilic addition, and indeed it does so, but not very readily. Many of the reactions of the C≡C bond proceed by nucleophilic addition — a type of mechanism that is rare with the C=C bond. Radical addition is also encountered, as it is with the C=C bond.

The reason for alkynes reacting less readily than alkenes towards electrophiles, but more readily towards nucleophiles, stems from the very short length of the C≡C bond:

$$H-C{\equiv}C-H \qquad \text{cf.} \qquad \underset{H}{\overset{H}{\diagdown}}C{=}C\underset{\diagdown H}{\overset{\diagup H}{}}$$

$$\underset{\substack{\longleftrightarrow \\ 0.120 \\ nm}}{} \qquad\qquad \underset{\substack{\longleftrightarrow \\ 0.134 \\ nm}}{}$$

In consequence, the two π bonds merge to give a stable cylinder of bonding around the central σ bond (Fig. 1.10). This is not readily polarised, even though it is accessible, and cannot serve easily as an electron source in forming bonds with electrophiles. Uncatalysed electrophilic addition is therefore uncommon.

Furthermore, the confinement of the six electrons of the $C\equiv C$ bond to a small region between the carbon atoms leaves those atoms open to the approach of other electrons. This shows itself in two ways.

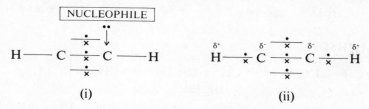

(i) nucleophilic attack can occur with strongly basic (i.e. electron donating) anions, such as RO^- and $RCOO^-$.
(ii) The shared pairs of electrons constituting the C—H bonds in ethyne are able to approach the carbon atoms more closely than usual so that the hydrogen atoms acquire an appreciable positive charge, i.e. they become acidic.

Electrophilic addition

Many electrophilic reagents, e.g. H_2SO_4 or HCl, which attack a $C=C$ bond, do not attack a $C\equiv C$ bond except in the presence of catalysts, usually salts of copper, mercury or nickel. These function by providing cations which are able to disrupt the cylinder of π bonding, with the formation of π-complexes, e.g.

$$\begin{array}{c} Hg^{2+} \\ \uparrow \\ CH\equiv CH \end{array}$$

$H_2SO_4 + H_2O$

The hydration of alkynes differs from that of alkenes, both in the conditions of the reaction and the chemical nature of the product.

When ethyne is treated with dilute sulphuric acid at 60 °C (333 K) in the presence of a little mercury(II) sulphate, it hydrates rapidly to give ethanal. The reaction, which at one time was used for the manufacture of ethanal, proceeds via an unstable enol, i.e. an unsaturated alcohol with a $-\overset{\displaystyle OH}{\underset{\displaystyle |}{C}}=\overset{\displaystyle |}{C}-$ grouping.

$$CH\equiv CH + H_2O \xrightarrow[\text{catalyst}]{HgSO_4 + H_2SO_4} CH_2=CHOH \xrightarrow{\text{rearrangement}} CH_3C\overset{\displaystyle O}{\underset{\displaystyle H}{\diagup}}$$

ethenol
(vinyl alcohol)

ethanal

The following mechanism has been proposed:

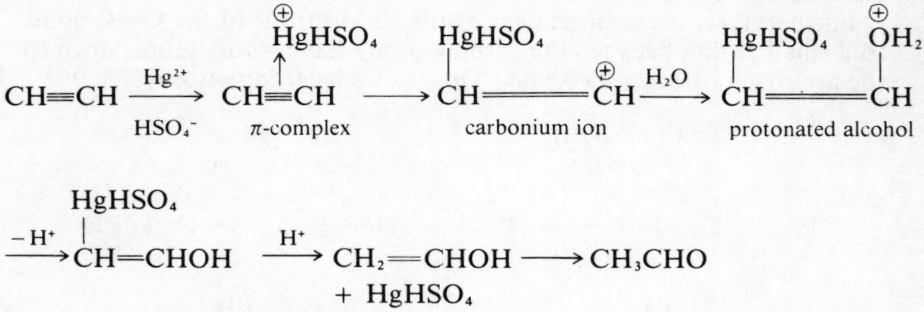

$$CH{\equiv}CH \xrightarrow[HSO_4^-]{Hg^{2+}} \underset{\pi\text{-complex}}{\overset{\overset{\oplus}{HgHSO_4}}{CH{\equiv}CH}} \longrightarrow \underset{\text{carbonium ion}}{\overset{HgHSO_4}{CH{=\!=\!=}\overset{\oplus}{CH}}} \xrightarrow{H_2O} \underset{\text{protonated alcohol}}{\overset{HgHSO_4 \quad \overset{\oplus}{OH_2}}{CH{=\!=\!=}CH}}$$

$$\xrightarrow[-H^+]{} \overset{HgHSO_4}{CH{=}CHOH} \xrightarrow{H^+} CH_2{=}CHOH \longrightarrow CH_3CHO + HgHSO_4$$

Other alkynes, upon hydration, give rise to ketones. Thus, propyne yields propanone, the hydrogen atom of the reagent joining the carbon atom with the greater number of hydrogen atoms already present:

$$CH_3C{\equiv}CH + H_2O \xrightarrow[H_2SO_4]{HgSO_4} CH_3\overset{\overset{\displaystyle OH}{|}}{C}{=}CH_2 \xrightarrow{\text{rearrangement}} CH_3COCH_3$$

Hydrogen halides

In their behaviour towards alkynes, hydrogen halides exhibit their usual order of reactivity in accordance with their acid strengths: $HI > HBr > HCl > HF$.

Hydriodic acid reacts readily with ethyne, especially in the presence of a catalyst, to give 1,1-diiodoethane:

$$CH{\equiv}CH \xrightarrow{HI} CH_2{=}CHI \xrightarrow[\text{(Markownikoff's rule)}]{HI} CH_3CHI_2$$

Hydrobromic acid acts in a similar fashion, but not hydrochloric acid, for the addition of HCl to a $C{=}C$ bond is difficult:

$$CH{\equiv}CH + HCl(aq) \xrightarrow[65\ °C\ (338\ K)]{HgCl_2} CH_2{=}CHCl$$
dilute

This reaction is adapted for the manufacture of chloroethene (vinyl chloride) for the plastics industry:

$$CH{\equiv}CH + HCl(g) \xrightarrow[150\ °C\ (423\ K)]{HgCl_2\ \text{on charcoal}} CH_2{=}CHCl$$

Gaseous hydrogen fluoride reacts only under pressure, but liquid anhydrous hydrogen fluoride is more reactive and will give 1,1-difluoroethane:

$$CH{\equiv}CH \xrightarrow{\text{HF}} CH_2{=}CHF \xrightarrow{\text{HF}} CH_3CHF_2$$

Halogens

The addition of halogens to alkynes, like the corresponding reactions with alkenes, may proceed by either an electrophilic or a free radical mechanism according to the conditions. In the presence of a 'halogen carrier', such as anhydrous aluminium chloride or iron(III) chloride, electrophilic addition occurs very readily but without explosion. Ethyne and chlorine react together to give the tetrachloride, although it is possible to isolate the dichloride:

$$CH{\equiv}CH \xrightarrow{\text{Cl}_2} \underset{\text{1,2-dichloroethene}}{CHCl{=}CHCl} \xrightarrow{\text{Cl}_2} \underset{\text{1,1,2,2-tetrachloroethane}}{CHCl_2\,CHCl_2}$$

1,1,2,2-Tetrachloroethane, a very toxic compound, is used for making 1,1,2-trichloroethene (trichloroethylene), a useful degreasing solvent. The reaction — a dehydrochlorination — can be brought about either by treating the tetrachloride with hot calcium hydroxide solution, or by heating it to about 500 °C (773 K) in the presence of barium chloride:

$$CHCl_2\,CHCl_2 \xrightarrow[\text{or heat with BaCl}_2]{\text{Ca(OH)}_2 \text{ solution}} CHCl{=}CCl_2 + HCl$$

The principal dry cleaning solvent now in use, hexachloroethane (perchloroethylene), is made from 1,1,2-trichloroethene as follows:

$$CHCl{=}CCl_2 \xrightarrow[\text{FeCl}_3]{\text{Cl}_2} CHCl_2\,CCl_3 \xrightarrow{\text{Ca(OH)}_2} CCl_2{=}CCl_2 \xrightarrow[\text{FeCl}_3]{\text{Cl}_2} CCl_3CCl_3$$

Liquid bromine reacts with ethyne to give $CHBr_2CHBr_2$, but with bromine water the reaction stops mostly at the first stage:

$$CH{\equiv}CH + Br_2(aq) = CHBr{=}CHBr$$

Iodine reacts only with difficulty, but in ethanolic solution $CHI{=}CHI$ can be obtained. (Iodine does not normally add to a $C{=}C$ bond.)

Cl_2 in aqueous solution

A molecule of ethyne can accept, in effect, two molecules of hypochlorous acid. Subsequent loss of water leads to the formation of a dichloroaldehyde:

$$CH{\equiv}CH \xrightarrow{\text{2HClO}} CHCl_2\,CH(OH)_2 \longrightarrow CHCl_2C{\overset{\displaystyle O}{\underset{\displaystyle H}{\big\langle}}} + H_2O$$

dichloroethanal

O_3

Alkynes (like alkenes) react with ozone to give ozonides, which on hydrolysis yield 1,2-dicarbonyl compounds:

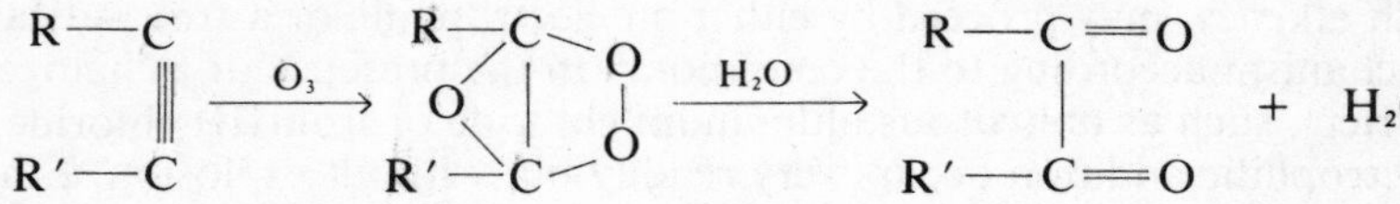

(R, R' = a hydrocarbon radical or H)

HCN

The catalysed addition of hydrogen cyanide to ethyne gives propene-nitrile (acrylonitrile):

$$CH{\equiv}CH + HCN \xrightarrow[80\ °C\ (353\ K)]{CuCl\ and\ NH_4Cl(aq)} CH_2{=}CHCN$$

Propenenitrile is used in making poly(propenenitrile) ('Acrilan'). It is no longer manufactured in this way but from propene, by oxidation in the presence of ammonia (Table 7.2).

Polymerisation

Alkynes will polymerise. Any resemblance to the polymerisation of alkenes is no more than superficial, for alkynes do not respond to peroxide catalysts or even Ziegler catalysts. Instead, they polymerise in the presence of the metal salts that catalyse so many of their other reactions. Long chain polymers are not obtained.

Ethyne, for example, polymerises in the presence of ammoniacal copper(I) chloride to give the dimer, 1-buten-3-yne (vinylacetylene) and the trimer, 1,5-hexadien-3-yne (divinylacetylene):

$$2CH{\equiv}CH \xrightarrow{[Cu(NH_3)_2]^+} \overset{4}{C}H{\equiv}\overset{3}{C}\overset{2}{C}H{=}\overset{1}{C}H_2$$
1-buten-3-yne

$$CH{\equiv}CCH{=}CH_2 + CH{\equiv}CH \xrightarrow{[Cu(NH_3)_2]^+} CH_2{=}CHC{\equiv}CCH{=}CH_2$$
1,5-hexadien-3-yne

Hydrogen chloride adds to the $C{\equiv}C$ bond of 1-buten-3-yne in accordance with Markownikoff's rule to give 2-chloro-1,3-butadiene, commonly known as chloroprene:

$$CH{\equiv}CCH{=}CH_2 + HCl = CH_2{=}CClCH{=}CH_2$$

On polymerisation, chloroprene gives the synthetic rubber, neoprene.

In the presence of nickel cyanide ethyne gives a cyclic tetramer, cyclooctatetraene. Unlike benzene, this compound is not aromatic, for the molecule possesses alternating double and single bonds. The reason is that the molecule is not planar, as a result of which the p orbitals which do not contribute to the σ-framework lie in different planes from one another. Because of this, any such orbital can overlap with only *one* of its neighbours, and delocalisation of π electrons is impossible.

$$4CH{\equiv}CH =$$

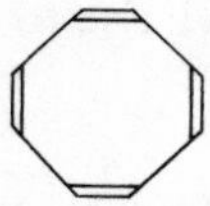

Radical addition

Halogens
Chlorine and bromine will add to a C≡C bond in the absence of a catalyst. The reactions are accelerated by light, and recent work has shown that they proceed by a free radical mechanism. (Ultraviolet light favours the formation of free radicals.) **The direct chlorination of ethyne is dangerous, for the reaction mixture may explode with the formation of carbon and hydrogen chloride.**

Polymerisation
Benzene is formed when ethyne is passed through a red hot tube:

$$3CH{\equiv}CH =$$

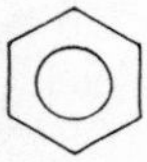

The yield of benzene is low, and the reaction cannot be regarded as a useful method of preparation. Nevertheless, it is of academic interest in that it is one of the few ways of converting an aliphatic compound into an aromatic one.

Nucleophilic addition
Alcohols, phenols and carboxylic acids add to the C≡C bond because they give rise to anions which can donate electrons to one of the carbon atoms of the triple bond. An alcohol, for example, attacks through its alkoxide ion:

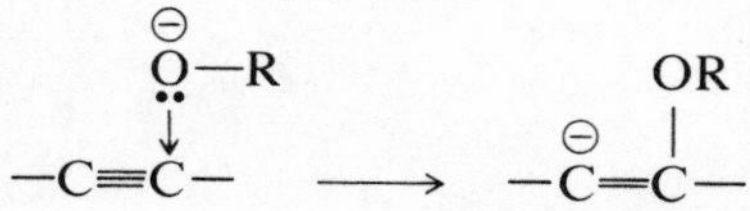

The carbanion (i.e. ion in which a negative charge resides on a carbon atom) then attracts the proton from a second molecule of alcohol:

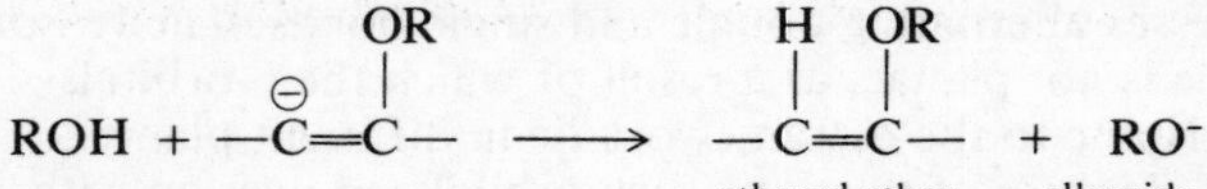

$$\text{ROH} + \overset{\overset{\displaystyle \text{OR}}{\displaystyle |}}{\underset{\ominus}{-\text{C}}}=\text{C}- \longrightarrow \overset{\overset{\displaystyle \text{H}}{\displaystyle |}\,\overset{\displaystyle \text{OR}}{\displaystyle |}}{-\text{C}=\text{C}-} + \text{RO}^-$$

ethenyl ether   alkoxide ion

An alkoxide ion is thus re-formed, and enables the reaction to continue.

Addition reactions with carboxylic acids give rise to ethenyl esters, e.g.

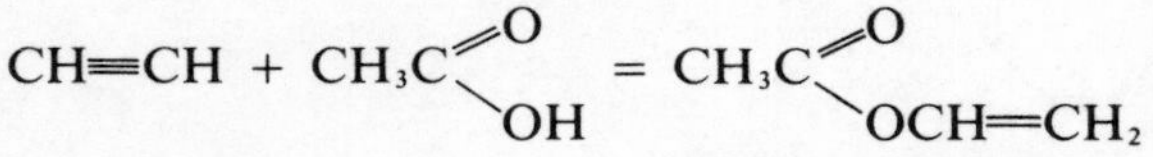

$$\text{CH}{\equiv}\text{CH} + \text{CH}_3\text{C}\overset{\displaystyle \nearrow\text{O}}{\underset{\displaystyle \searrow\text{OH}}{}} = \text{CH}_3\text{C}\overset{\displaystyle \nearrow\text{O}}{\underset{\displaystyle \searrow\text{OCH}{=}\text{CH}_2}{}}$$

ethenyl ethanoate
(vinyl acetate)

The addition may be conducted in the presence of mercury(II) salts, in which case it proceeds by an electrophilic mechanism.

Large quantities of ethenyl ethanoate are manufactured for the plastics industry. On emulsion polymerisation it gives a material (PVA) which forms the basis of many household emulsion paints.

Hydrogenation
The hydrogenation of alkynes to give, first, alkenes and then alkanes is described in §7.2.

Chapter 9

The C━━━C bond in arenes

Data on the C┄┄C bond

covalent bond length/nm	0.139
bond dissociation enthalpy/kJ mol^{-1} at 298 K	518

C┄┄C is the symbol used to represent the delocalised carbon–carbon double bond in the benzene ring. The bond is most simply studied in the *arenes,* i.e. aromatic hydrocarbons such as benzene and methylbenzene.

9.1 Aromatic compounds

Early in the last century, certain naturally occurring substances were discovered which were fundamentally different from aliphatic compounds, in that they contained at least six carbon atoms and could not readily be degraded to compounds with less than six carbon atoms. Such compounds were often fragrant (e.g. benzaldehyde — 'oil of almonds') and were called *aromatic compounds.*

The six carbon atoms forming the basis of the structure are in the form of a regular, planar hexagon known as a *benzene ring.* Aromatic compounds therefore comprise benzene and other compounds with at least one benzene ring. However, the special characteristics of benzenoid aromatic compounds are shared by certain other ring compounds possessing $4n + 2$ delocalised electrons (where n is 0, 1, 2, 3 etc.) and these compounds, too, are also classed as 'aromatic'. Three well known examples are as follows.

	pyridine	pyrrole	thiophene
Formula	CH / CH CH / CH CH / N	CH═══CH / CH CH / N H	CH═══CH / CH CH / S
Symbol	(ring with N)	(ring with N H)	(ring with S)
Number of delocalised electrons $4n + 2$	1 from each C atom + 1 from N atom = 6 — 6	1 from each C atom + 2 (lone pair) from N atom = 6 — 6	1 from each C atom + 2 (lone pair) from S atom = 6 — 6

These compounds are described as *heterocyclic,* because they are ring compounds in which at least one of the atoms of the ring is different from the others. By no means all heterocyclic compounds are aromatic; only those which conform to the $4n + 2$ rule are classed as such. The study of these compounds lies outside the scope of this book.

The structure of benzene has already been discussed in Chapter 1. It is important to bear in mind that the ring does not possess three double bonds alternating with single bonds as proposed by Kekulé. The π electrons of the double bonds are not concentrated in three particular localities, but are delocalised, i.e. uniformly distributed around the ring. All six carbon–carbon bonds of the ring are identical to one another, as indicated by X-ray diffraction measurements of covalent bond length. This dispersal of electrons leads to great stability; see the calculation of stabilisation energy (§1.3). The benzene ring, once formed, is very difficult to disrupt, even at high temperatures or by powerful oxidising agents. Indeed, aromatic compounds are often formed at high temperatures. Because of the reduced concentration of electrons, the C⋯C bond is less susceptible than the alkenic C═C bond to electrophilic reagents in addition reactions.

9.2 Nomenclature of arenes

The following notes apply only to aromatic hydrocarbons. The naming of substituted arenes is covered in other chapters.

Monocyclic arenes

Monosubstituted compounds are named systematically as alkylbenzenes, although a few have trivial names which are still recognised by IUPAC, e.g.

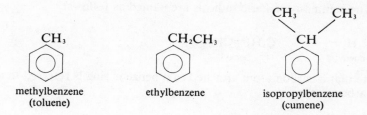

methylbenzene
(toluene)

ethylbenzene

isopropylbenzene
(cumene)

Disubstituted arenes are named by numbering the positions of the benzene ring 1—6. Thus, compound I (below) is called 1,2-diethyl-benzene, II is 1,3-diethylbenzene and III is 1,4-diethylbenzene. As always, the numbering is kept as low as possible: compound I, for instance, cannot be named as, say, 5,6-diethylbenzene.

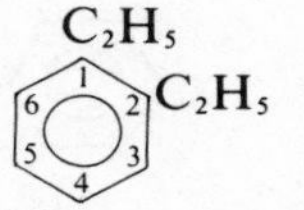

I

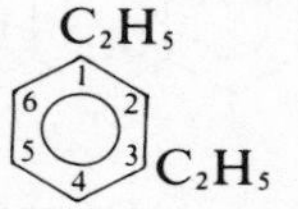

II

III

According to an older system, 1,2-compounds are named as *ortho* (*o*) compounds, 1,3- as *meta* (*m*) and 1,4- as *para* (*p*). Thus, I, II and III become, respectively, *o*-diethylbenzene, *m*-diethylbenzene and *p*-diethylbenzene.

The trivial name of 'xylene' is usually used for a dimethylbenzene.

If the substituent groups are different from each other, and one of them forms part of a trivially named compound, then that group is used as a reference for naming purposes, e.g.

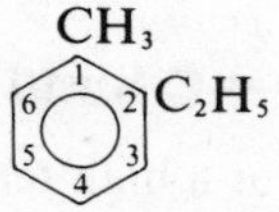

2-ethyltoluene (or *o*-ethyltoluene)

If neither group forms part of a trivially named compound, the latter group alphabetically is used as the reference, e.g.

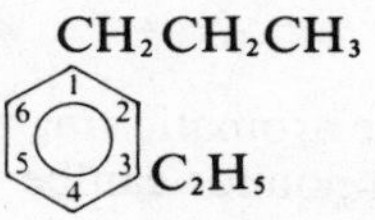

3-ethylpropylbenzene

The common aromatic radicals are named as follows:

$$C_6H_5—$$
phenyl

$$\overset{\alpha}{C_6H_5—CH_2—}$$
benzyl

Note that a carbon atom attached to a benzene ring is referred to as an α-carbon atom.

Polycyclic arenes

Arenes with more than one benzene ring may have fused or non-fused ring systems. Among the fused polycyclic systems, naphthalene, anthracene and phenanthrene are the simplest and best known.

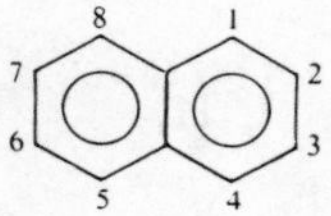

naphthalene 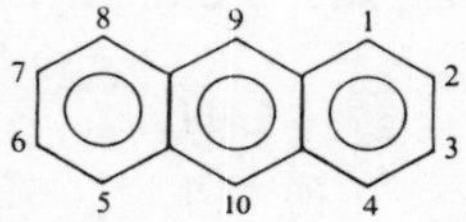   anthracene 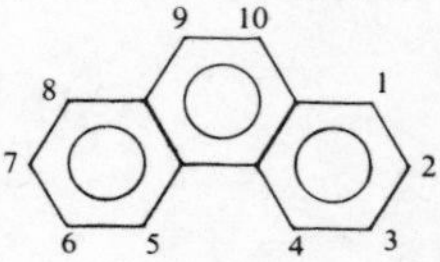   phenanthrene

Non-fused polycyclic systems which are joined to one another by single or double bonds are called *ring assemblies*. The commonest is biphenyl.

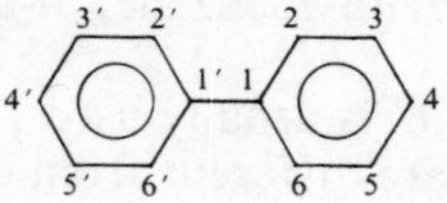

biphenyl

9.3 Formation of the C⚌C bond

The benzene ring is formed at a high temperature by the dehydrogenation of *alicyclic compounds,* i.e. cyclic aliphatic compounds. A simple example is the dehydrogenation of cyclohexane:

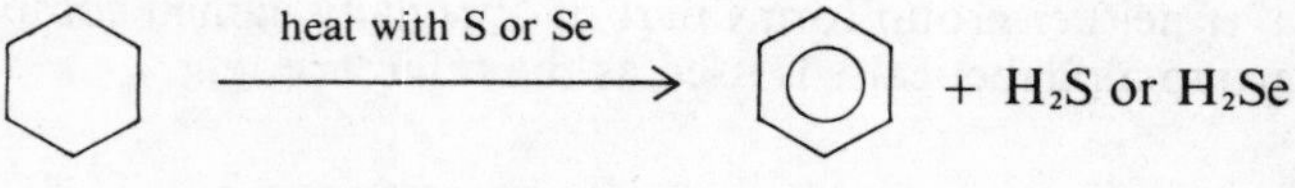

The tendency of a stable aromatic ring to be formed is so strong that various other alicyclic compounds can be used as well as cyclohexane and its derivatives, e.g.

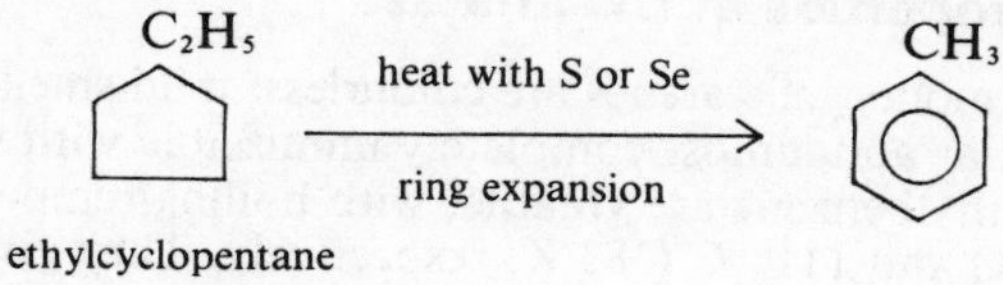

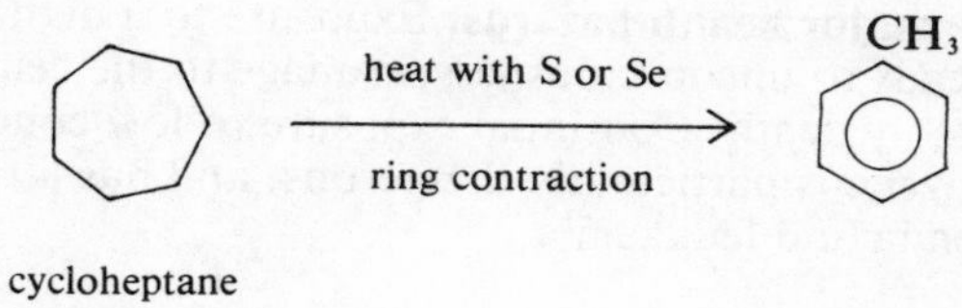

ethylcyclopentane → (heat with S or Se, ring expansion) → methylbenzene (CH_3)

cycloheptane → (heat with S or Se, ring contraction) → methylbenzene (CH_3)

Acyclic (open-chain) alkanes can also be employed, for at high temperatures and in the presence of suitable catalysts they undergo cyclisation to alicyclic hydrocarbons and then dehydrogenation to arenes, e.g.

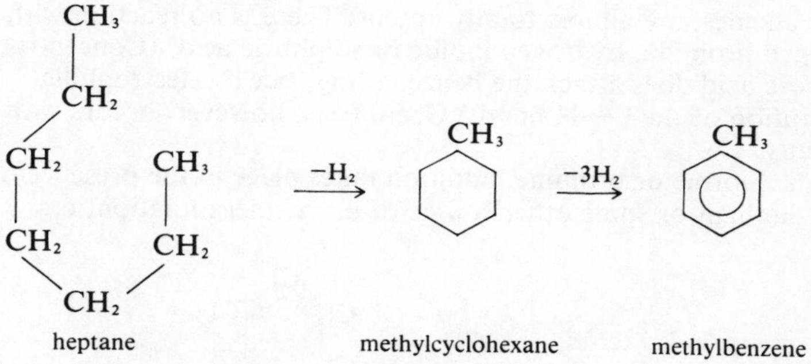

heptane $\xrightarrow{-H_2}$ methylcyclohexane $\xrightarrow{-3H_2}$ methylbenzene

This principle is utilised in the manufacture of arenes by the *platform-ing* (i.e. platinum reforming) of naphtha. The naphtha is first fractionally distilled so that it contains only C_6-C_9 hydrocarbons, and treated to remove catalyst poisons, notably sulphur. It is then mixed with hydrogen and passed over a catalyst of platinum on a silica–alumina base. (The purpose of the hydrogen is to avoid deposition of carbon on the platinum catalyst.) The reactor is maintained at a temperature of 450–530 °C (723–803 K) and a pressure of up to 13 atm (1 300 kPa). After cooling, the product is passed to a liquid-gas separator. Most of the hydrogen-rich gas is recycled, while the liquid is fractionally distilled. The fraction containing arenes may be either blended into motor fuel or treated (e.g. by solvent extraction) to isolate the aromatics. Further fractionation gives (i) benzene, (ii) methylbenzene, and (iii) a mixture of dimethylbenzenes and ethylbenzene.

9.4 General properties of the arenes

All the common monocyclic arenes are colourless, mild smelling liquids. They are non-polar and almost completely immiscible with water. Benzene and methylbenzene are volatile, with boiling temperatures of 80.1 °C (353.3 K) and 111 °C (383 K) respectively. They give rise to flammable vapours which burn with a yellow, sooty flame.

The arenes are major health hazards. Exposure to a high concentration of vapour leads to unconsciousness, damage to the central nervous system, and possibly death. Continual exposure to low concentrations is also risky. Benzene is particularly dangerous, and has a reputation for causing anaemia and leukaemia.

9.5 Chemical properties of the C⚌C bond

The delocalised carbon–carbon double bond is far less reactive than the alkenic double bond. Indeed, the principal reactions of the benzene ring are electrophilic *substitution* reactions of the C—H bond (Chapter 10).

Electrophilic addition reactions to the double bond, so characteristic of the alkenes, are almost totally absent. There is no reaction with hydrogen bromide, hydrogen iodide or sulphuric acid. (Concentrated sulphuric acid does attack the benzene ring, but by electrophilic substitution of the C—H bond.) Ozonolysis, however, occurs with difficulty.

With chlorine or bromine, addition takes place in the presence of direct sunlight or some other source of ultraviolet radiation, e.g.

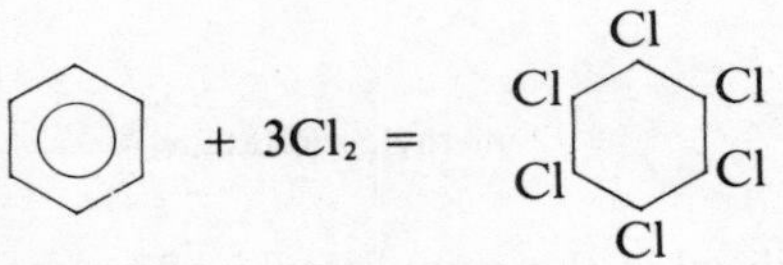

1,2,3,4,5,6-hexachlorocyclohexane
(benzene hexachloride)

This is a free radical chain reaction, initiated by the homolytic fission of a chlorine molecule:

$$Cl_2 \xrightarrow[\text{light}]{\text{ultraviolet}} 2Cl\cdot$$

The benzene ring is then attacked in two stages:

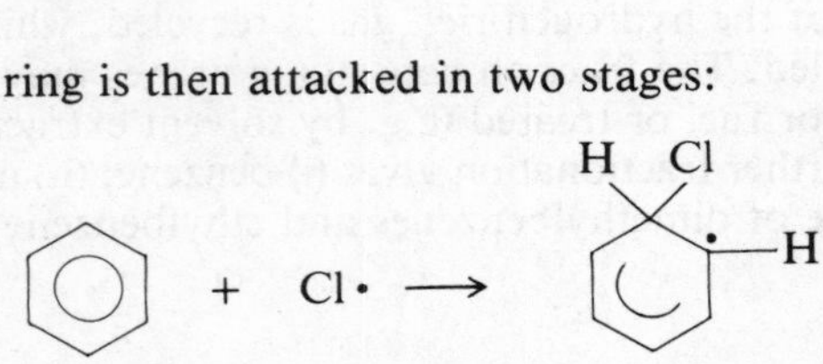

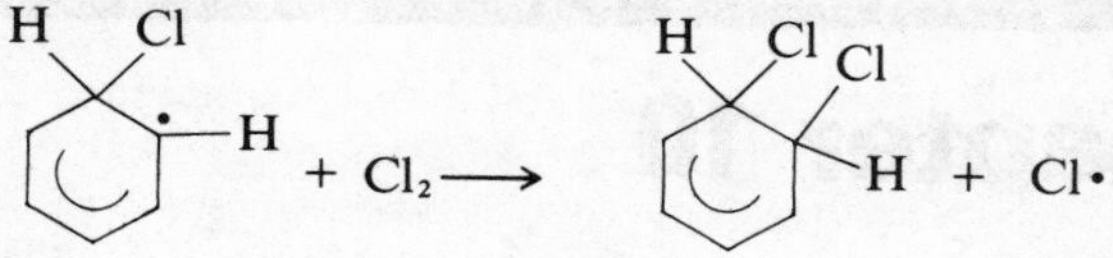

It will be seen that the second of these steps releases a chlorine free radical that can propagate the chain. Addition continues until six atoms of chlorine have been introduced, after which the chain reaction is terminated in the usual way by the coupling of radicals.

In the presence of a halogen carrier, chlorine or bromine attacks a C—H bond of the benzene ring by electrophilic substitution (§10.2).

There is little reaction between benzene and iodine. Fluorine, however, reacts with benzene in the vapour phase and in the presence of cobalt(III) fluoride or silver fluoride. The principal product is dodecafluorocyclohexane, C_6F_{12}, formed by addition plus substitution:

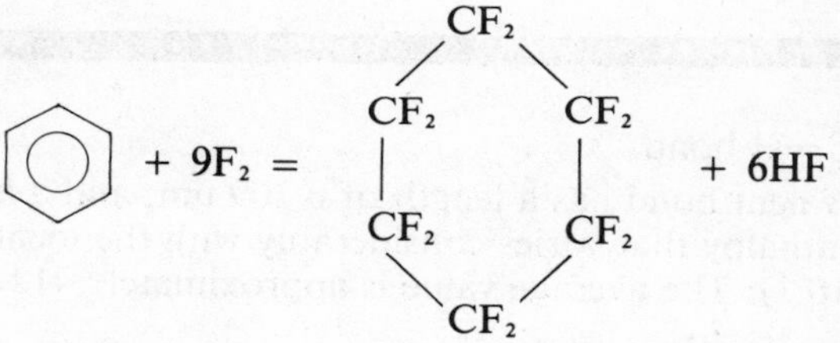

Various fluorides of low relative molecular mass, e.g. CF_4, are also formed, but there are no aromatic fluorine products.

The hydrogenation of benzene occurs much less readily than that of the alkenes. With Adams' catalyst (finely divided platinum) at 25 °C (298 K), the reaction takes about 25 hours:

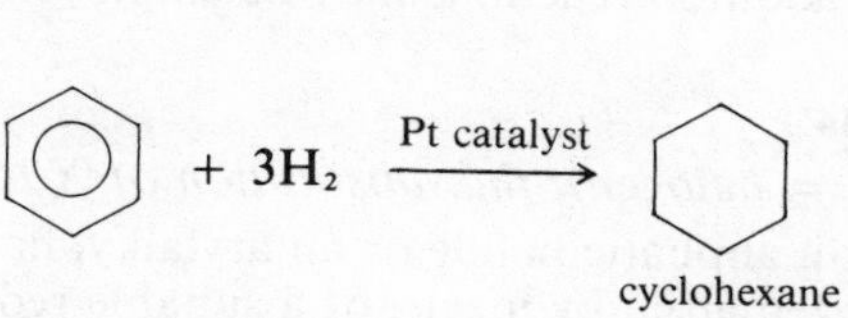

With nickel as the catalyst, a temperature of 200 °C (473 K) and a pressure of 200 atm (20 000 kPa) are required.

In a similar manner, phenol can be hydrogenated to give cyclohexanol:

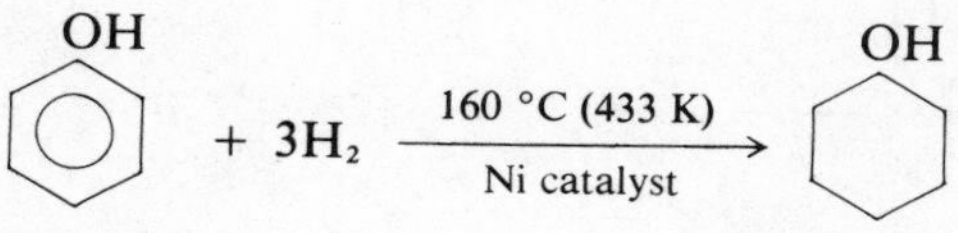

Chapter 10

The C─H bond in alkanes, arenes and C─H acidic compounds

Data on the C—H bond

The C—H covalent bond has a length of 0.109 nm, and a bond dissociation enthalpy that varies considerably with the location of the bond (Table 10.1). The average value is approximately 412 kJ mol^{-1} at 298 K.

With the exception of fully halogenated compounds, nearly all organic compounds possess the C—H bond. The formation and properties of the bond depend very much on its location, and in particular whether it forms part of an aliphatic or aromatic system.

10.1 Formation of the C — H bond

Both substitution and addition reactions may be employed.

Substitution methods

From C—X (where X = halogen); the substitution of X by H (§11.5)
The halogen atom of an aliphatic halide or an arylalkyl halide can be substituted by a hydrogen atom by means of a suitable reducing agent, e.g. metal + protic solvent, or lithium tetrahydridoaluminate(III), e.g.

$$CH_3CH_2CH_2Cl \xrightarrow{\text{Zn + HCl}} CH_3CH_2CH_3$$

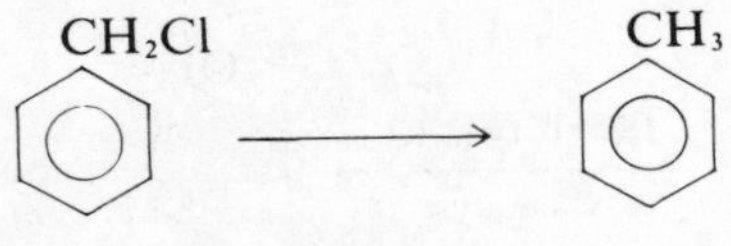

The corresponding reduction of an aryl halide is complicated by the low reactivity of a halogen atom when it is attached directly to a benzene ring, and by the fact that drastic reducing conditions could result in the hydrogenation of the benzene ring. Good results may be achieved by the use of a nickel–aluminium alloy in the presence of alkali, e.g.

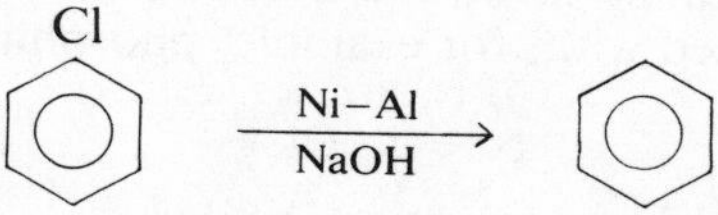

From C=O; the substitution of O by 2H

The conversion of $\diagdown$C=O (the carbonyl group of an aldehyde or ketone) to $\diagdown$CH$_2$ requires a powerful reductant; otherwise an alcohol is formed instead $\left(\diagdown\text{C=O} \rightarrow \diagdown\text{C} \begin{smallmatrix} \diagup\text{H} \\ \diagdown\text{OH} \end{smallmatrix} \right)$. Perhaps the best known reagent is zinc amalgam and boiling concentrated hydrochloric acid; this is the *Clemmensen reduction*. The mechanism is not well understood. The only certain fact is that the reaction does not proceed via an alcohol, for alcohols cannot be reduced to alkanes by means of this reagent. For example

$$\left.\begin{array}{l} \text{CH}_3\text{CH}_2\text{CHO} \\ \\ \text{or} \\ \\ \text{CH}_3\text{COCH}_3 \end{array}\right\} \xrightarrow{\text{Zn–Hg + conc. HCl}} \text{CH}_3\text{CH}_2\text{CH}_3$$

An alternative reducing agent for the conversion of $\diagdown$C=O to $\diagdown$CH$_2$ is a mixture of concentrated hydriodic acid and red phosphorus.

From C—OH; the substitution of OH by H

This is a difficult substitution to achieve in one stage. The usual reductants of the type metal + protic solvent are ineffective, but phenols can be successfully reduced by distilling them in the dry state with zinc dust, e.g.

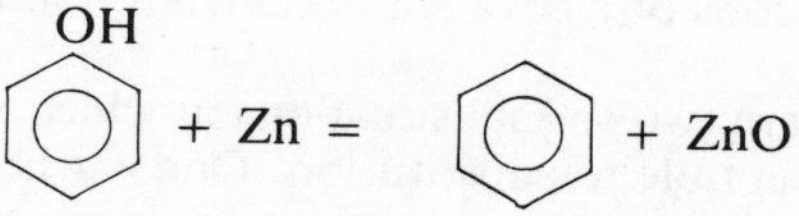

Alcohols, in contact with heated zinc, suffer cracking rather than reduction. They are best converted to alkanes by treatment with concentrated hydriodic acid and a little red phosphorus at 150 °C (423 K):

$$ROH \xrightarrow[\text{substitution}]{\text{HI}} RI \xrightarrow[\text{reduction}]{\text{HI}} RH$$

From C—NH₂; the substitution of NH₂ by H (§14.6)

Direct reduction is impossible. However, aromatic primary amines can be diazotised with sodium nitrite and hydrochloric acid, and the resulting diazonium salt reduced with, for example, phosphinic acid:

$$\underset{\text{phenylamine (aniline)}}{\overset{NH_2}{\bigcirc}} \xrightarrow{NaNO_2 + HCl} \underset{\text{benzenediazonium chloride}}{\overset{\overset{\oplus}{N} \equiv N \ Cl^-}{\bigcirc}} \xrightarrow{HPH_2O_2} \bigcirc + N_2(g) + HCl$$

Addition methods

The hydrogenation of C=C and C≡C bonds, by hydrogen in the presence of a nickel, platinum or palladium catalyst, leads to the formation of C—H bonds (§7.4).

10.2 Chemical properties of the C—H bond

The conditions under which a C—H bond reacts, and the mechanisms of the reactions, vary enormously with the location of the bond. The electronegativity difference between carbon and hydrogen is only 0.4 (C = 2.5, H = 2.1 on the Pauling scale), which means that although there is a certain amount of charge separation in the C—H bond, $\overset{\delta^-}{C}-\overset{\delta^+}{H}$, it is normally insufficient to permit electrophilic attack at the carbon atom.

With alkanes, and the alkyl groups of alkylaromatic hydrocarbons such as methylbenzene, substitution proceeds by a free radical mechanism, whereby ultraviolet light or visible light at the blue end of the spectrum is used to split the reagent into free radicals which are highly reactive and capable of attacking the hydrocarbon. Ultraviolet light is potentially able to break many covalent bonds and bring about many reactions. This is because its frequency is such that it can be absorbed and utilised in exciting the electrons of a covalent bond from bonding to anti-bonding orbitals.

However, there are two special situations in which the hydrogen atom is particularly susceptible to substitution. One is where the hydrogen atom is directly bonded to a benzene ring. Aromatic compounds, because of their delocalised π electrons, behave quite differently from alkenes. Electrophilic addition does not occur readily, as this would destroy the aromatic ring with its stable π bonding. Instead, the principal reactions are electrophilic substitutions.

The other special situation involves C—H acidic compounds. The hydrogen atom of a C—H bond is not normally acidic, but the presence of certain groupings, notably C=O and C≡C, can cause acidity. Aldehydes, ketones, carboxylic acid derivatives and certain alkynes all show this effect and are therefore known as *C—H acidic compounds*.

C—H bonds are also involved in elimination reactions, namely dehydrohalogenation (§11.5), dehydration (§12.5) and dehydrogenation (§6.4).

The C—H bond in alkanes and alkylaromatic hydrocarbons
Halogenation

In general, alkanes react rapidly with halogens either in ultraviolet light or on heating to give a mixture of compounds. Methane and chlorine, for example, give a mixture of mono-, di-, tri- and tetrachloromethanes:

$$CH_4 + Cl_2 = CH_3Cl + HCl$$
$$CH_3Cl + Cl_2 = CH_2Cl_2 + HCl$$
$$CH_2Cl_2 + Cl_2 = CHCl_3 + HCl$$
$$CHCl_3 + Cl_2 = CCl_4 + HCl$$

It is impossible to stop the chlorination at any of the intermediate stages, although the composition of the product can be controlled to some extent by adjusting the methane : chlorine ratio.

From ethane a mixture of nine chlorides is obtained, and the higher alkanes give even more complex mixtures. Their separation on anything but a small scale by chromatographic techniques is virtually impossible, and the reactions are thus of little use.

Alkylaromatic hydrocarbons undergo side chain halogenation at the α-position in essentially the same manner as alkanes, except that the substitution is *stepwise* and can be stopped at any of the intermediate stages. Methylbenzene, for example, can be chlorinated at its boiling temperature in the presence of ultraviolet light to give three chlorides:

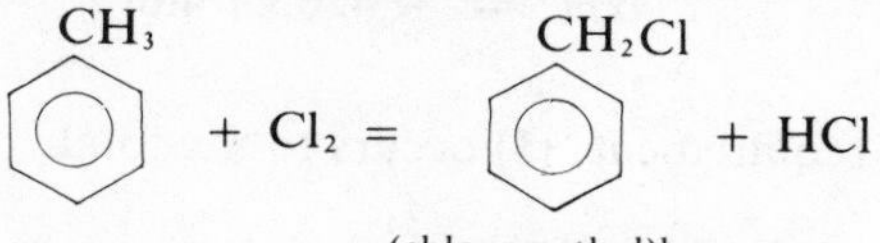

(chloromethyl)benzene

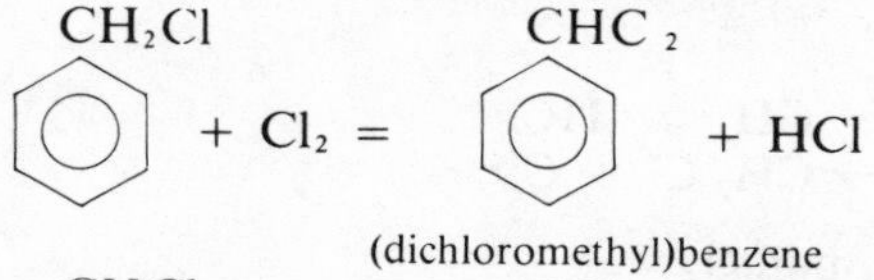

(dichloromethyl)benzene

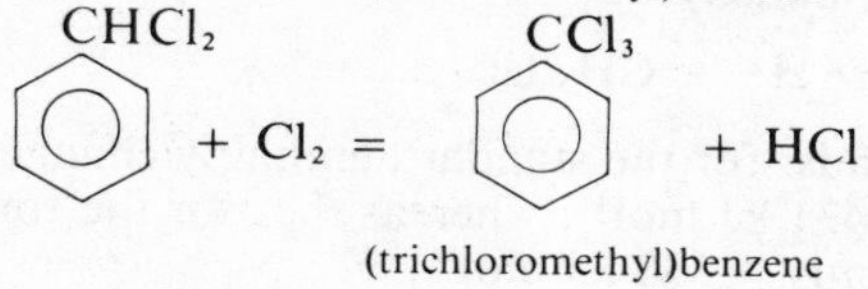

(trichloromethyl)benzene

By stopping the chlorination when the increase in mass corresponds to that which can be calculated from the equation, a good yield of (chloromethyl)benzene or (dichloromethyl)benzene can be obtained.

Different halogens have widely different reactivities, the order being $F_2 > Cl_2 > Br_2 > I_2$. Fluorine reacts violently to give highly fluorinated products, some of which arise through the cleavage of carbon–carbon bonds. The reaction is more controllable if it is conducted in the presence of copper turnings, and if the fluorine is diluted with nitrogen, but even so fluorinated hydrocarbons are usually prepared indirectly. Chlorination can be carried out safely and smoothly in the presence of diffused sunlight or at about 400 °C (673 K) in the dark. Bromine is much less reactive than chlorine, both in ultraviolet light and at elevated temperatures, and is capable of attacking only tertiary hydrogen atoms (see below). Iodine does not react.

Different substrates, too, have different reactivities. 2-Methylpropane (isobutane) and 2-methylbutane (isopentane), for example, are particularly reactive because they possess tertiary hydrogen atoms. Hydrogen atoms are classified as primary, secondary or tertiary, according to whether they are attached to primary, secondary or tertiary carbon atoms. (A *primary carbon atom* is attached to one hydrocarbon radical, a *secondary carbon atom* to two, and a *tertiary carbon atom* to three.) Tertiary hydrogen atoms are the most readily substituted. The order is tertiary > secondary > primary, although, as we shall see, the relative reactivity varies considerably from one type of halogenation to another.

All these facts can be explained in terms of a free radical chain reaction. Let us consider first the chlorination of methane.

Initiation Two reactions are conceivable:

$$Cl\!\overset{\frown}{\underset{\smile}{\cdot}}\!Cl \xrightarrow[\text{or heat}]{\text{ultraviolet light}} 2Cl\cdot \qquad \Delta H^{\ominus} = +242 \text{ kJ mol}^{-1} \qquad (1)$$

$$\overset{H}{\underset{H}{\big|}}\!\!\!\!\overset{\frown}{H\!-\!C\!\overset{\cdot}{\underset{\smile}{\cdot}}\!H} \to CH_3\cdot + H\cdot \qquad \Delta H^{\ominus} = +426 \text{ kJ mol}^{-1} \qquad (2)$$

Because of the lower enthalpy requirement, (1) occurs to the total exclusion of (2).

Propagation

Reaction 1 $Cl\cdot + CH_4 \longrightarrow CH_3\cdot + HCl$
Reaction 2 $CH_3\cdot + Cl_2 \longrightarrow CH_3Cl + Cl\cdot$

An alternative to reaction 1, namely

$$Cl\cdot + CH_4 \longrightarrow H\cdot + CH_3Cl$$

is precluded on enthalpy grounds, for the standard enthalpy change on forming the H—Cl bond is -431 kJ mol^{-1}, whereas that for the formation of the C—Cl bond is only -338 kJ mol^{-1}.

The chlorine free radical formed in reaction 2 can attack either another molecule of methane, to give more CH_3Cl, or a molecule of CH_3Cl to give CH_2Cl_2: consequently, the product is inevitably a mixture of chlorinated methanes.

Termination By the combination of any available pairs of free radicals:

$$2Cl\cdot \longrightarrow Cl_2$$
$$CH_3\cdot \; + \quad Cl\cdot \longrightarrow CH_3Cl$$
$$2CH_3\cdot \longrightarrow CH_3CH_3$$

The presence of traces of ethane in the product provides powerful evidence to support this theory.

The rapidity of the reaction is due to the fact that chlorine free radicals are highly reactive, and that every time one of them is consumed another is formed. Bromination, in contrast, is relatively slow, because bromine free radicals are less reactive than chlorine free radicals.

The radical chlorination of methylbenzene resembles that of methane, with propagation represented by the following equations:

benzyl free radical

However, the α C—H bonds in (chloromethyl)benzene and (dichloromethyl)benzene are much less reactive than those in methylbenzene; consequently, the chlorination of (chloromethyl) benzene does not commence until that of methylbenzene is complete. When radical bromination of methylbenzene is carried out, (tribromomethyl)benzene is not formed.

The likelihood of a free radical chain reaction taking place can be gauged quite accurately by considering the overall enthalpy change, ΔH, i.e. the algebraic sum of the individual enthalpy changes involved. (Strictly, ΔG should be considered rather than ΔH.) For example, we can calculate the standard enthalpy change for the conversion of methane to chloromethane, using the bond dissociation enthalpies shown in Table 10.1. To break the Cl—Cl and CH_3—H bonds *requires* (242 + 426) kJ mol^{-1}, i.e. $\Delta H^{\ominus} = +668$ kJ mol^{-1}. To form the Cl—H and C—Cl bonds *liberates* (431 + 338) kJ mol^{-1}, i.e. $\Delta H^{\ominus} = -769$ kJ mol^{-1}. The net enthalpy change for the reaction is thus

Table 10.1 Bond dissociation enthalpies/kJ mol^{-1}

F—F	158	CH$_3$—H	426
Cl—Cl	242	CH$_3$CH$_2$—H	410
Br—Br	193	$\begin{array}{c}CH_3 \\ CH_3\end{array}$>CH—H	393
I—I	151	(CH$_3$)$_3$C—H	376
F—H	562	C—F*	484
Cl—H	431	C—Cl*	338
Br—H	366	C—Br*	276
I—H	299	C—I*	238

*Denotes average bond dissociation enthalpy.

$-769 + 668 = -101$ kJ mol^{-1}. The reaction is exothermic, and capable of proceeding.

By performing similar calculations, we can see why different halogens have different reactivities, and why the reactivity of a hydrogen atom varies with its molecular environment. We shall discuss each aspect in turn.

Reactivity of the halogens

The high reactivity of fluorine can be ascribed to the relatively low bond dissociation enthalpy of the F—F bond, coupled with high values for the F—H and C—F bonds. A repeat of the previous calculation for

$$CH_4 + F_2 = CH_3F + HF$$

gives $\Delta H^{\ominus} = -462$ kJ mol^{-1}. Iodine, on the other hand, forms weak bonds (see I—H and C—I), so that even though the bond dissociation enthalpy for I—I is low, the overall standard enthalpy change for the reaction

$$CH_4 + I_2 = CH_3I + HI$$

can be calculated to be $+40$ kJ mol^{-1}. This would therefore be an endothermic reaction, and unlikely to occur of its own accord. We can pinpoint where the difficulty arises. Obviously it is not in forming iodine free radicals; but consider the first of the propagation reactions:

$$I\cdot + CH_3—H \rightarrow H—I + CH_3\cdot$$

$\Delta H^{\ominus} = DH^{\ominus}(CH_3—H) - DH^{\ominus}(I—H) = 426 - 299 = +127$ kJ mol^{-1} (endothermic). The iodine free radical is thus incapable of attacking the C—H bond in methane, and a chain reaction cannot be propagated.

Reactivity of the alkanes

The differences between the reactivities of primary, secondary and tertiary hydrogen atoms stem from differences in bond dissociation enthalpies; see Table 10.1 and compare the values for C—primary H in ethane, C—secondary H in propane and C—tertiary H in 2-methylpropane. The variation reflects the relative stabilities of primary, secondary and tertiary free radicals. Thus, we can argue that a tertiary hydrogen atom is particularly reactive because it is readily lost to yield a relatively stable tertiary free radical.

If we compare $\Delta H^{\ominus}$ for the chlorination of methane (-101 kJ mol^{-1}) with that for the chlorination of the tertiary hydrogen atom in 2-methylpropane (-151 kJ mol^{-1}), we see that the difference is relatively small compared with that between $\Delta H^{\ominus}$ for the chlorination of methane and for its fluorination (-462 kJ mol^{-1}). In other words, the effect on reactivity of changing from chlorine to fluorine is far greater than the effect of changing from one type of hydrogen atom to another. Consequently, fluorine is not very *selective* in what it attacks: primary, secondary and tertiary hydrogen atoms are attacked with almost equal ease, the reactivity difference between them being in the ratio 1 : 1.2 : 1.4 at 27 °C (300 K).

However, when we turn from the chlorination of alkanes to their bromination, we find that the change in $\Delta H^{\ominus}$ is relatively small ($\Delta H^{\ominus}$ for the bromination of methane is -23 kJ mol^{-1}). The reactivity differences between the three types of hydrogen atom now become highly significant. Bromine is a most selective reagent, and the ratio of the reactivities of primary, secondary and tertiary hydrogen atoms towards bromine has been estimated at 1 : 32 : 1 600 at 127 °C (400 K). It is a general rule that the higher the reactivity of a reagent, the lower is its selectivity.

Chlorosulphonation

Sulphur dioxide and chlorine, acting together on an alkane in the presence of ultraviolet light, bring about the formation of a sulphonyl chloride, RSO_2Cl, in a reaction known as *chlorosulphonation*. (Sulphonyl chlorides are acid chlorides of sulphonic acids, RSO_2OH.) The chain reaction is initiated by the homolysis of molecular chlorine, and is propagated by the following three reactions:

$$Cl\cdot + RH \rightarrow R\cdot + HCl$$

$$R\cdot + SO_2 \rightarrow R\text{—}\overset{\displaystyle O}{\underset{\displaystyle O}{\overset{\cdot}{S}}}$$

$$R\text{—}\overset{\displaystyle O}{\underset{\displaystyle O}{\overset{\cdot}{S}}} + Cl_2 \rightarrow R\text{—}\overset{\displaystyle O}{\underset{\displaystyle O}{S}}\text{—}Cl + Cl\cdot$$

Nitration

Nitration means a reaction with nitric acid in which a hydrogen atom becomes substituted by a nitro group, NO_2:

$$RH + HNO_3 = RNO_2 + H_2O$$

nitroalkane

Alkanes may be nitrated either in the liquid phase, with concentrated nitric acid under pressure at 140 °C (413 K), or in the vapour phase at 420 °C (693 K). The latter process is worked commercially with propane to produce nitroalkanes for use as fuels and solvents:

$$CH_3CH_2CH_3 \xrightarrow[\substack{420\ °C \\ (693\ K)}]{HNO_3} \begin{cases} CH_3CH_2CH_2NO_2 & \text{1-nitropropane} \\ CH_3CH(NO_2)CH_3 & \text{2-nitropropane} \\ CH_3CH_2NO_2 & \text{nitroethane} \\ CH_3NO_2 & \text{nitromethane} \end{cases}$$

Because of the relatively high reactivity of secondary hydrogen atoms, the principal product of this reaction is 2-nitropropane. The nitroethane and nitromethane arise through cracking.

Aliphatic nitration is believed to have a free radical mechanism. Unlike the nitration of benzene, it is not an electrophilic substitution and cannot be brought about by a mixture of concentrated nitric acid and concentrated sulphuric acid.

Autoxidation (oxidation with O_2)

It is well known that aliphatic ethers form peroxides (i.e. compounds with the peroxo group, —O—O—) when they are allowed to stand in contact with air and light. Ethoxyethane (diethyl ether), for example, gives dihydroxyethyl peroxide, $CH_3CH(OH)—O—O—CH(OH)CH_3$. **These peroxides explode with extraordinary violence on heating, and any ether that has not been standing over sodium or sodium hydroxide must be tested for peroxides before distillation.** Test strips, impregnated with iodide, are specially manufactured for the purpose. If no precautions are taken, the involatile peroxides remain in the distillation flask and may explode when the temperature rises towards the end of the distillation.

Oxidation with molecular oxygen, known as *autoxidation*, is not confined to ethers. It is a reaction of the C—H bond, and is thus widespread in organic chemistry. Although the final products may vary, the initial reaction always involves the formation of a hydroperoxide by a free radical chain reaction:

$$RH + O_2 = ROOH$$

Initiation Under the action of ultraviolet light or heat, free radicals are formed by homolytic fission of C—H bonds in the compound concerned. (In difficult cases free radicals can be supplied by means of a peroxide catalyst.)

$$RH \rightarrow R\cdot + H\cdot$$

Propagation

Reaction 1 $R\cdot + \dot{O}{=}\dot{O} \rightarrow R{-}O{-}O\cdot$
Reaction 2 $R{-}O{-}O\cdot + RH \rightarrow R{-}O{-}O{-}H + R\cdot$

Molecular oxygen is able to combine with a radical in the first propagation reaction because it is a *diradical*, i.e. it has two unpaired electrons.

Termination By the combination of free radicals, e.g.

$$R{-}O{-}O\cdot + R\cdot \rightarrow R{-}O{-}O{-}R$$

Certain autoxidations are of commercial importance. For example, the most economic way of manufacturing phenol involves the autoxidation of isopropylbenzene (§12.3). The lower aliphatic carboxylic acids are made by the aerial oxidation of $C_5{-}C_7$ alkanes (§13.4), and benzaldehyde and benzoic acid are obtained by a similar process from methylbenzene (§13.4).

The hardening of unsaturated *drying oils*, such as linseed oil, and of resins made from them is a polymerisation process which begins by autoxidation. Autoxidation also occurs when rubber ages and when fats and oils become rancid.

Another well known example is the aerial oxidation of aldehydes to carboxylic acids. Benzaldehyde, for instance, in a partly filled bottle, will soon deposit crystals of benzoic acid:

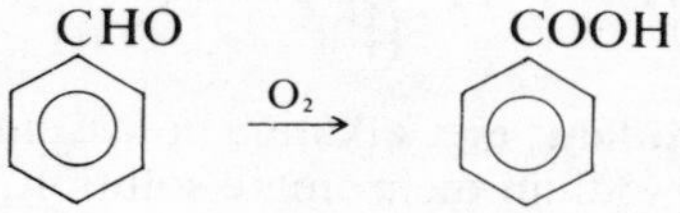

The aldehyde is first oxidised to a peroxoacid in accordance with the mechanism described above:

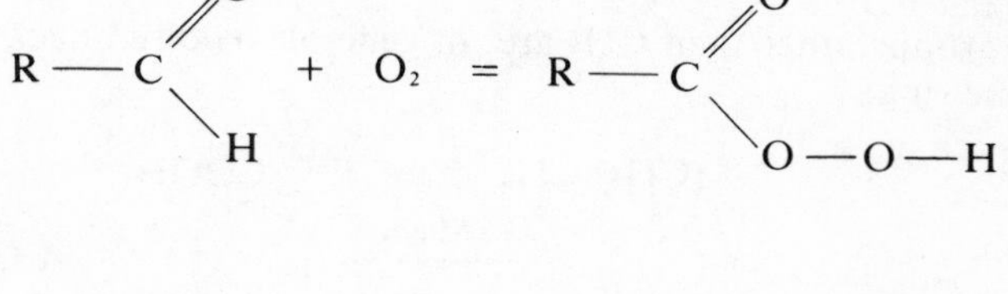

Oxidation–reduction then occurs between the peroxoacid and another molecule of aldehyde to give two molecules of carboxylic acid:

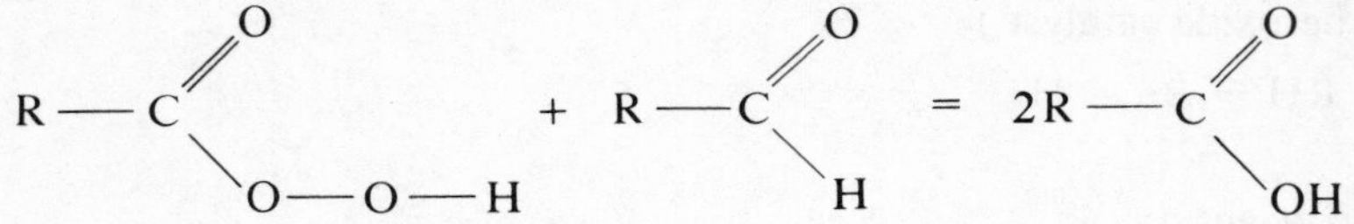

$$R - C{\Large<}^{O}_{O-O-H} \quad + \quad R - C{\Large<}^{O}_{H} \quad = \quad 2R - C{\Large<}^{O}_{OH}$$

Note that the oxidation of aldehydes *in aqueous solution* proceeds by an entirely different mechanism (§13.7).

Oxidation

Alkanes are relatively resistant to attack by oxidising agents, but do respond to acidified permanganate or dichromate solutions on heating. In a free radical chain reaction, closely allied to autoxidation, an alkane is oxidised to the corresponding alcohol:

$$RH \rightarrow ROH$$

In practice, the situation is complicated by the oxidation of the alcohol to give an aldehyde or a ketone. Any aldehyde so formed oxidises to a carboxylic acid. In addition, cleavage of C—C bonds occurs to give products of low relative molecular mass, and rearrangement reactions may also take place. The oxidation of alkanes is thus of little value for preparing definite compounds.

Alkyl groups which are attached to a benzene ring are much more susceptible to attack, and the oxidation of alkylaromatic hydrocarbons is of both industrial and laboratory importance. Methyl groups are oxidised according to the general scheme

$$-CH_3 \rightarrow -CH_2OH \rightarrow -C{\Large<}^{O}_{H} \rightarrow -C{\Large<}^{O}_{OH}$$

With a powerful oxidant, e.g. alkaline potassium permanganate solution, or acidified sodium dichromate solution, or dilute nitric acid, oxidation is complete, e.g.

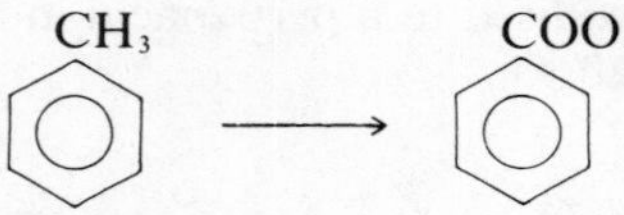

Alkyl groups other than CH_3 are, in general, oxidised back to the benzene ring:

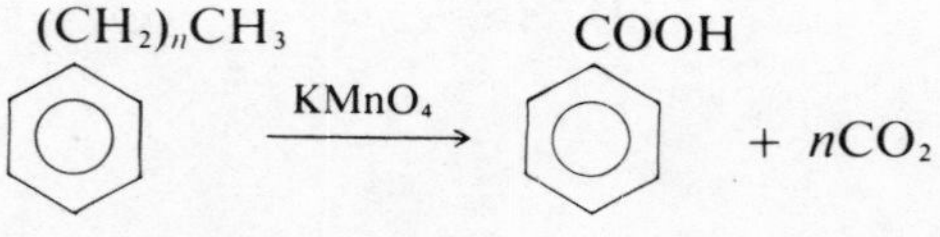

If more than one methyl group is present, both are oxidised by permanganate or dichromate solutions, but only one by dilute nitric acid, e.g.

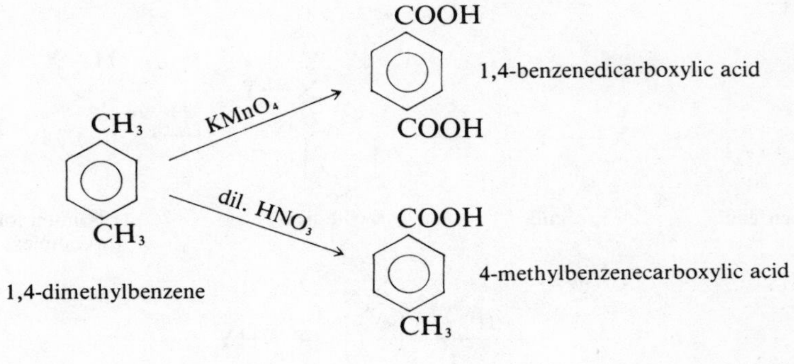

1,4-dimethylbenzene → (KMnO₄) → 1,4-benzenedicarboxylic acid (COOH/COOH)

1,4-dimethylbenzene → (dil. HNO₃) → 4-methylbenzenecarboxylic acid (COOH/CH₃)

With certain less powerful oxidants, intermediate oxidation products can be obtained, e.g.

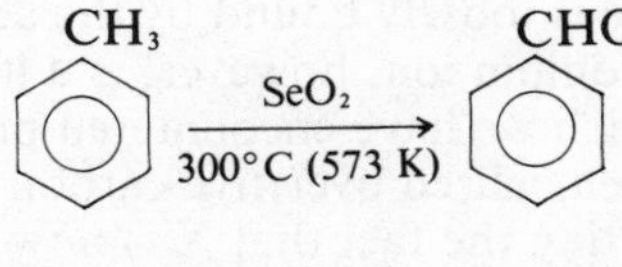

$$CH_3 \xrightarrow[300°C\ (573\ K)]{SeO_2} CHO$$

To some extent, benzaldehyde is made commercially by the liquid phase oxidation of methylbenzene with manganese(IV) oxide and dilute sulphuric acid.

The C—H bond in the benzene ring

The π electrons of the benzene ring, like those of the C=C bond in alkenes, attract electrophilic reagents. However, while an alkene undergoes electrophilic *addition*, an aromatic compound suffers electrophilic *substitution*. The two reactions are not as different as they may appear, and it is useful to begin by revising the subject of electrophilic addition:

alkene + electrophile → π-complex

slow → carbonium ion + Y^- → fast → product

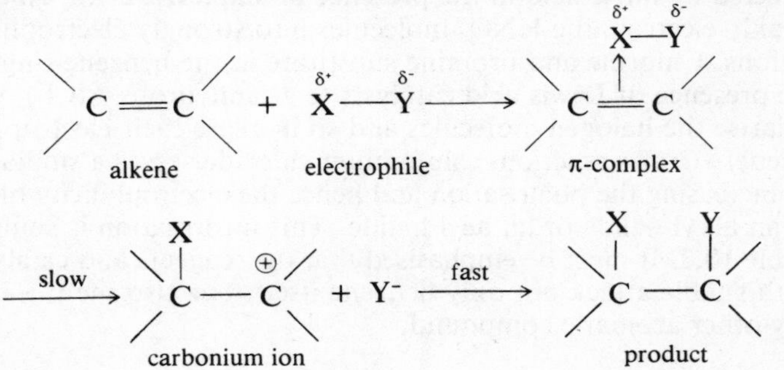

Electrophilic substitution in aromatic compounds is very similar to this, the only real difference being that in the final stage of the reaction the anion Y^-, instead of adding to the carbonium ion, abstracts a proton from the carbonium ion so as to re-form the very stable aromatic ring:

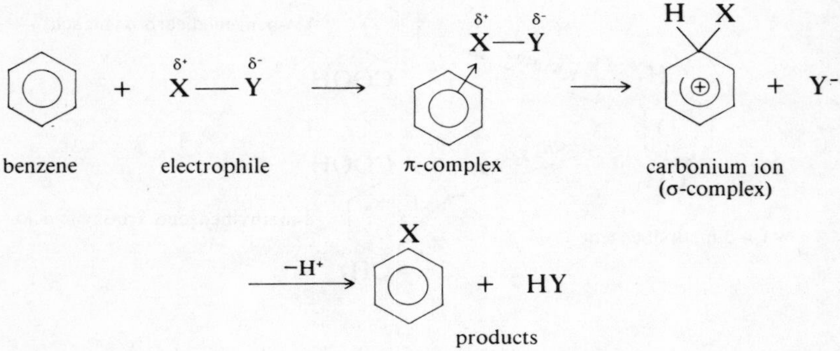

The π-complex is similar to that which is formed with alkenes, in that the electrophile is loosely bound by the acceptance of a pair of π electrons. The carbonium ion, however, is a little different from the carbonium ions which we have encountered previously, in that the positive charge is delocalised over five carbon atoms. It is usually called a σ-*complex*, reflecting the fact that X is now joined to a carbon atom by a σ bond. The slowest and hence the rate determining stage of the reaction is usually the formation of the σ-complex (cf. addition to alkenes).

Because of the delocalisation of π electrons, the benzene ring is less susceptible than the alkenic double bond to electrophilic attack. Hydrogen bromide and hydrogen iodide, well known reagents for alkenes, do not attack benzene. With the exception of sulphuric acid, the few electrophiles which do react with the C—H bonds of the benzene ring require the presence of catalysts to increase their *electrophilicity*, i.e. their electrophilic character. Aromatic nitration, for example, is effected by nitric acid in the presence of sulphuric acid, which converts weakly electrophilic HNO_3 molecules into strongly electrophilic NO_2^+ cations. Chlorine and bromine substitute in the benzene ring only in the presence of Lewis acid catalysts (e.g. anhydrous $AlCl_3$), which polarise the halogen molecules and so increase their electrophilicity. In Friedel—Crafts reactions, aluminium chloride serves a similar purpose, by increasing the polarisation and hence the electrophilicity of molecules of an alkyl halide or an acyl halide. This information is summarised in Table 10.2. It must be emphasised that the reagents and catalysts shown in this table attack not only benzene itself, but also the C—H bonds of any other aromatic compound.

Table 10.2 Catalysts for electrophilic substitution in the benzene ring

Process	Reagent	Catalyst	Electrophile
nitration	HNO_3	H_2SO_4	NO_2^+
chlorination	Cl_2	anhydrous $AlCl_3$ or $FeCl_3$	$\overset{\delta^+}{Cl}-\overset{\delta^-}{Cl}\cdots\cdots AlCl_3$*
bromination	Br_2	anhydrous $AlBr_3$ or $FeBr_3$	$\overset{\delta^+}{Br}-\overset{\delta^-}{Br}\cdots\cdots AlBr_3$*
Friedel–Crafts alkylation	RX	anhydrous $AlCl_3$	$\overset{\delta^+}{R}-\overset{\delta^-}{X}\cdots\cdots AlCl_3$*
Friedel–Crafts acylation	RCOX	anhydrous $AlCl_3$	$\begin{matrix}R\\ \diagdown\\ \diagup\\ X\end{matrix}\overset{\delta^+}{C}=\overset{\delta^-}{O}\cdots\cdots AlCl_3$*
sulphonation	H_2SO_4	none required	SO_3, i.e. $\overset{\overset{\delta^-}{O}}{\underset{\delta^-O\diagup\diagdown O\delta^-}{\overset{\delta^+}{S}}}$

*These species act as electrophiles because they possess atoms with a partial positive charge.

Some other reagents, notably nitrous acid and iodine, although incapable of attacking benzene, will react with certain aromatic compounds in which the ring is activated by substituent groups, such as HO, NH_2, NHR and NR_2, which can exert a mesomeric effect (§1.4). The interaction of a lone pair of electrons (on O or N) with the π electrons of the benzene ring increases the electron density of the ring and hence its attraction for electrophiles.

Nitration
Nitric acid can participate in aromatic substitution reactions because the benzene ring is highly resistant to oxidation. (Any attempt at a similar reaction with an alkene would lead to oxidative cleavage.) The electrophile is the *nitryl cation*, NO_2^+, formed by the protonation of nitric acid:

$$H^+ + \overset{..}{\underset{H}{O}}-NO_2 \rightleftharpoons \overset{H}{\underset{H}{\overset{\diagdown}{O}}}\overset{\oplus}{}-NO_2 \rightleftharpoons H_2O + NO_2^+$$

Nitric acid is capable of self-protonation,

$$2HNO_3 \rightleftharpoons \overset{H}{\underset{H}{\diagdown}} \overset{\oplus}{O} - NO_2 + NO_3^-$$

$$\Updownarrow$$

$$H_2O + NO_2^+$$

but the change occurs only to a small extent and nitric acid is thus a weak nitrating agent when used alone.

Nitric acid is best protonated by a strong acid. Concentrated sulphuric acid serves the purpose, and is the usual catalyst for aromatic nitration:

$$HNO_3 + H_2SO_4 \rightleftharpoons H_2NO_3^+ + HSO_4^-$$
$$H_2NO_3^+ + H_2SO_4 \rightleftharpoons H_3O^+ + NO_2^+ + HSO_4^-$$

In accepting a proton, the nitric acid molecule behaves as a base!

Evidence in support of this theory is provided by the depression of freezing temperature that is observed when nitric acid is dissolved in sulphuric acid. Values which are four times those expected on the basis of a covalent structure are found, suggesting that one molecule of nitric acid gives rise to four ions. Salts containing the NO_2^+ ion are known.

A mixture of concentrated nitric acid and concentrated sulphuric acid, known as 'nitrating acid', converts benzene into nitrobenzene, a yellow oily liquid with a boiling temperature of 211 °C (484 K):

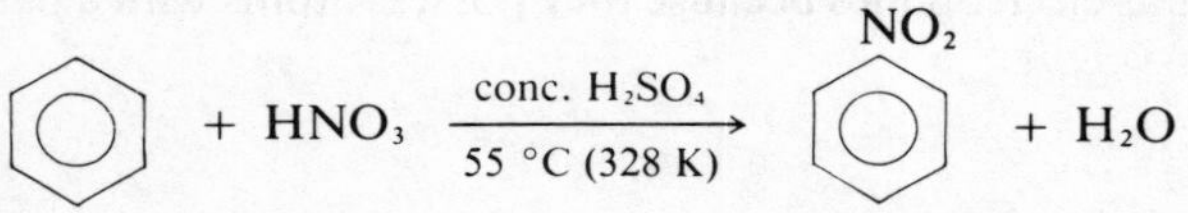

In detail:

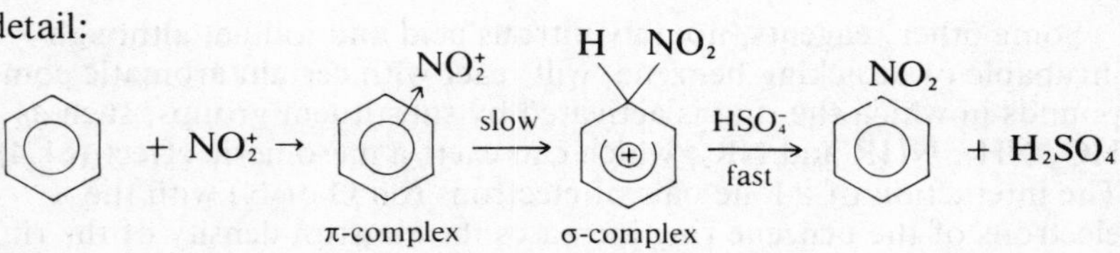

π-complex   σ-complex

Further nitration may occur to give 1,3-dinitrobenzene, a yellow solid with a melting temperature of 90 °C (363 K). The nitro group already in position is 1,3-directing, i.e. it directs all other incoming groups into the 3-position. Dinitration is relatively difficult, for NO_2, like all 1,3-directing groups, deactivates the benzene ring by an electron withdrawing effect.

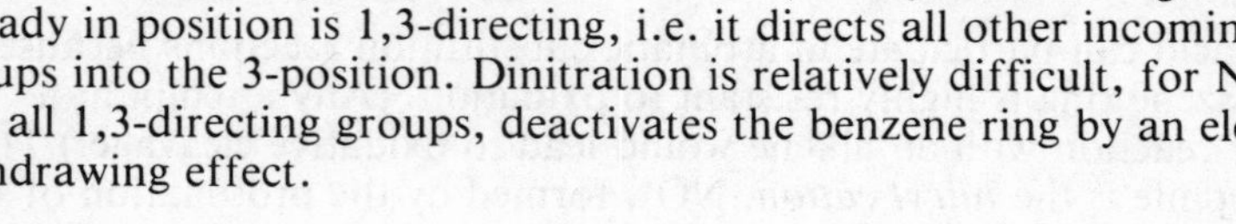

Only a little 1,3,5-trinitrobenzene is formed.

Methylbenzene is easier to nitrate than benzene, because the methyl group has a positive inductive effect, i.e. an electron releasing effect. Because the methyl group is 1,2- and 1,4-directing (see below), the ultimate product is methyl-2,4,6-trinitrobenzene, a high explosive commonly called trinitrotoluene (TNT):

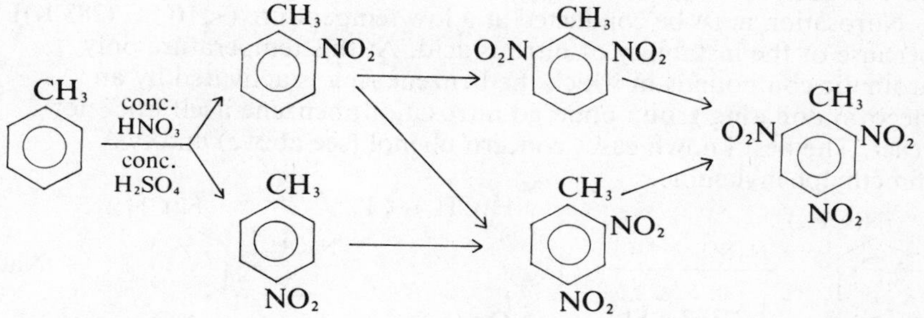

Phenol undergoes nitration with remarkable ease, for the HO group strongly activates the benzene ring by the mesomeric effect. Dilute nitric acid (without sulphuric acid) is sufficient to give a mixture of 2- and 4-nitrophenols:

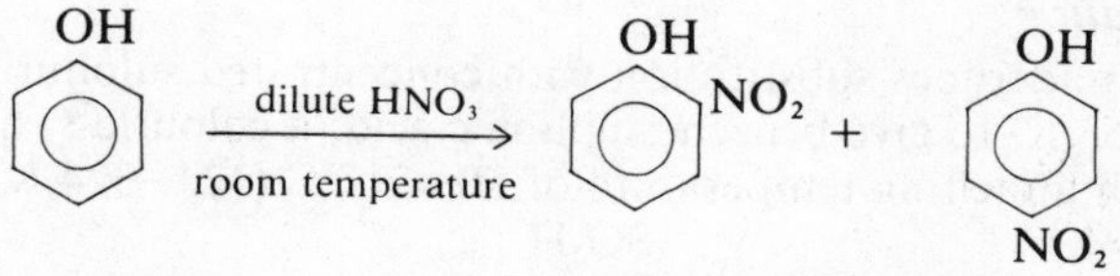

To some extent the phenol is nitrated by nitryl cations formed by the self-protonation of nitric acid, but it has been found that the main reaction is not nitration but *nitrosation* (see below), initiated by traces of nitrous acid in the nitric acid. This is followed by oxidation of the nitrosophenol, to give the nitrophenol and further nitrous acid:

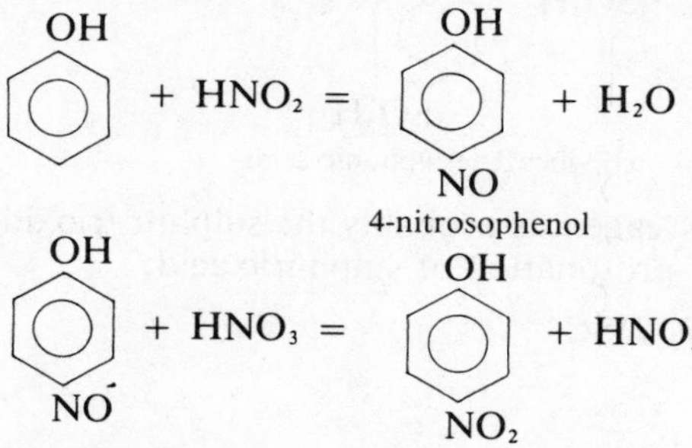

Nitrosation

In the same way that 'nitration' means the substitution of the nitro group, NO_2, by means of nitric acid, so 'nitrosation' means the substitution of the nitroso group, NO, by means of nitrous acid:

$$ArH + HNO_2 = ArNO + H_2O$$

And in the same way that nitration involves the nitryl cation, NO_2^+, formed by the protonation of nitric acid, so nitrosation requires the *nitrosyl cation*, NO^+, formed by the protonation of nitrous acid (§14.6).

Nitrosation must be conducted at a low temperature ($<10\ °C$ (283 K)) because of the instability of nitrous acid. At this temperature only aromatic compounds in which the benzene ring is activated by an electron donating group undergo nitrosation; benzene itself does not react. The best known cases concern phenol (see above) and *N,N*-dimethylphenylamine:

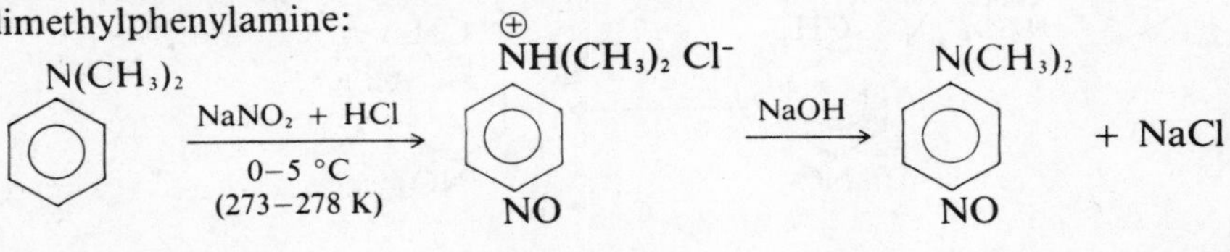

N,N-dimethylphenylamine
(*N,N*-dimethylaniline)

4-nitrosodimethylphenyl-
ammonium chloride
(yellow solid)

4-nitrosodimethylphenyl-
amine (green solid)

This reaction is of importance as a test for aromatic tertiary amines.

Sulphonation

Benzene undergoes substitution with concentrated sulphuric acid at 80 °C (353 K) to give benzenesulphonic acid, a colourless, highly soluble solid with a melting temperature of 50–51 °C (323–324 K).

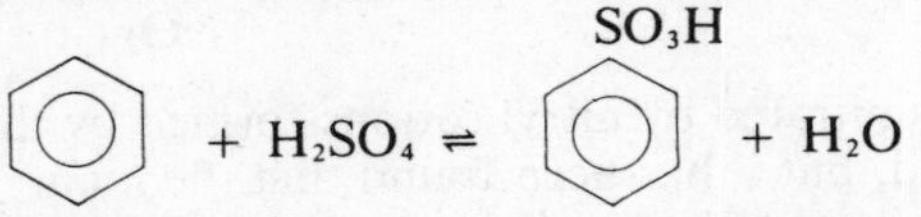

Methylbenzene and the other alkylbenzenes behave in a similar way:

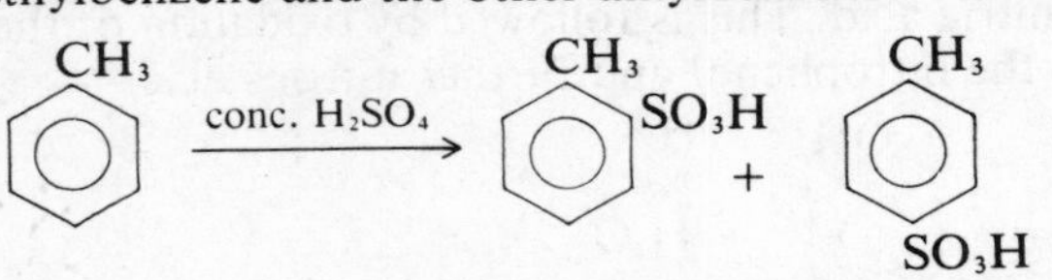

2- and 4-methylbenzenesulphonic acids

The principal electrophilic reagent is probably the sulphur trioxide molecule, formed by the self-protonation of sulphuric acid:

$$2H_2SO_4 \rightleftharpoons H_3SO_4^+ + HSO_4^-$$

$$\Updownarrow$$

$$H_3O^+ + SO_3$$

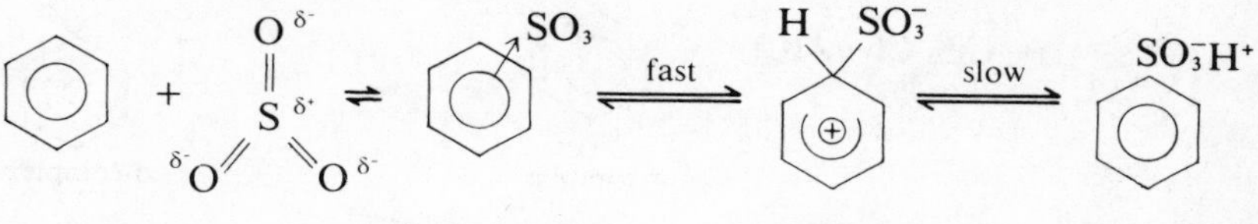

π-complex σ-complex

This mechanism is different from that of nitration in two respects. First, the slowest (hence, rate determining) stage is the abstraction of a proton. Second, the reaction is reversible. The reverse change, *desulphonation*, may be accomplished by refluxing the sulphonic acid with dilute sulphuric acid or hydrochloric acid, or by treating it with superheated steam.

The sulphonic acid group is 1,3-directing, and deactivating. Thus, di- and trisulphonic acids can be obtained only if sulphonation is conducted under vigorous conditions, e.g. with fuming sulphuric acid, or at a high temperature, or with a catalyst such as silver sulphate.

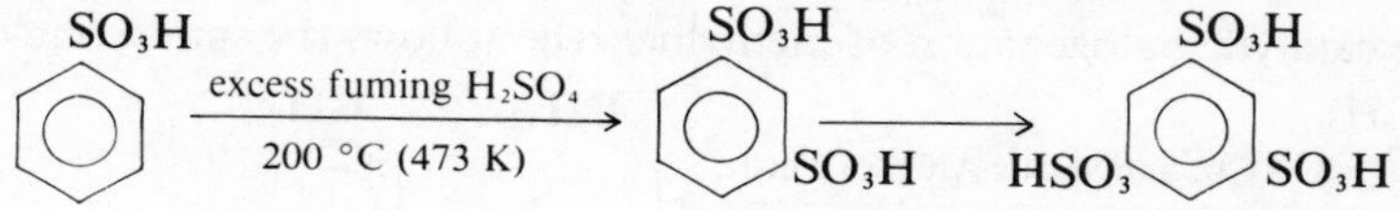

Halogenation

In the presence of ultraviolet light, benzene reacts with both chlorine and bromine in free radical addition reactions to give hexahalo-cyclohexanes (§9.5). Methylbenzene under these conditions becomes substituted in the side chain by a similar mechanism (see above).

However, in the *absence* of ultraviolet light, but in the presence of a suitable catalyst (anhydrous $AlCl_3$ or $FeCl_3$ for chlorination; $AlBr_3$ or $FeBr_3$ for bromination), benzene and other arenes undergo electrophilic substitution on the ring. Anhydrous aluminium chloride, like all Lewis acid catalysts, attracts a lone pair of electrons from the outer shell of a chlorine atom. This in turn polarises the chlorine molecule:

$$\overset{\delta+}{Cl} \rightarrow \overset{\delta-}{Cl} : \cdots \rightarrow \overset{\delta-}{AlCl_3}$$

The partial negative charge is distributed among the four chlorine atoms attached to the aluminium atom, although it is often written, for convenience, on the aluminium atom itself. This convention is adopted in the mechanism which follows, and also in Table 10.2.

The positively charged chlorine atom can attack the benzene ring in the usual manner to give chlorobenzene, a colourless liquid with a boiling temperature of 132 °C (405 K):

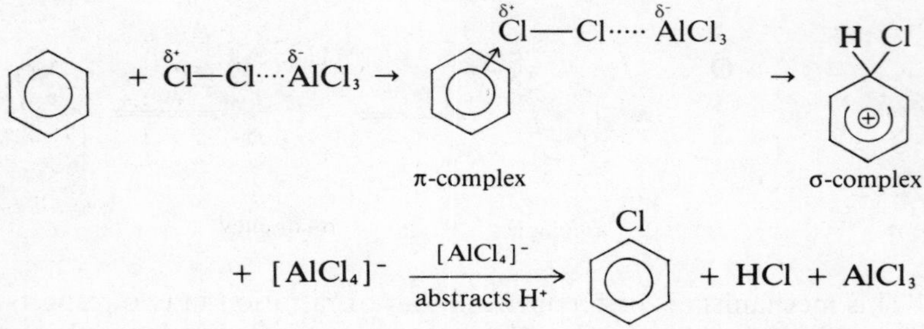

π-complex

σ-complex

$+ [AlCl_4]^- \xrightarrow[\text{abstracts H}^+]{[AlCl_4]^-}$ $+ HCl + AlCl_3$

Chlorination and bromination can be conducted at room temperature. To ensure that only monosubstitution occurs, equimolar proportions of benzene and the halogen are used: with excess reagent a mixture of 1,2- and 1,4-dihalobenzenes is formed:

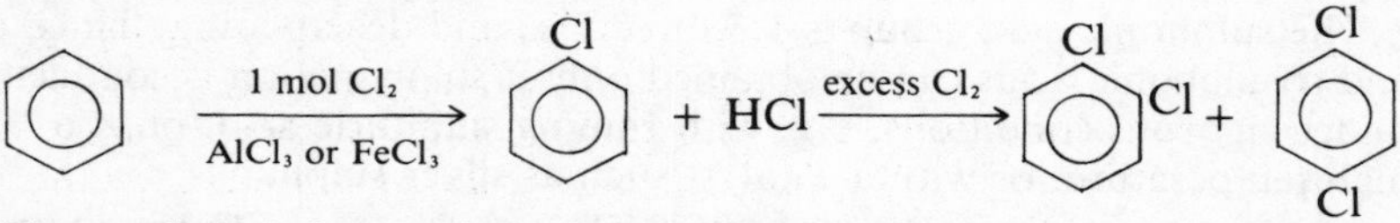

The catalytic halogenation of methylbenzene follows the same course:

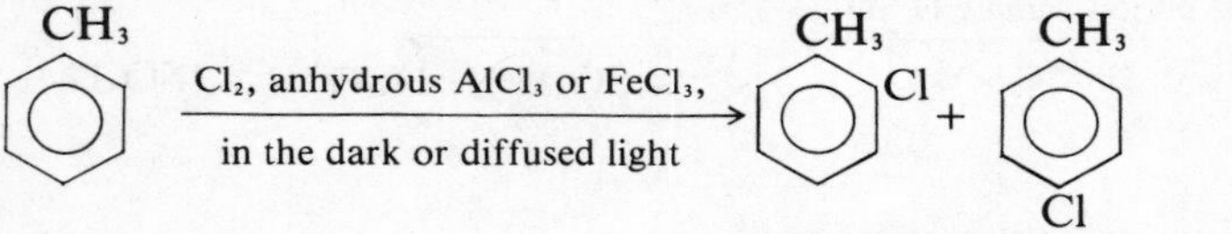

2- and 4-chloromethylbenzenes

Phenol and phenylamine, in which the benzene ring is activated by an electron donating group (HO or NH_2), undergo *tri*halogenation when they are treated with chlorine water or bromine water at room temperature, e.g.

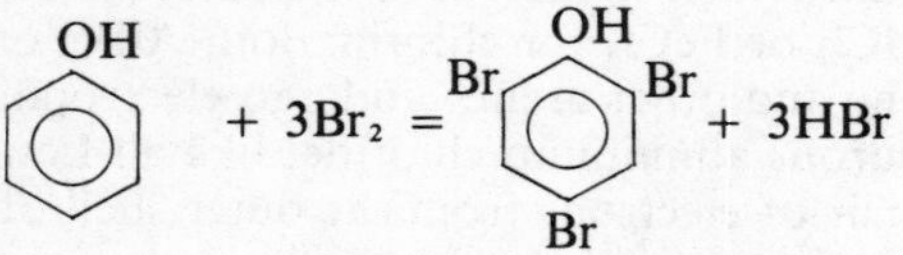

2,4,6-tribromophenol

Phenylamine, likewise, gives 2,4,6-tribromophenylamine. The products precipitate as white crystalline solids immediately the reactants are mixed together. No catalyst is required. The activating influence of HO and NH_2 groups is such that it is extremely difficult to get other than tribromo- or trichloro-derivatives.

The iodination of benzene produces only a small amount of iodobenzene:

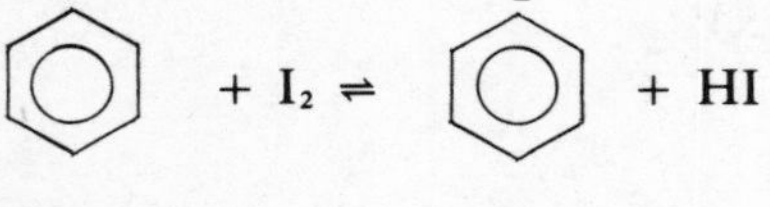

The reaction is reversible, and equilibrium lies well to the left-hand side. However, in the presence of a compound which reacts with HI, such as an oxidising agent or a mild alkali, equilibrium can be destroyed and a good yield of the iodide obtained. Thus, the iodination of benzene in the presence of nitric acid (an oxidant) gives iodobenzene in 87 per cent yield. Similarly, phenylamine can be iodinated to give 4-iodophenyl-amine in the presence of sodium hydrogencarbonate.

The fluorination of benzene (§9.5) produces no aromatic fluorine compounds.

Friedel–Crafts alkylation

The introduction of an alkyl group into the benzene ring by means of an alkyl halide or other alkylating agent is referred to as a *Friedel–Crafts reaction*. A Lewis acid catalyst is needed and, while many compounds have been suggested for this purpose, aluminium chloride usually gives the best yield. The reaction is often conducted in the presence of excess hydrocarbon which acts as a solvent but, should additional solvent be required, carbon disulphide or nitrobenzene may be used. At the optimum temperature of 20 °C (293 K) a reaction time of about 12 hours is needed, e.g.

$$\text{benzene} + CH_3Cl \xrightarrow[\text{20 °C (293 K)}]{AlCl_3} \text{toluene} + HCl$$

The mechanism of the reaction is similar to that of chlorination. The function of the catalyst is to enhance the polarisation and thus the electrophilicity of the alkyl halide.

$$\text{benzene} + \overset{\delta+}{R}\rightarrow\overset{\delta-}{Cl}\cdots AlCl_3 \rightarrow \underset{\pi\text{-complex}}{\boxed{\overset{\overset{\delta+}{R}-Cl\cdots\overset{\delta-}{AlCl_3}}{}}} \rightarrow \underset{\sigma\text{-complex}}{\boxed{\overset{H\quad R}{\oplus}}}$$

$$+ [AlCl_4]^- \xrightarrow[\text{abstracts } H^+]{[AlCl_4]^-} \overset{R}{\boxed{}} + HCl + AlCl_3$$

Other alkylating agents include alcohols and alkenes. Aluminium chloride is again the preferred catalyst, although it is less effective with alcohols and more of it is needed, e.g.

$$\text{benzene} \underset{C_2H_5OH + 1 \text{ mol } AlCl_3}{\overset{CH_2=CH_2 + 0.2 \text{ mol } AlCl_3}{\rightleftharpoons}} \overset{C_2H_5}{\boxed{}}$$

Limitations of the Friedel–Crafts reaction

Friedel–Crafts alkylation is of only limited importance in the laboratory as it suffers from a number of disadvantages, as follows.

(i) It is often difficult to prevent continued substitution, because one alkyl group will activate the ring towards further alkylation. With benzene and iodomethane, for instance, unless there is a large excess of benzene, the product is likely to be not methylbenzene but 1,3,5-trimethylbenzene (mesitylene).

(ii) When straight chain alkyl groups are being introduced, there is a marked tendency for them to become branched, e.g.

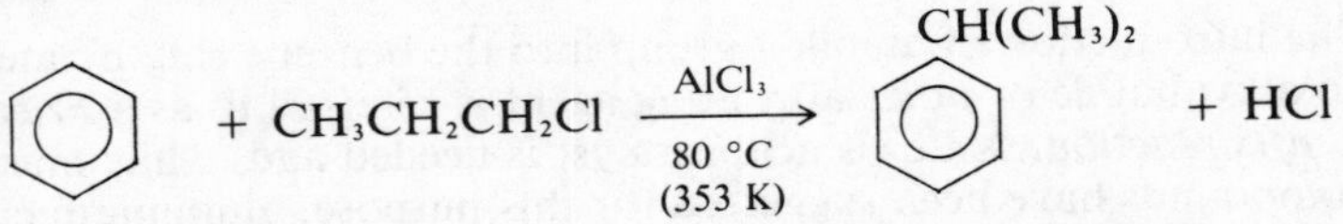

$$\text{C}_6\text{H}_6 + CH_3CH_2CH_2Cl \xrightarrow[\substack{80\ °C \\ (353\ K)}]{AlCl_3} \text{C}_6\text{H}_5CH(CH_3)_2 + HCl$$

The reason is that the alkyl halide–halogen complex tends to ionise, to give a carbonium ion which functions as an electrophilic reagent. Chain branching occurs because a primary carbonium ion largely rearranges to a more stable secondary carbonium ion before it attacks the benzene ring:

$$CH_3CH_2CH_2Cl \cdots AlCl_3 \rightleftharpoons CH_3CH_2\overset{\oplus}{C}H_2 + [AlCl_4]^-$$

$$CH_3CH_2\overset{\oplus}{C}H_2 \rightleftharpoons CH_3\overset{\oplus}{C}HCH_3$$

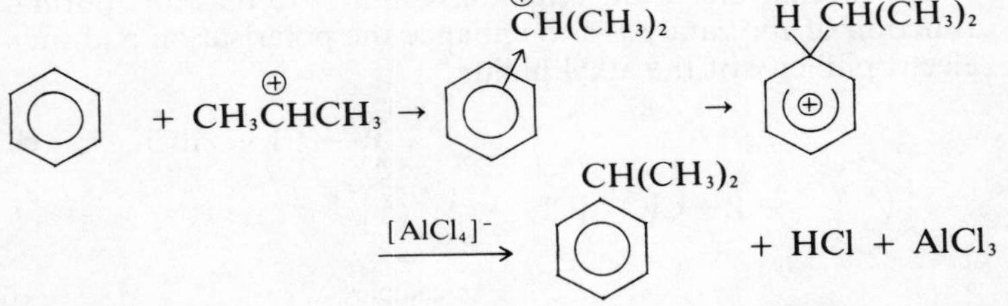

$$\text{C}_6\text{H}_6 + CH_3\overset{\oplus}{C}HCH_3 \rightarrow \cdots \rightarrow \cdots$$

$$\xrightarrow{[AlCl_4]^-} \text{C}_6\text{H}_5CH(CH_3)_2 + HCl + AlCl_3$$

(iii) The reaction must be carefully controlled to get the desired orientation of a disubstituted product. Suppose, for instance, that we wished to prepare 1,2- and 1,4-dimethylbenzenes by the methylation of methylbenzene — a reasonable objective since the CH_3 group is 1,2- and 1,4-directing:

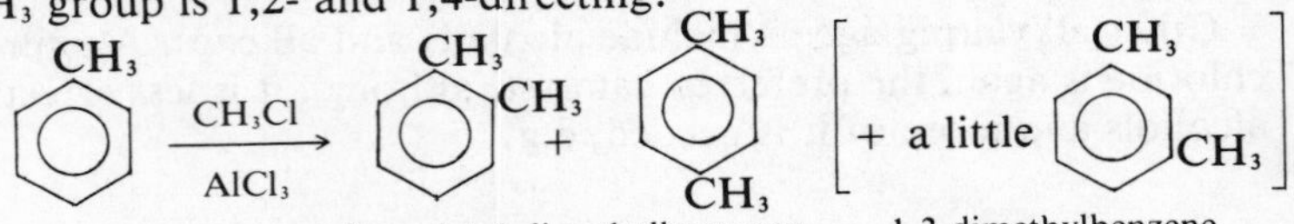

1,2- and 1,4-dimethylbenzenes   1,3-dimethylbenzene

To achieve this aim the reaction would have to be stopped after a short time, otherwise 1,3-dimethylbenzene would be the principal product. The reason is that Friedel–Crafts catalysts promote both alkylation and *dealkylation*, i.e. the reverse reaction. 1,2- and 1,4-Dimethylbenzenes are dealkylated in preference to the thermodynamically more stable 1,3-isomer, which thus accumulates and accounts for an ever-increasing proportion of the product.

Friedel–Crafts acylation

An acyl group, $R—C{\overset{O}{\diagup}}$ or $Ar—C{\overset{O}{\diagup}}$, can be introduced into the benzene ring by means of an acyl chloride (often called an 'acid chloride') or an acid anhydride. As in Friedel–Crafts alkylations, aluminium chloride is usually employed as a catalyst, and an excess of the hydrocarbon or carbon disulphide may be used as a solvent. The reaction, which provides the principal route to aromatic ketones, proceeds best at room temperature, although a reaction time of about 12 hours is necessary in order to ensure a good yield, e.g.

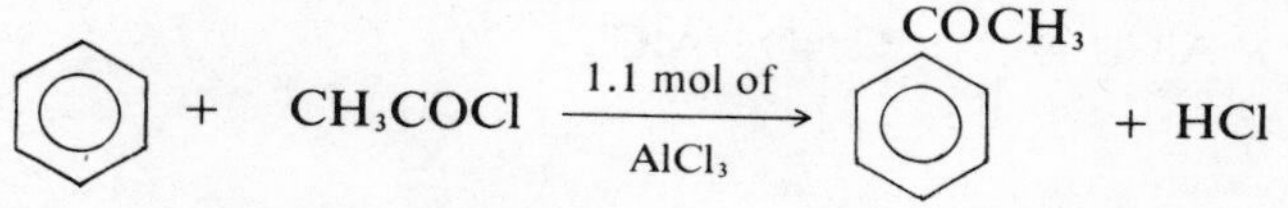

| | + CH_3COCl | $\xrightarrow[\text{AlCl}_3]{\text{1.1 mol of}}$ | | + HCl |

ethanoyl chloride   acetophenone
(formed as a complex with AlCl$_3$)

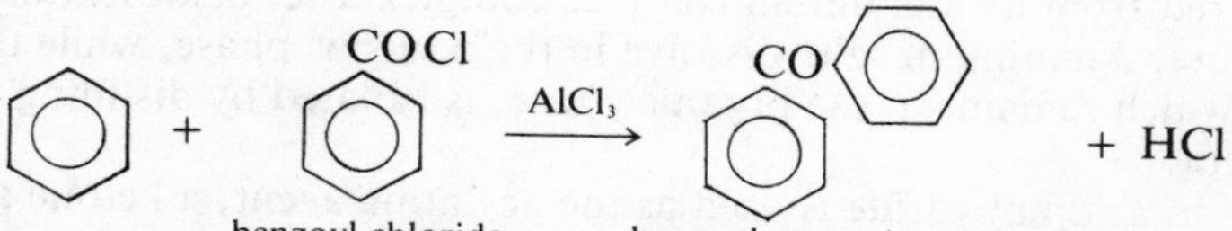

benzoyl chloride   benzophenone (as a complex)

Although the mechanism is similar to that of the Friedel–Crafts alkylation, there is some uncertainty about the origin of the electrons which are attracted by the catalyst. Clearly, the chlorine atom could donate a lone pair of electrons from its outer shell, as in the previous case, but oxygen donates electrons more readily than chlorine and it is considered likely that electrons of the C=O bond are involved:

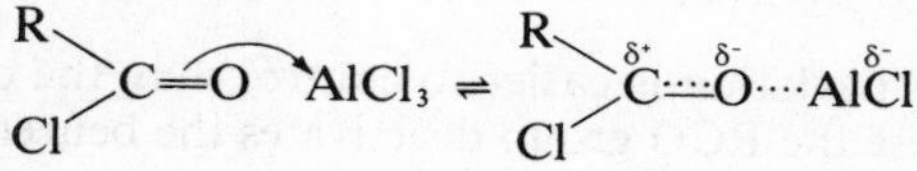

As before, it is conventional to write δ^- on the aluminium atom. Electrophilic substitution then occurs in the usual way:

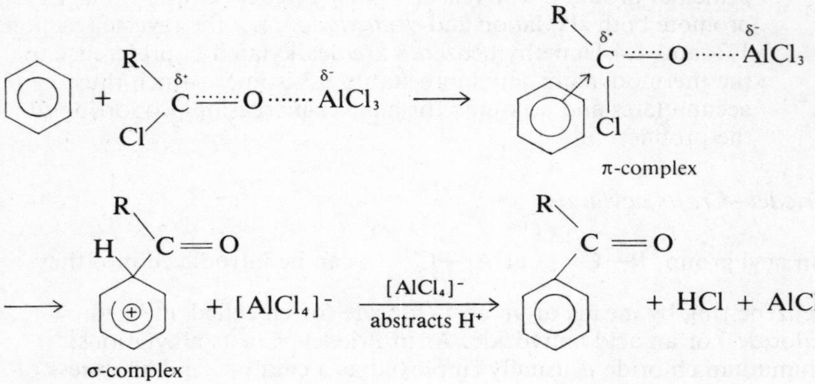

π-complex

σ-complex

The ketone, like the reagent, coordinates to the aluminium chloride:

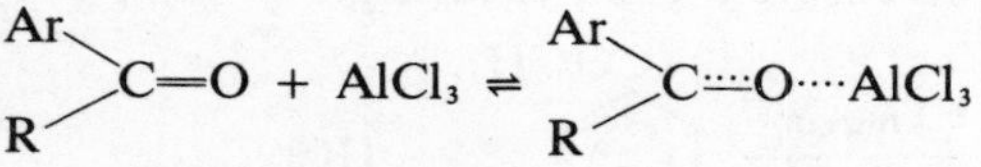

This means that an unusually large amount of catalyst is needed, since up to 1 mol of aluminium chloride per mole of acyl chloride is gradually removed during the reaction. When the reaction is complete, the ketone is recovered from its aluminium chloride complex after acidification of the mixture. Aluminium salts dissolve in the aqueous phase, while the ketone, which remains in the organic phase, is isolated by distilling off the solvent.

When an acid anhydride is used as the acylating agent, a ketone and a carboxylic acid are formed. Because both of these compounds react with 1 mol of aluminium chloride, at least 2 mol of the catalyst must be employed, e.g.

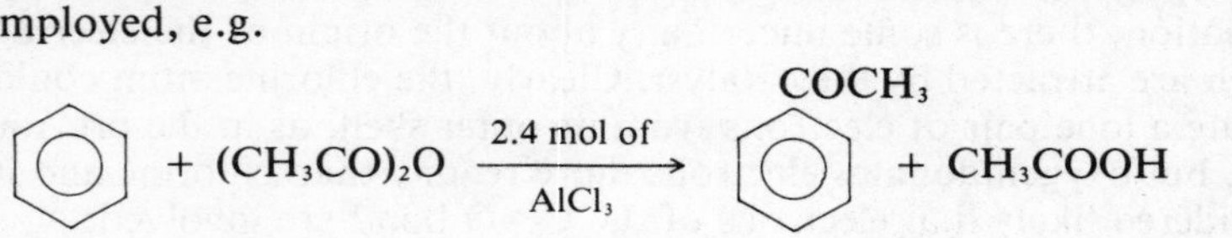

ethanoic anhydride

Friedel−Crafts acylation is easier to control than the corresponding alkylation. Because the RCO group deactivates the benzene ring, further reaction, leading to disubstituted products, does not occur.

Aromatic disubstitution

The reactions described above show that disubstitution yields predominantly either a 1,3-compound or a mixture of 1,2- and 1,4-isomers. It is the group already present which influences disubstitution; not the incoming group. 1,3-Directing groups always deactivate the ring, i.e. they render further electrophilic substitution difficult, while most 1,2- and 1,4-directing groups activate the ring and make further substitution easier. An important exception to this rule concerns the halogen atoms, Cl, Br and I: they are 1,2- and 1,4-directing but deactivating.

Although various rules have been proposed to help in predicting the directive influence of a substituent group, none is entirely satisfactory. The best approach lies in classifying substituent groups according to their structure.

Class (i)   Those in which the substituent atom (i.e. the one which is attached to the ring) is further joined by a multiple bond to a more electronegative atom, e.g.

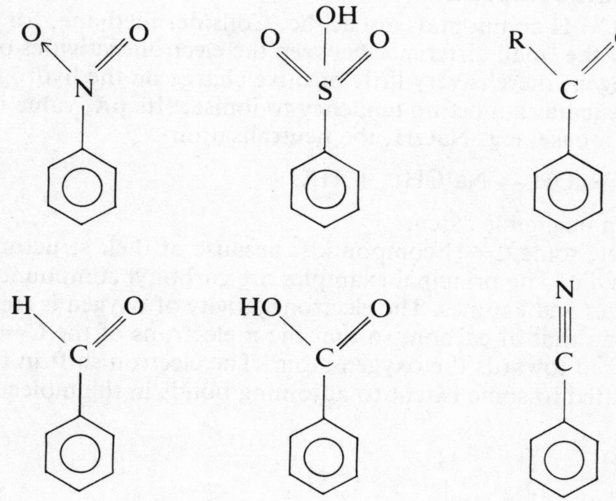

All such groups are 1,3-directing.

Class (ii)   Those which consist of a single atom, or groups XY, XY_2, etc, in which there is no multiple bonding, e.g.

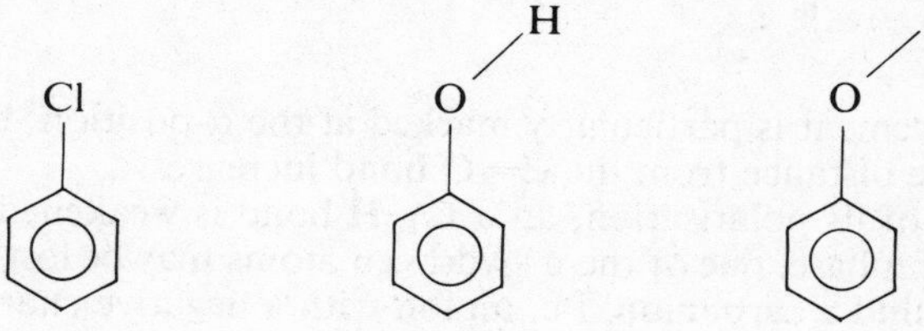

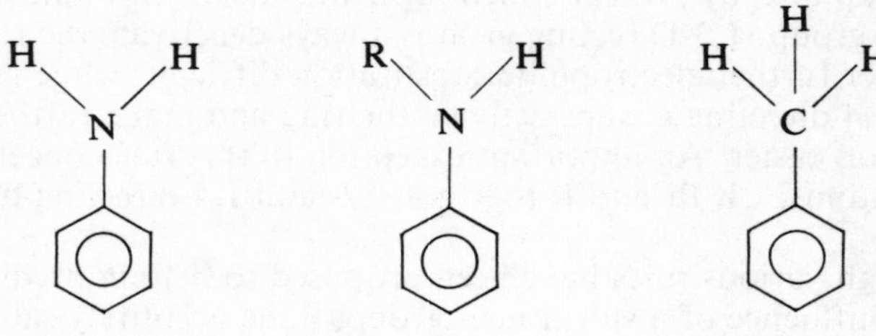

All the groups illustrated here are both 1,2- and 1,4-directing.

However, a few such groups, notably CCl_3, $\overset{\oplus}{N}H_3$ and $\overset{\oplus}{N}R_3$, are 1,3-directing.

C—H acidic compounds

Very few C—H compounds are acidic. Consider methane, for example. Because of the small difference between the electronegativities of carbon and hydrogen, there is very little positive charge on the hydrogen atoms and methane has almost no tendency to ionise. (Its pK_a value is over 40.) With a base, e.g. NaOH, the neutralisation

$$CH_4 + NaOH \rightleftharpoons Na^+CH_3^- + H_2O$$

occurs to a negligible extent.

However, some C—H compounds, because of their structures, are slightly acidic. The principal examples are carbonyl compounds, such as aldehydes and ketones. The electronegativity of oxygen is considerably greater than that of carbon, so that the π electrons of the C=O bond are displaced towards the oxygen atom. The electron shift in this bond is transmitted to some extent to adjoining bonds in the molecule, thus:

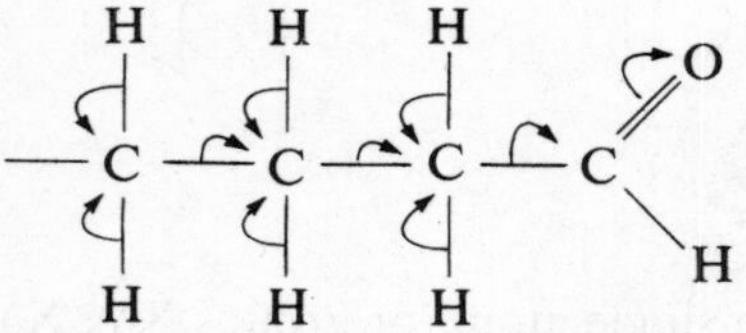

Note Curved arrows show the displacement of bonding pairs of electrons.

Positions   γ   β   α

The displacement is particularly marked at the α-position, but dies away as the distance from the C=O bond increases.

Because of its polarisation, an α C—H bond is weakened. In the presence of a base, one of the α-hydrogen atoms may be lost as a proton, leaving behind a carbanion, i.e. an ion with a negative charge on a carbon atom:

$$RCH_2C{\overset{\displaystyle O}{\underset{\displaystyle H}{<}}} + HO^- = R\overset{\ominus}{C}HC{\overset{\displaystyle O}{\underset{\displaystyle H}{<}}} + H_2O$$

Derivatives of carboxylic acids, notably acid chlorides and esters, act in the same way, as do nitriles and nitro compounds. Carboxylic acids themselves, however, behave somewhat differently, for when they are treated with a base neutralisation occurs in such a way that negative charge resides on an *oxygen* atom:

$$R-C{\overset{\displaystyle O}{\underset{\displaystyle OH}{<}}} + HO^- = R-C{\overset{\displaystyle O}{\underset{\displaystyle O^\ominus}{<}}} + H_2O$$

Other groupings which increase the acidity of a hydrogen atom are C=C and C≡C; the rule is that acidity increases in the following order:

$$\underset{\diagup}{\overset{\diagdown}{C}}-\underset{\diagdown}{\overset{\diagup}{C}}-H \quad < \quad \underset{\diagup}{\overset{\diagdown}{C}}=C{\overset{H}{\underset{\diagdown}{<}}} \quad < \quad -C\equiv C-H$$

Nevertheless, alkenes in general undergo addition reactions almost entirely to the exclusion of substitution, and the situation with alkynes is similar except that *metalation* occurs readily, with the substitution of hydrogen atoms in ≡C—H groupings by atoms of certain metals, notably sodium, copper and silver.

We shall now discuss the more important C—H acidic compounds, except that their reactions with carbonyl compounds will be left until Chapter 13.

Aldehydes and ketones

As a result of the acidity of their α-hydrogen atoms, aldehydes and ketones are readily halogenated at the α-position under alkaline conditions. Chlorine, bromine and even iodine give trisubstituted products, e.g.

$$CH_3COCH_3 + 3Br_2 + 3HO^- = CBr_3COCH_3 + 3Br^- + 3H_2O$$

propanone   1,1,1-tribromopropanone

The reaction is an electrophilic substitution, comparable with electrophilic substitution in benzene except that a catalyst is not required. The bromine molecule becomes polarised as it approaches the carbanion:

$$\underset{\substack{\delta^- \quad \delta^+ \\ Br-Br}}{} \quad \underset{\substack{| \\ H}}{H-\overset{\ominus}{\underset{|}{\ddot{C}}}-\overset{\overset{O}{\parallel}}{C}-CH_3} \rightarrow H-\overset{\overset{\uparrow Br}{}}{\underset{\underset{H}{|}}{C}}-\overset{\overset{O}{\parallel}}{C}-CH_3 + Br^-$$

The monobrominated ketone is more susceptible to bromination than the original compound, because the bromine atom exerts a $-I$ effect. This enhances the acidity of the remaining α-hydrogen atoms and, in addition, stabilises the resultant carbanion by delocalising its negative charge. As a result, the reaction continues to give the dibromo- and ultimately the tribromoketone.

Trihalogenated aldehydes and ketones are difficult to isolate because they readily undergo alkaline cleavage to give haloforms, i.e. trihalomethanes (see below).

In the presence of acid, the halogenation of aldehydes and ketones proceeds by a different mechanism, and this leads to a reversal of two of the principles just described. α-Hydrogen atoms are attacked, as before, but this time the monosubstituted product can be isolated as it is *less* readily halogenated than the original compound, e.g.

$$CH_3COCH_3 + Br_2 \xrightarrow{H^+ \text{ catalyst}} CH_2BrCOCH_3 + HBr$$

1-bromopropanone

When other methyl ketones are used, the methylene group (carbon 3) is attacked preferentially; not the methyl group:

$$\overset{1}{C}H_3\overset{2}{C}O\overset{3}{C}H_2R + Br_2 = CH_3COCHBrR + HBr$$

Haloform reaction

Halogenation under alkaline conditions is used in the *haloform reaction* for preparing the haloforms, namely: chloroform, $CHCl_3$; bromoform, $CHBr_3$; and iodoform, CHI_3. Three reactants are required.

(i) Ethanal or a methyl ketone. The $CH_3C{\overset{\nearrow O}{\searrow}}$ group in an aldehyde or ketone is essential.

(ii) The appropriate halogen, i.e. chlorine, bromine or iodine.

(iii) An alkali, e.g. sodium hydroxide.

The reaction proceeds in two stages.

Substitution The $CH_3C{\overset{\displaystyle O}{\diagup}}_{\diagdown}$ grouping becomes halogenated to

$CX_3C{\overset{\displaystyle O}{\diagup}}_{\diagdown}$, where X = Cl, Br or I, e.g.

$$CH_3CHO + 3Cl_2 + 3NaOH = CCl_3CHO + 3NaCl + 3H_2O$$

In the case of propanone (see above) only *one* CH_3 group becomes substituted.

Alkaline cleavage The trihalogenated aldehyde or ketone is immediately attacked by the alkali to give a haloform and a salt of a carboxylic acid, e.g.

$$CCl_3CHO + NaOH = CHCl_3 + HCOONa$$

This unlikely looking reaction — it involves the breaking of a C—C bond — becomes understandable when it is realised that the carbonyl carbon atom is quite strongly positively charged, because the usual charge separation of the C=O bond is reinforced by the inductive effect of *three* halogen atoms. Thus, a nucleophilic substitution (§11.5) can occur, in which an HO^- ion joins the carbonyl carbon atom and displaces a CCl_3^- anion:

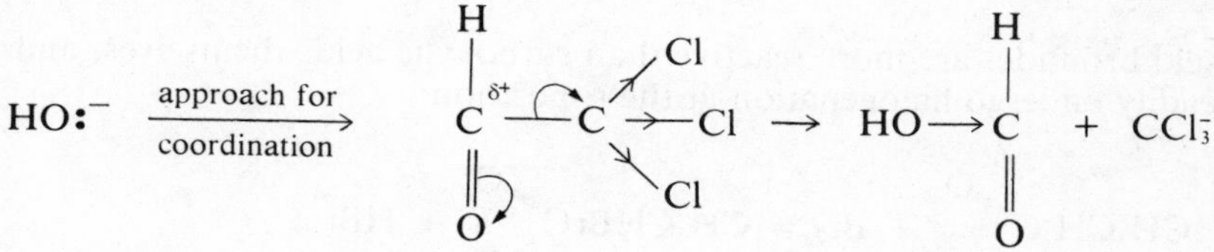

The transfer of a proton from $HCOOH$ to CCl_3^- completes the reaction:

$$HCOOH + CCl_3^- \rightarrow HCOO^- + CHCl_3$$

Ethanol and secondary alcohols possessing a methyl group also yield haloforms upon treatment with halogen and alkali, not because they themselves undergo the haloform reaction, but because they become oxidised by the halogen or the hypohalite ion to aldehydes or methyl ketones:

$$C_2H_5OH + NaClO = CH_3CHO + NaCl + H_2O$$
$$CH_3CH(OH)R + NaClO = CH_3COR + NaCl + H_2O$$

These principles are utilised in the *iodoform test* in qualitative analysis. The test is carried out by placing approximately 1 cm³ of the unknown compound in a test tube and then adding about 3 cm³ of a solution of iodine in aqueous potassium iodide. This is followed by sodium hydroxide solution, dropwise, until the brown colour of the iodine almost disappears. Finally, the solution is cooled. A yellow crystalline precipitate of iodoform is positive for the presence of ethanal, a methyl ketone, ethanol, or a secondary alcohol with a methyl group.

Carboxylic acids

Carboxylic acids do not react with halogens under alkaline conditions because a carbanion is not formed (see above), but they do so, largely at the α-position, in the presence of red phosphorus, e.g.

$$CH_3CH_2COOH + Br_2 \xrightarrow{\text{red phosphorus}} CH_3CHBrCOOH + HBr$$

propanoic acid   2-bromopropanoic acid

The purpose of the red phosphorus is to convert some of the acid into its acid bromide (acyl bromide):

$$2P + 3Br_2 = 2PBr_3$$

$$3CH_3CH_2C\overset{O}{\underset{OH}{\big\langle}} + PBr_3 = 3CH_3CH_2C\overset{O}{\underset{Br}{\big\langle}} + H_2PHO_3$$

Acid bromides are more reactive than carboxylic acids themselves, and readily undergo halogenation at the α-position:

$$CH_3CH_2C\overset{O}{\underset{Br}{\big\langle}} + Br_2 = CH_3CHBrC\overset{O}{\underset{Br}{\big\langle}} + HBr$$

Finally, the brominated acid bromide and carboxylic acid react together with an exchange of functional groups:

$$CH_3CHBrC\overset{O}{\underset{Br}{\big\langle}} + CH_3CH_2C\overset{O}{\underset{OH}{\big\langle}} = CH_3CHBrC\overset{O}{\underset{OH}{\big\langle}}$$

$$+ CH_3CH_2C\overset{O}{\underset{Br}{\big\langle}}$$

The acid bromide once again becomes substituted, and so the reaction continues.

Further halogenation may occur, but in a stepwise manner, because the halogenation of carboxylic acids increases their resistance to attack under these conditions. The chlorination of ethanoic acid at its boiling temperature is particularly well known.

$$CH_3COOH + Cl_2 \xrightarrow{\text{red phosphorus}} CH_2ClCOOH + HCl$$

$$CH_2ClCOOH + Cl_2 = CHCl_2COOH + HCl$$
$$CHCl_2COOH + Cl_2 = CCl_3COOH + HCl$$

Each stage is complete before the next commences.

The reaction can be stopped at either of the intermediate stages with a good yield of chloroethanoic acid or dichloroethanoic acid. Contrast this with the chlorination of methane, which proceeds by a free radical mechanism and invariably gives a mixture of products.

Alkynes

1-Alkynes, containing the grouping $\equiv CH$, are weak protic acids. Ethyne (acetylene) has a pK_a value of ~ 26, which shows that it is a weaker acid than ethanol ($pK_a \approx 16$) but much stronger than methane ($pK_a > 40$). Such acidity is remarkable in a hydrocarbon, and results from the relative remoteness of the electrons of the $C \equiv C$ bond from those of

the C—H bond (§8.4). This leads to charge separation, $\equiv \overset{\delta^-}{C} \overset{\cdot\cdot}{\underset{\times}{\cdot}} \overset{\delta^+}{H}$, and the loss of the hydrogen atom to a suitable acceptor.

Sodium

Ethyne reacts with sodium to give monosodium acetylide, $NaC \equiv CH$, or disodium acetylide, $NaC \equiv CNa$, depending on the proportions of the reactants. (With a limited amount of sodium the monosodium derivative predominates.) Potassium behaves similarly. Reaction can be brought about either by bubbling ethyne through a solution of sodium in liquid ammonia — in effect, by reaction with sodium amide, $NaNH_2$ — or by passing the gas over warm sodium:

$$CH \equiv CH + Na^+NH_2^- = NaC \equiv CH + NH_3$$
$$2CH \equiv CH + 2Na = 2NaC \equiv CH + H_2$$

Note that the NH_2^- ion is strongly basic and accepts a proton.

Sodium acetylides are electrovalent and are decomposed by water, cf. calcium acetylide.

$$NaC \equiv CH + H_2O = CH \equiv CH + NaOH$$

They will also react with alkyl halides to form higher alkynes:

$$NaC \equiv CH + RX = RC \equiv CH + NaX$$

The other hydrogen atom can now be replaced by sodium and the process repeated — not necessarily with the same alkyl halide — to give an alkyne $RC \equiv CR$.

Copper

A reddish brown precipitate of copper(I) acetylide can be obtained by passing ethyne through an ammoniacal solution of copper(I) chloride:

$$CH \equiv CH + 2[Cu(NH_3)_2]^+ = CuC \equiv CCu + 2NH_4^+ + 2NH_3$$

Ammonia acts as the proton acceptor.

The general reaction is:

$$RC{\equiv}CH \xrightarrow{\text{ammoniacal CuCl}} RC{\equiv}CCu(s)$$

This is used in the laboratory as a test for the presence of the ${\equiv}CH$ grouping.

Copper(I) acetylide has a covalent structure and is unaffected by water, but it can be decomposed by acids to give pure ethyne. **It detonates in the dry state when it is struck or heated.**

Silver

Silver acetylide, which has properties similar to those of copper(I) acetylide, can be prepared by the action of ethyne on ammoniacal silver nitrate. It forms as a white precipitate:

$$CH{\equiv}CH + 2[Ag(NH_3)_2]^+ = AgC{\equiv}CAg + 2NH_4^+ + 2NH_3$$

For a general case,

$$RC{\equiv}CH \xrightarrow{\text{ammoniacal AgNO}_3} RC{\equiv}CAg(s)$$

Chapter 11

The C—halogen bond

Data on C—halogen bonds

	C—F	C—Cl	C⋯Cl (C_6H_5Cl)	C—Br	C—I
covalent bond length/nm	0.138	0.177	0.169	0.193	0.214
bond dissociation enthalpy/ kJ mol^{-1} at 298 K	484	338	457	276	238

Carbon—halogen bonds are encountered in a wide variety of organic halides, as follows:

(i) *alkyl halides,* i.e. haloalkanes, e.g. chloromethane, CH_3Cl;

(ii) *aryl halides,* i.e. aromatic halides in which a halogen atom is attached directly to the benzene ring, e.g. chlorobenzene, C_6H_5Cl;

(iii) *arylalkyl halides,* i.e. aromatic halides in which one or more halogen atoms appear in an aliphatic side chain, e.g. (chloromethyl)benzene, $C_6H_5CH_2Cl$;

(iv) *1,1-dihalides,* i.e. dihalides in which the two halogen atoms reside on the same carbon atom, e.g. 1,1-dichloroethane, CH_3CHCl_2;

(v) *1,2-dihalides,* i.e. dihalides in which the two halogen atoms are attached to adjoining carbon atoms, e.g. 1,2-dichloroethane, CH_2ClCH_2Cl;

(vi) *acyl halides,* otherwise known as *acid halides,* e.g. ethanoyl chloride, CH_3COCl;

(vii)*polyhalides,* e.g. trichloromethane, $CHCl_3$.

11.1 Nomenclature of organic halides

The commonest types of systematic names listed by IUPAC are as follows.

(i) *Substitutive names,* which are based on the names of the alkanes or arenes from which the halides are derived. For example, the compound of formula CH_3Cl is derived from methane by the substitution of an atom of chlorine for one of hydrogen, and is given the substitutive name of chloromethane.

(ii) *Radicofunctional names,* which are drawn from the names of the hydrocarbon radicals present in the molecules. On this system the compound CH_3Cl is called methyl chloride. Table 11.1 shows some common radicofunctional names, together with the equivalent substitutive names.

Table 11.1 Nomenclature of organic halides

Formula	Substitutive name	Radicofunctional name
CH_3Cl	chloromethane	methyl chloride
CH_3CH_2Cl	chloroethane	ethyl chloride
$CH_3CH_2CH_2Cl$	1-chloropropane	propyl chloride
$CH_3CHClCH_3$	2-chloropropane	isopropyl chloride
$CH_2{=}CHCl$	chloroethene	vinyl chloride
$CH_2{=}CHCH_2Cl$	1-chloro-2-propene	allyl chloride
CH_2Cl_2	dichloromethane	methylene dichloride
CH_3CHCl_2	1,1-dichloroethane	ethylidene dichloride
CH_3CCl_3	1,1,1-trichloroethane	ethylidyne trichloride
 C_6H_5Cl (Cl-benzene)	chlorobenzene	phenyl chloride*
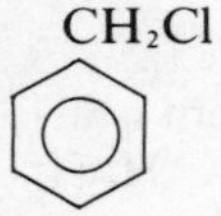 $C_6H_5CH_2Cl$ (CH₂Cl-benzene)	(chloromethyl)benzene	benzyl chloride
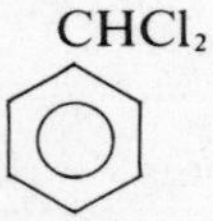 $C_6H_5CHCl_2$ (CHCl₂-benzene)	(dichloromethyl)benzene	benzylidene dichloride
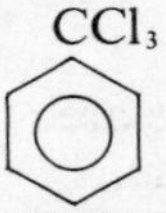 $C_6H_5CCl_3$ (CCl₃-benzene)	(trichloromethyl)benzene	benzylidyne trichloride*

*These names are not in current use. For $C_6H_5CCl_3$, the trivial name of benzotrichloride is widely used, although this is not recognised by IUPAC.

For the *haloforms,* i.e. trihalomethanes, CHX_3, the trivial names of fluoroform, chloroform, bromoform and iodoform are retained by IUPAC.

For the naming of acyl halides, see §13.1.

11.2 Formation of the C-halogen bond

Both substitution and addition reactions may be employed.

Substitution methods

From C—H; the substitution of H by Cl, Br or I (§10.2)
Alkanes undergo free radical chlorination or bromination in ultraviolet light at room temperature, or in the dark on heating to about 400 °C (673 K), e.g.

$$CH_4 + Cl_2 = CH_3Cl + HCl$$

Similar reactions occur when methylbenzene is treated with chlorine or bromine under these conditions.

The substitution of H by halogen occurs more readily, but by different mechanisms, in aldehydes, ketones and carboxylic acids, where the C—H bond is activated by a neighbouring C=O bond, e.g.

$$CH_3COCH_3 + 3Cl_2 + 3NaOH = CCl_3COCH_3 + 3NaCl + 3H_2O$$

The C—H bond in the benzene ring is subject to electrophilic substitution with chlorine or bromine in diffused light and the presence of a halogen carrier, e.g.

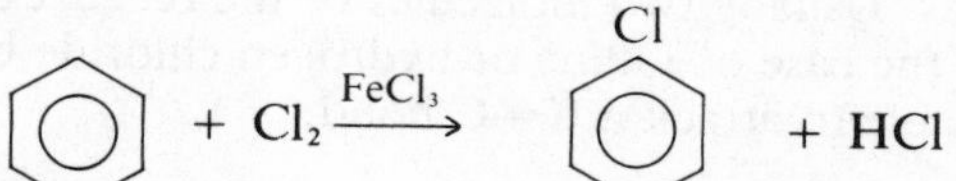

From C—OH; the substitution of HO by Cl, Br or I
Alcohols can be converted into alkyl halides by treatment with a covalent inorganic halide, usually a hydrogen halide or a phosphorus halide, e.g.

$$C_2H_5OH + PCl_5 = C_2H_5Cl + POCl_3 + HCl \qquad (\S 12.5)$$

Phenols do not react with hydrogen halides, and even with phosphorus halides there is little aryl halide formation (§12.5).

The HO group in carboxylic acids responds to phosphorus halides but not to hydrogen halides, e.g.

$$CH_3C\begin{matrix}{=O}\\{\diagdown OH}\end{matrix} + PCl_5 = CH_3C\begin{matrix}{=O}\\{\diagdown Cl}\end{matrix} + POCl_3 + HCl \qquad (\S 13.8)$$

From C—NH₂; the substitution of NH₂ by Cl, Br or I (§14.6)

Aromatic (but not aliphatic) primary amines can be diazotised with nitrous acid at 5 °C (278 K). Suitable treatment of the diazonium salt will result in the substitution of fluorine, chlorine, bromine or iodine atoms.

From C=O; the substitution of O by 2Cl or 2Br

Aldehydes and ketones react with phosphorus pentachloride to give 1,1-dichlorides, e.g.

$$CH_3C{\overset{O}{\underset{H}{\diagdown}}} + PCl_5 = CH_3CHCl_2 + POCl_3$$

Yields are quite good from aldehydes, but less so from ketones.

Phosphorus pentabromide can be used to prepare 1,1-dibromides, but has a tendency to dissociate into phosphorus tribromide and bromine, and the latter may bring about substitution in the akyl group of the carbonyl compound.

Addition methods

From C=C (§7.4)

Alkenes undergo electrophilic addition with chlorine or bromine to give 1,2-dihalides and with hydrogen halides to give alkyl halides.

From C≡C (§8.4)

Alkynes, like alkenes, respond to the addition of halogens and hydrogen halides. Usually two molecules of the reagent are involved, but only one in the case of iodine or hydrogen chloride because these reagents are unable to attack a C=C bond.

11.3 Manufacture of organic halides

Table 11.2 shows how some of the lower aliphatic halides are manufactured. Several of these compounds are in common use as solvents, but are toxic. **Tetrachloromethane in particular represents a major health hazard, for its vapour, even at low concentrations, can cause damage to the liver.**

11.4 General properties of organic halides

The introduction of a halogen atom into a hydrocarbon raises its relative molecular mass and hence its boiling temperature. Of the alkyl halides, only three (CH_3Cl, CH_3Br and C_2H_5Cl) are gaseous at ordinary temperatures. The rest, below C_{16}, are liquids. Because of the relationship between relative molecular mass and boiling temperature, the iodides have higher boiling temperatures than the bromides, which in turn are higher boiling than the chlorides.

Table 11.2 Industrial chemistry of some aliphatic halides

Halide	Manufacture	Uses
chloromethane	$CH_3OH + HCl \xrightarrow[\text{catalyst}]{ZnCl_2} CH_3Cl + H_2O$ or $CH_4 + Cl_2 \xrightarrow[\text{(673 K)}]{400°C} CH_3Cl + HCl$	for the manufacture of tetramethyllead, silicone polymers and methylcellulose
bromomethane	$CH_3OH + HBr = CH_3Br + H_2O$	as a fumigant for grain
chloroethane	$CH_3CH_3 + Cl_2 \xrightarrow[\text{(673 K)}]{400°C} CH_3CH_2Cl + HCl$ $CH_2{=}CH_2 + HCl \xrightarrow[\substack{\text{(from} \\ \text{above)}}]{\substack{35°C\ (308\ K) \\ 30\ atm \\ (3\ 000\ kPa) \\ AlCl_3\ catalyst}} CH_3CH_2Cl$	for the manufacture of tetraethyllead and ethylcellulose; as a local anaesthetic
dichloromethane	chlorination of methane	as a solvent for plastics; in paint strippers
1,2-dichloroethane	$CH_2{=}CH_2 + Cl_2 \xrightarrow[\text{catalyst}]{FeCl_3} CH_2ClCH_2Cl$	for the manufacture of chloroethene; as a solvent, e.g. for degreasing metals
trichloromethane (chloroform)	chlorination of methane	for the manufacture of poly(tetrafluoroethene), see Table 7.2
tetrachloromethane (carbon tetrachloride)	chlorination of methane or carbon disulphide	for the manufacture of CCl_2F_2 (Freon), used as a refrigerant and an aerosol propellant
hexachloroethane (perchloroethylene)	from ethyne, via 1,1,2-trichloroethene (see p. 137), or by the pyrolysis of CCl_4	as a safe solvent for 'drycleaning' processes

Although the halides are strongly polar compounds — the negative inductive effect of the halogen atom underlies much of their chemistry — they are immiscible with water because of the absence of hydrogen bonding. Their densities are mostly greater than 1 g cm^{-3} (1 kg dm^{-3}), and with water they form a separate *lower* phase.

All organic halides are fire resistant, although they decompose when heated in air. Tetrachloromethane, for instance, gives a certain amount of carbonyl chloride — a highly toxic gas. Certain halides, notably bromochlorodifluoromethane, are used in fire extinguishers.

11.5 Chemical properties of the C-halogen bond

The bond is most easily studied in the alkyl halides. These compounds undergo nucleophilic substitution reactions with a wide range of reagents, and elimination reactions (dehydrohalogenation) to give alkenes. Furthermore, they react with magnesium to form Grignard reagents.

Nucleophilic substitution reactions

The carbon atom of the C—halogen bond carries a partial positive charge, due to the $-I$ effect of the halogen atom (§1.4). The carbon atom therefore attracts nucleophilic reagents, such as the HO^- ion. As the ion approaches the carbon atom, the electrons of the C—X bond (where X = Cl, Br or I) are repelled towards the halogen atom. The bond becomes increasingly polarised, i.e. less covalent and more electrovalent. Eventually, the HO^- ion coordinates on to the carbon atom, through a lone pair of electrons in the outer shell of the oxygen atom, and at about the same time the electrons of the C—X bond move completely to the halogen atom to give a separate halide ion, X^-.

HO⁻ ion approaching C atom

HO⁻ ion close to C; C—X bond strongly polarised

HO⁻ ion coordinated on to C atom; X⁻ ion leaving

In this way the HO group replaces the halogen atom in a *nucleophilic substitution reaction*.

Alkyl halides vary considerably in their reactivity towards nucleophiles. Besides the decrease in reactivity which is always experienced on ascending a homologous series, there are two other factors which play an important role.

(i) The nature of the halogen. The reactivity order of organic halides in general is

iodides > bromides > chlorides

At first sight we might expect chlorides to be the most reactive. The C—Cl bond is more polarised than C—Br or C—I, because of the relatively high electronegativity of chlorine, and this leads to a strong attraction for nucleophiles. However, and this is the more important consideration, the C—I bond is the most easily polarised *at the approach of nucleophiles*. Consequently, the bond is easily broken with the loss of an iodide ion. (Note that the C—I bond has a low bond dissociation enthalpy.) I⁻ is said to be a 'good leaving group'.

(ii) The class of halide. Like alcohols, alkyl halides are classified as primary, secondary or tertiary compounds:

RCH_2X	R_2CHX	R_3CX
primary	*secondary*	*tertiary*

For a given halogen atom, the reactivity order is

$$\text{tertiary} > \text{secondary} > \text{primary}$$

The cause is attributed to the $+I$ effect of alkyl groups. In a molecule of a tertiary alkyl halide there are three such groups attached to the carbon atom of the C—halogen bond. Each exerts its electron releasing effect. This increases the polarisation of the C—halogen bond and therefore weakens it:

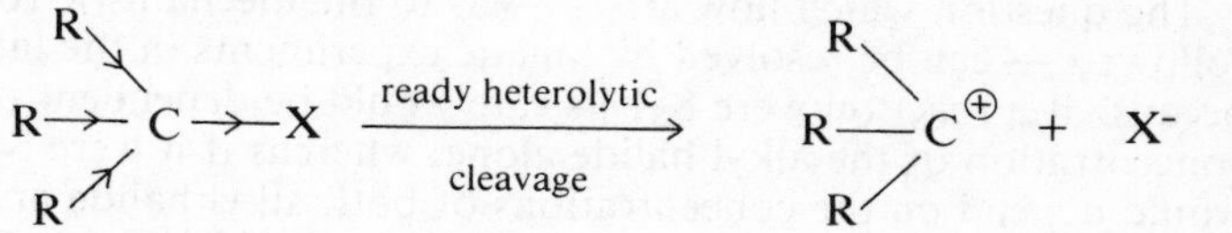

In secondary and primary alkyl halides there are fewer alkyl groups and the effect is correspondingly smaller.

S_N1 and S_N2 reactions

In a reaction between an alkyl halide and a hydroxide, it is conceivable that a halide ion X^- could leave before or after the HO^- ion joins the carbon atom. If X^- should leave *before* the HO^- ion joins, a carbonium ion will exist for a short time as shown in route 1 below. If, on the other hand, X^- should leave *after* the HO^- ion has joined, there will exist momentarily a *transition state* in which X and HO (each with a partial negative charge) lie at right angles to a plane in which the other atoms or groups lie at 120° to one another; see route 2.

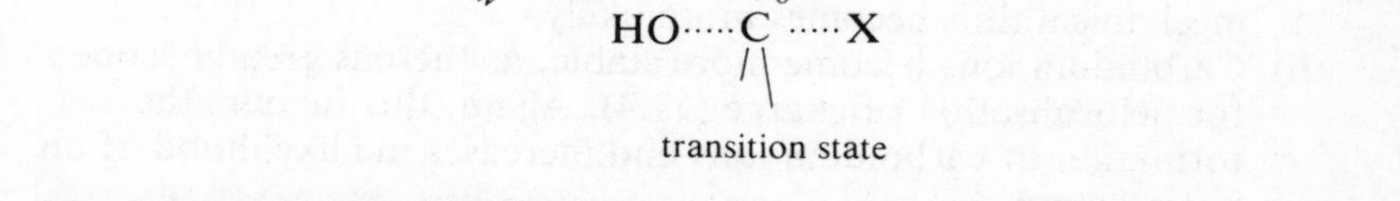

transition state

The first stage of both routes, involving the attainment of an unstable intermediate, would probably be slow and hence rate determining. (In contrast, the second stage, involving a highly reactive carbonium ion or transition state, would occur very quickly.) The formation of a carbonium ion (route 1) concerns only the alkyl halide:

$$RX \rightarrow R^+ + X^-$$

and since there is only one species undergoing change in the rate determining step, route 1 is described as S_N1. (S = substitution,

N = nucleophilic, 1 = unimolecular.) However, the formation of the transition state (route 2) involves both the alkyl halide and the hydroxide ion:

$$RX + HO^- \rightarrow \overset{\delta^-}{H}\overset{|\overset{\delta^+}{}}{O}\cdots\overset{}{\underset{/\,\backslash}{R}}\cdots\overset{\delta^-}{X}$$

Because there are two species taking part in the rate determining step the reaction is bimolecular and said to be S_N2.

The question which now arises — as to the mechanistic route actually followed — can be resolved by kinetic experiments in the laboratory, because if a reaction were S_N1 its rate would be dependent on the molar concentration of the alkyl halide alone, whereas if it were S_N2 its rate would depend on the concentrations of both alkyl halide and alkali.

It is found that the hydrolysis of tertiary alkyl halides is first order, i.e.

rate $\propto$ [RX]   (independent of [HO$^-$])

Therefore only one species (the alkyl halide) is undergoing change in the rate determining stage and such reactions have an S_N1 mechanism.

The hydrolysis of primary alkyl halides, however, is second order:

rate $\propto$ [RX] [HO$^-$]

Here, two species are undergoing change in the rate determining step and the reaction must be S_N2.

The rates at which secondary alkyl halides hydrolyse depend on the concentrations of the halides and, to a lesser extent, the alkali, which suggests that the reactions are partly S_N1 and partly S_N2, i.e. some molecules react by one route and some by the other.

The variation of behaviour is due mainly to the + I effect of alkyl groups. On progressing from primary to secondary to tertiary alkyl halides, the number of alkyl groups exerting this effect increases. This has the following consequences.

(i) The C—halogen bond becomes increasingly polarised, so that carbonium ions can be formed more readily (see above). An S_N1 mechanism thus becomes more likely.

(ii) Carbonium ions become more stable, as there is greater scope for delocalisation of charge (§7.4). Again, this favours the formation of carbonium ions and increases the likelihood of an S_N1 reaction.

(iii) The partial positive charge on the carbon atom of the C—halogen bond becomes increasingly neutralised, so that coordination of

nucleophiles is less likely. For example, $\overset{\delta^+}{C}H_3 \rightarrow \underset{\underset{\overset{\uparrow}{\overset{\delta^+}{C}H_3}}{}}{\overset{\overset{\overset{\delta^+}{C}H_3}{\downarrow}}{C}} - \overset{\delta^-}{X}$ is less susceptible than

$$H-\underset{\underset{\displaystyle H}{|}}{\overset{\overset{\displaystyle H}{|}}{C}}-X$$

to the coordination of HO⁻ ions. An

S_N2 mechanism is thus increasingly unlikely.

Steric (i.e. spatial) factors also need to be taken into account. For example, we can argue that a tertiary alkyl halide is more likely to form a planar carbonium ion than an overcrowded transition state, in which five bulky groups are attached to a single carbon atom.

By reasoning along these lines we can account for the experimental findings (Fig. 11.1) that there is a gradual transition from S_N2 to S_N1 mechanism associated with increasing substitution in the alkyl halides.

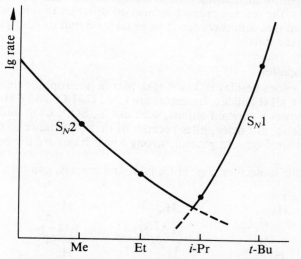

Me = methyl, Et = ethyl, *i*-Pr = isopropyl, *t*-Bu = *tert*-butyl

Fig. 11.1 Rates of hydrolysis of alkyl halides.

Stereochemistry of nucleophilic substitution
In all S_N2 reactions, a nucleophile joins one side of the substrate molecule while another atom or group of atoms leaves from the other side, e.g.

$$HO^- + \underset{\underset{\displaystyle R'}{\diagdown}}{\overset{\overset{\displaystyle R}{|}}{\underset{H}{C}}}-X \longrightarrow HO-\underset{\underset{\displaystyle R'}{\diagdown}}{\overset{\overset{\displaystyle R}{|}}{\underset{H}{C}}} + X^-$$

The spatial arrangement of atoms or groups about the functional carbon atom is therefore changed to a mirror image form, and we say that there is an *inversion of configuration*. The effect is of no consequence unless the carbon atom concerned is asymmetric, and even in such cases the inversion of configuration does not *necessarily* produce an opposite rotation of plane polarised light. If there were no change of functional group we should certainly observe a change from dextro-rotation to laevo-rotation (or vice versa), but the substitution of one functional group by another can prevent this.

In S_N1 reactions we might logically expect to obtain a racemic mixture of dextro and laevo forms, since a trigonal planar carbonium ion stands an equal chance of attack from either side. In practice, however, carbonium ions are so reactive that they are often attacked by the nucleophile before the leaving group has had time to get well away. Consequently, the leaving group hinders coordination to one side of the carbonium ion, and there tends to be an inversion of configuration as with S_N2 reactions.

Other nucleophiles

Many other anions besides HO^- will take part in nucleophilic substitution reactions with alkyl halides. Examples are CN^-, $C_2H_5O^-$ and CH_3COO^-. There are, however, certain anions, such as Cl^-, SO_4^{2-}, CO_3^{2-} and PO_4^{3-}, which do not act as nucleophiles because of their reluctance to donate lone pairs of electrons. In general, strong Lewis bases are the best nucleophiles.

Nucleophilic molecules, e.g. H_2O, NH_3 and amines, can take part in

these reactions, although not as readily as nucleophiles with a full negative charge. Water, for example, reacts with alkyl halides in essentially the same manner as alkalis,

$$RX + H_2O = ROH + HX$$
$$RX + HO^- = ROH + X^-$$

but the reactions take a great deal longer. This can be shown by refluxing bromoethane with aqueous sodium hydroxide for a few minutes, and then adding dilute nitric acid and aqueous silver nitrate. A turbidity results, indicating the presence of bromide ions. If, however, the experiment is repeated with water instead of sodium hydroxide solution, there is no noticeable turbidity with silver nitrate until the mixture has been heated for several hours.

The conditions under which certain well known nucleophiles react with alkyl halides and the importance of these reactions in the laboratory will now be discussed.

CN^-

Organic cyanides, known as *nitriles*, are formed when alkyl halides are refluxed with sodium cyanide or potassium cyanide in aqueous ethanol:

$$RX + K^+ (C\equiv N)^- = R-C\equiv N + K^+ X^-$$
$$\text{e.g. } CH_3CH_2Br + KCN = CH_3CH_2CN + KBr$$
propanenitrile

Sodium cyanide and potassium cyanide are extremely poisonous.

This reaction is useful in organic syntheses, for it represents one of the very few ways of increasing the length of a carbon chain. Nearly always, a reaction product contains the same number of carbon atoms as the starting material, but by reacting potassium cyanide with an alkyl halide an extra carbon atom can be introduced. For example, an alcohol, ROH, could be converted to a higher homologue, RCH_2OH, by one of the following routes:

$$ROH \xrightarrow{HBr} RBr \xrightarrow{KCN} RCN \begin{array}{c} \xrightarrow{\text{reduction}} RCH_2NH_2 \xrightarrow{HNO_2} \\ \\ \xrightarrow{\text{hydrolysis}} RCOOH \xrightarrow{Li[AlH_4]} \end{array} RCH_2OH$$

As this scheme suggests, nitriles are important intermediates in the synthesis of primary amines and carboxylic acids (Chapter 15).

If an alkyl halide is allowed to react with silver cyanide, which is a covalent compound, an *isocyanide* is formed:

$$RX + AgCN = R-N\equiv C + AgX \qquad (\S15.2)$$

RO^-

Ethers are obtained, in a reaction known as the *Williamson synthesis*, e.g.

$$C_2H_5Br + C_2H_5O^-Na^+ \xrightarrow[\text{ethanolic solution}]{\text{reflux in}} C_2H_5-O-C_2H_5 + NaBr$$
sodium ethoxide ethoxyethane

The sodium ethoxide is prepared by dissolving sodium in ethanol.

The product of this particular reaction is a *simple ether*, so called because the two hydrocarbon radicals are identical. The Williamson synthesis is most valuable for the preparation of *mixed ethers*, in which the radicals are dissimilar, because there is no other means of making such compounds, e.g.

$$CH_3CHBrCH_3 + C_2H_5O^-Na^+ = (CH_3)_2CH-O-C_2H_5 + NaBr$$

2-bromopropane   2-ethoxypropane

RCOO⁻

RCOO⁻

Alkyl halides react rapidly with the silver salts of carboxylic acids to give esters, e.g.

$$C_2H_5Br + CH_3C\!\!\begin{array}{c}O\\O^-Ag^+\end{array} \xrightarrow[\text{ethanolic solution}]{\text{reflux in}} CH_3C\!\!\begin{array}{c}O\\OC_2H_5\end{array} + AgBr$$

ethyl ethanoate

The cheaper sodium salts can be used instead, but silver salts are preferred because the silver ion, Ag^+, catalyses the reactions.

A variation on the more usual esterification between carboxylic acids and alcohols, this method has the merit of giving almost 100 per cent yield of ester and not just an equilibrium mixture.

Metals

Metals

Sodium, zinc and other metals which lie towards the top of the electrochemical series behave very much like nucleophilic reagents, because of the ease with which they can ionise and donate electrons (§13.7). In the presence of protic solvents, e.g. dilute HCl or H_2SO_4, such metals can attack alkyl halides and reduce them to alkanes, e.g.

$$CH_3CH_2Br \xrightarrow{Zn + \text{dilute HCl}} CH_3CH_3$$

An alternative reducing agent is lithium tetrahydridoaluminate(III) in dry ether.

NH₃ (ammonolysis)

NH₃ (ammonolysis)

Ammonia is a weaker nucleophile than anions such as HO^- and CN^-, and reactions with alkyl halides occur only on heating in a sealed tube to a temperature of 100 °C (373 K). Amine salts are formed, in equilibrium with primary amines and hydrogen halides:

$$H_3N\!: + \begin{array}{c}|\\C\!-\!\!\underset{x}{\cdot}\!X\\/\quad\backslash\end{array} \longrightarrow H_3N\!:\begin{array}{c}|\\C\!-\!\!\underset{x}{\cdot}\!X\\/\quad\backslash\end{array} \longrightarrow \left[H_3N\!-\!\!\begin{array}{c}|\\C\\/\quad\backslash\end{array}\right]^+ X^-$$

e.g. $C_2H_5Br + NH_3$ (alcoholic) $\xrightarrow[(373\text{ K})]{100\ °C} C_2H_5NH_3^+\ Br^-$ ethylammonium bromide

$$\updownarrow$$

$$C_2H_5NH_2 + HBr$$

ethylamine

This preparation, like all sealed tube reactions, is hazardous because of the risk of explosion. A safety shield is essential.

The product when cooled to room temperature is entirely an amine salt, from which the free amine can be liberated by double decomposition with a base, e.g.

$$C_2H_5NH_3^+\ Br^- + NaOH = C_2H_5NH_2 + NaBr + H_2O$$

Good yields of amines can be obtained only from primary alkyl halides: secondary and tertiary halides (except for isopropyl halides) undergo an elimination reaction to give alkenes.

The ammonolysis of alkyl halides is known as the *Hofmann method* of preparing amines. It suffers from the disadvantage that it invariably gives a mixture of four compounds, namely a primary amine, a secondary amine, a tertiary amine and a quaternary ammonium salt, e.g.

$$C_2H_5Br \xrightarrow{NH_3} C_2H_5NH_2 \quad (C_2H_5)_2NH \quad (C_2H_5)_3N \quad \text{and} \quad [(C_2H_5)_4N]^+\ Br^-$$

|ethylamine | diethylamine | triethylamine | tetraethylammonium bromide |

The reason for this is that amines, like ammonia, can act as nucleophiles and react with the alkyl halide, e.g.

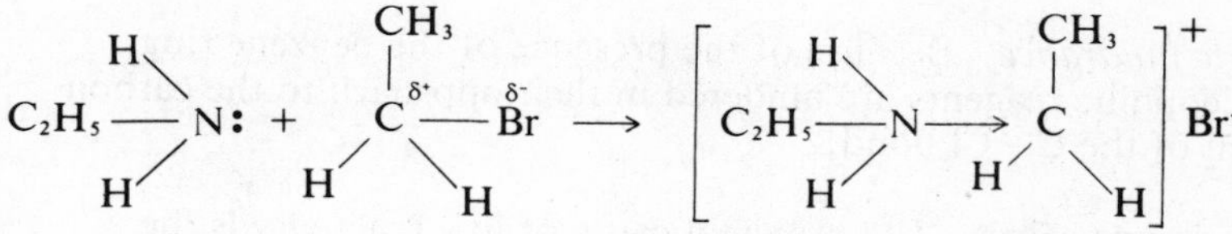

i.e. $[(C_2H_5)_2NH_2]^+Br^-$

The composition of the product can be controlled to some extent by adjusting the ratio of alkyl halide to ammonia. (A large excess of ammonia favours the primary amine.) Primary, secondary and tertiary amines can be separated by fractional distillation (unless their boiling temperatures are very close together) or by chromatographic techniques.

To convert an alkyl halide to a primary amine alone, it is better to react it with potassium phthalimide rather than ammonia in the *Gabriel phthalimide synthesis*:

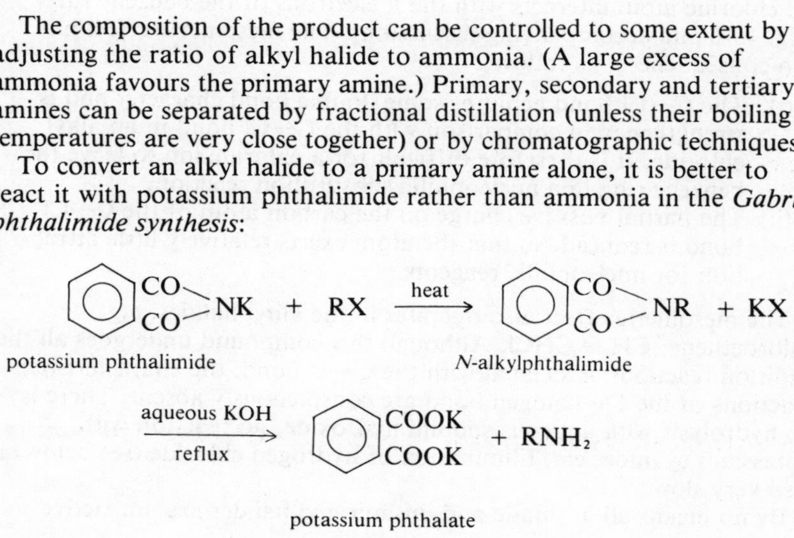

potassium phthalimide   *N*-alkylphthalimide

potassium phthalate

Special cases

Aryl halides and unsaturated halides

Aryl halides, such as chlorobenzene, bromobenzene and iodobenzene, are remarkably unreactive compounds. For example, from our knowledge of the chemistry of alkyl halides we should expect chlorobenzene to be hydrolysed to phenol on refluxing with aqueous alkali:

Cl $\quad\quad\quad\quad\quad\quad\quad\quad$ OH

$$\bigcirc + NaOH(aq) = \bigcirc + NaCl$$

However, this reaction does not occur, except at high temperature (300 °C; 573 K) and high pressure (150–200 atm; 15 000–20 000 kPa). Likewise, the reaction with ammonia, to give phenylamine, requires a temperature of 200 °C (473 K), a pressure of 50 atm (5 000 kPa), and the presence of a copper(I) oxide catalyst. With many other nucleophiles that react readily with alkyl halides, e.g. CN^-, RO^- and $RCOO^-$, there is no reaction at all.

The halogen atom in an aryl halide is said to be *non-labile,* i.e. not readily substituted. The cause lies partly in the geometry of the molecule and partly in the nature of the bonding.

Steric hindrance   Because of the presence of the benzene ring, nucleophilic reagents are hindered in their approach to the carbon atom of the C—Cl bond.

Mesomeric effect   The principal cause of low reactivity is the mesomeric effect, whereby a lone pair of electrons in the outer shell of the chlorine atom interacts with the π electrons of the benzene ring (Fig. 1.16 and accompanying discussion). The mesomeric effect has two consequences, as follows.

(i) The C—Cl bond acquires some double bond character and is strengthened in comparison with the C—Cl bond in an alkyl chloride. It is therefore difficult for a chloride ion to leave the benzene ring in a nucleophilic substitution reaction.

(ii) The partial positive charge on the carbon atom of the C—Cl bond is reduced, so that the atom exerts relatively little attraction for nucleophilic reagents.

The mesomeric effect also operates in the vinyl halides, e.g. chloroethene, $CH_2{=}CHCl$. Although this compound undergoes all the addition reactions associated with the C$=$C bond, the characteristic reactions of the C—halogen bond are conspicuously absent. There is no hydrolysis with aqueous sodium hydroxide, no reaction with potassium cyanide, etc. Elimination of hydrogen chloride (see below) is also very slow.

By no means all aromatic and unsaturated halides are unreactive.

Both (chloromethyl)benzene, $C_6H_5CH_2Cl$, and 3-chloro-1-propene (allyl chloride), $CH_2{=}CHCH_2Cl$, possess labile halogen atoms because in these compounds the chlorine atom and the π-system are prevented from interacting by an intervening saturated carbon atom. A halogen atom is non-labile only if it is attached to a carbon atom which is doubly bonded or which forms part of a benzene ring.

There are some other aromatic compounds, besides the aryl halides, which exhibit the mesomeric effect and are therefore reluctant to undergo nucleophilic substitution reactions. The principal examples are phenol and phenylamine.

1,1-Dihalides, e.g. 1,1-dichloroethane, CH_3CHCl_2

These compounds are generally made by the action of phosphorus pentahalides on the corresponding aldehydes and ketones, and we might expect them to hydrolyse to aldehydes or ketones rather than alcohols. This they do via haloalcohols. Substitution of one halogen atom by an HO group is followed by the elimination of a molecule of HX, e.g.

$$CH_3CHCl_2 \xrightarrow{\text{NaOH(aq)}} CH_3CH(OH)Cl \xrightarrow{-HCl} CH_3C{\overset{\textstyle O}{\underset{\textstyle H}{\diagup}}}$$

Likewise, (dichloromethyl)benzene hydrolyses to give benzaldehyde.

1,2-Dihalides, in contrast, behave as difunctional alkyl halides and give 1,2-diols on hydrolysis, e.g.

$$\begin{array}{c} CH_2Cl \\ | \\ CH_2Cl \end{array} + 2NaOH(aq) = \begin{array}{c} CH_2OH \\ | \\ CH_2OH \end{array} + 2NaCl$$

1,2-Diols are stable compounds, unlike 1,1-diols which spontaneously eliminate water to give aldehydes or ketones.

Acyl halides, e.g. ethanoyl chloride, $CH_3C{\overset{\textstyle O}{\underset{\textstyle Cl}{\diagup}}}$

The contrast between acyl halides and alkyl halides is obvious as soon as the stopper is removed from a bottle of ethanoyl chloride, for fumes of hydrogen chloride immediately appear in moist air,

$$CH_3C{\overset{\textstyle O}{\underset{\textstyle Cl}{\diagup}}} + H_2O = CH_3C{\overset{\textstyle O}{\underset{\textstyle OH}{\diagup}}} + HCl$$

The extreme lability of the chlorine atom indicates that the carbon atom to which it is attached has a particularly high partial positive charge. This is because the polarisation of the C—Cl bond is reinforced by a similar effect involving the π electrons of the C$=$O bond:

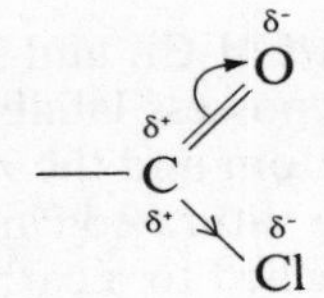

$$\begin{array}{c} \overset{\delta^-}{O} \\ \parallel \\ \underset{\delta^+}{C}\overset{\delta^+}{\diagdown} \\ \underset{Cl}{\overset{\delta^-}{\diagdown}} \end{array}$$

Reactions with nucleophiles thus occur very readily, but they are not necessarily nucleophilic substitution reactions. Addition—elimination reactions are also possible, i.e. nucleophilic addition across the C=O bond, followed by the elimination of HCl, and with most reagents the mechanism is of this kind. It is therefore customary to regard acyl halides primarily as carbonyl compounds, and we shall study their chemistry in Chapter 13.

Elimination of hydrogen halide

When an alkyl halide is boiled with an ethanolic solution of potassium hydroxide, it undergoes a certain amount of dehydrohalogenation (elimination of HX) to give an alkene, e.g.

$$CH_3CH_2Br + HO^-(alcoholic) = CH_2{=}CH_2 + H_2O + Br^-$$

H^+ and X^- are nearly always lost from adjoining carbon atoms. The process is known as *1,2-elimination* or *β-elimination*; X^- is said to leave the α (or 1) carbon atom and H^+ the β (or 2) carbon atom. The numbers (1 and 2) do not necessarily correspond to those used in the naming.

The mechanism of elimination is very similar to that of substitution, the main difference being that HO^- donates a lone pair of electrons to a β-hydrogen atom instead of a carbon atom. HO^-, therefore, does not behave as a nucleophile in this reaction, but acts as a Lewis base. Other Lewis bases, such as $C_2H_5O^-$, NH_3 and tertiary amines, can also be used.

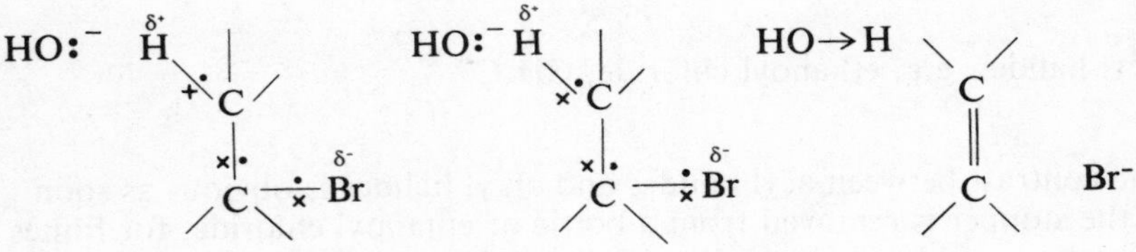

HO⁻ ion approaches the alkyl halide. Note that the inductive effect of the halogen atom, transmitted through the molecule, causes β-hydrogen atoms to acquire a partial positive charge.

HO⁻ ion reaches a β-hydrogen atom; C—Br and C—H bonds are strongly polarised.

HO⁻ ion coordinates to H⁺ to give a molecule of water; Br⁻ ion leaves.

A closer study of reaction mechanisms emphasises the similarity between elimination and substitution. Tertiary alkyl halides, which undergo unimolecular substitution reactions (S_N1), take part in unimolecular elimination reactions (E1). In both cases a carbonium ion is formed as an intermediate. Primary alkyl halides, whose substitution reactions are bimolecular (S_N2), take part in elimination reactions which are also bimolecular (E2). Both reactions proceed via a transition state:

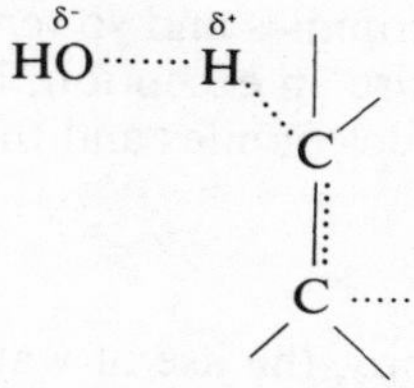

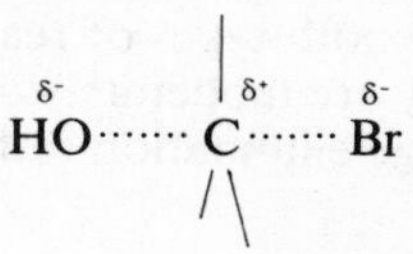

transition state for an elimination   *transition state for a substitution*

Only one alkene (ethene) can be obtained from the ethyl halides and each of the propyl halides (propene), but from many of the higher halides there are two possible products. 2-Chlorobutane, for instance, could conceivably give 1-butene or 2-butene:

$$CH_3CHClCH_2CH_3 \longrightarrow CH_2{=}CHCH_2CH_3 \quad \text{1-butene}$$
$$\searrow CH_3CH{=}CHCH_3 \quad \text{2-butene}$$

In fact the product is mainly 2-butene. This is because the transition state leading to the formation of 2-butene is relatively stable and easy to attain, compared with that leading to 1-butene. It is a general rule — *Saytzeff's rule* — that the hydrogen atom is eliminated from the carbon atom which bears the fewer hydrogen atoms.

Substitution versus elimination

It often happens that when an alkyl halide is treated with a reagent possessing a lone pair of electrons there are two competing reactions:

(i) substitution, e.g.

$$CH_3CH_2Br + HO^- = CH_3CH_2OH + Br^-$$

Here, the HO^- ion behaves as a nucleophile, coordinating to a carbon atom.

(ii) elimination of hydrogen halide, e.g.

$$CH_3CH_2Br + HO^- = CH_2{=}CH_2 + H_2O + Br^-$$

Here, the HO^- ion behaves as a base, abstracting a proton from the β-position.

The relative importance of substitution to elimination is not fixed, but depends very much on the conditions, namely the nature of the reagent, the nature of the solvent, the concentration of the reagent, and the temperature. We shall discuss each in turn.

Reagent

In theory, strong bases favour elimination while strong nucleophiles favour substitution. In practice, however, species, such as HO^- and RO^-, which are strong bases are also strong nucleophiles and so tend to promote both types of reaction. Tertiary amines are an exception, in that they are moderately strong bases but weak nucleophiles and thus encourage elimination rather than substitution.

Solvent

For tertiary halides, which react via carbonium ions, the use of water favours substitution. Carbonium ions are stabilised by water, a strongly polar solvent, but in less polar solvents, such as ethanol, carbonium ions tend to decompose with the formation of alkenes.

For primary and secondary halides, which react via transition states, the influence of the solvent is less clear.

An objection to the use of ethanolic alkali for elimination lies in the production of ethyl ethers, C_2H_5—O—R. These compounds are formed by neutralisation of the alcohol by the alkali, to give a metallic alkoxide, followed by a Williamson synthesis with the alkyl halide, e.g.

$$C_2H_5OH + NaOH \rightleftharpoons C_2H_5O^-Na^+ + H_2O$$
$$C_2H_5O^-Na^+ + C_2H_5Cl = C_2H_5-O-C_2H_5 + NaCl$$

Chloroethane with ethanolic alkali gives almost no ethene, and while the yield is a little better with the bromide and the iodide, the main product in all cases is ethoxyethane. The higher members of the homologous series give a much more satisfactory yield of alkene, e.g. 80 per cent yield of propene from the isopropyl halides.

Concentration of reagent

Dilute alkali favours substitution, while concentrated alkali promotes elimination. This is the result of a change of mechanism. At low concentrations the base does not participate in the first stage of the reaction and the mechanism is $S_N1/E1$, but at higher concentrations the base does take part and the mechanism changes to $S_N2/E2$. Corresponding to the change in mechanism there is an increase in the ratio of elimination to substitution.

Temperature

At low temperatures only substitution can occur. Elimination reactions are prevented from taking place because the activation energy is relatively high. An increase in temperature, provided that it is sufficient to supply this activation energy, will have a proportionately greater effect on the rate of elimination than substitution.

To summarise, an ideal reagent for substitution is dilute aqueous calcium hydroxide (a relatively weak base), at the lowest temperature consistent with a reasonable rate of reaction. Another reagent, frequently recommended, is wet silver oxide. A suitable reagent for elimination is a hot, concentrated, ethanolic solution of potassium hydroxide.

Grignard reagents

Alkyl halides and aryl halides (except for chlorobenzene) react with metallic magnesium suspended in dry ether to give a *Grignard reagent,* i.e. an alkyl- or arylmagnesium halide, e.g.

$$C_2H_5Br + Mg \xrightarrow{\text{dry ether}} C_2H_5MgBr$$

ethylmagnesium bromide

In consequence of the large electronegativity difference between carbon and magnesium, Grignard reagents are strongly polar compounds, $\overset{\delta^-}{R}-\overset{\delta^+}{Mg}-\overset{\delta^-}{X}$. They are stable only in the presence of an ether, two molecules of which coordinate to the magnesium atom to give a distorted tetrahedral structure, e.g.

Grignard reagents are valuable in the laboratory as many kinds of compounds can be made from them. In all their reactions they behave as bases. This may seem surprising, for they do not possess lone pairs of electrons except in the outer shell of the halogen atom. However, in situations which increase the charge separation of the C—Mg bond, Grignard reagents can serve as a source of carbanions, R^-. Such ions do possess lone pairs of electrons and can act as Lewis bases.

Their principal reactions are with the following substances.

Compounds possessing active hydrogen atoms

Water, alcohols, primary amines and secondary amines, all of which contain *active hydrogen atoms* (i.e. hydrogen atoms with a partial positive charge), react with Grignard reagents to give hydrocarbons, e.g.

$$CH_3MgI + H_2O = CH_4 + Mg^{2+} + HO^- + I^-$$
$$CH_3MgI + C_2H_5OH = CH_4 + Mg^{2+} + C_2H_5O^- + I^-$$

The reaction can be regarded as a neutralisation, in which the methanide ion, CH_3^-, coordinates to a proton from the active hydrogen compound:

$$\overset{\bullet\bullet}{\underset{\bullet\bullet}{C}}H_3 \overset{2+}{Mg} \bar{I} \quad base$$
$$\downarrow$$
$$H^+ HO^- \quad acid$$

Alkyl halides

Hydrocarbons are formed in a nucleophilic substitution reaction, e.g.

$$CH_3MgI + CH_3CH_2I = CH_3CH_2CH_3 + MgI_2$$

Essentially, a carbanion from the Grignard reagent coordinates to the carbon atom of the C—halogen bond:

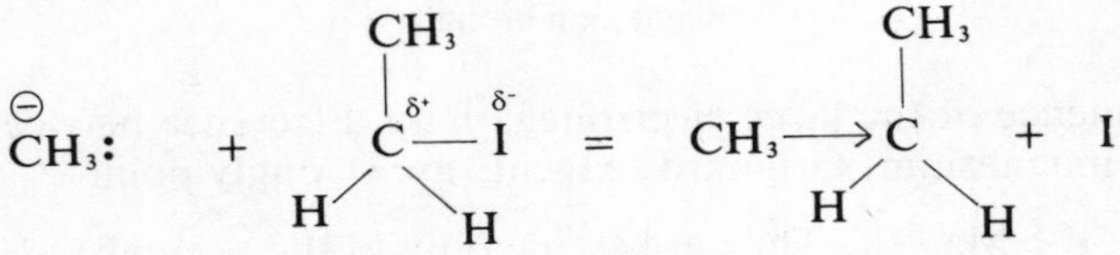

Carbonyl compounds

Carbonyl compounds, i.e. compounds which possess a C=O bond, undergo nucleophilic addition reactions with Grignard reagents. We shall discuss the mechanism in detail later (§13.7). Essentially, the positively charged magnesium atom becomes attached to the negatively charged oxygen atom, while the negatively charged alkyl group, R, joins the carbon atom:

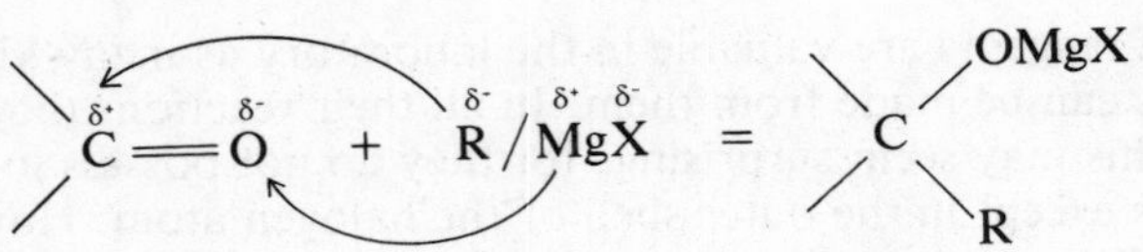

The adducts are complex magnesium salts. Although they possess a fair degree of ionic character, they are essentially covalent and can be hydrolysed to give various products depending on the type of carbonyl compound.

In all these conversions the following principles are observed.

(i) The Grignard reagent takes the carbonyl compound to the level of oxidation immediately below it. Ketones, for example, give rise to tertiary alcohols, aldehydes give secondary alcohols, and derivatives of carboxylic acids give (in the first instance) aldehydes and ketones.

(ii) All the hydrocarbon radicals (or hydrogen atoms) belonging to the Grignard reagent and the carbonyl compound appear in the hydrolysis product.

We shall now study some examples.

Ketones

The simplest example involves propanone and methylmagnesium iodide. The adduct, when hydrolysed under acidic conditions, gives 2-methyl-2-propanol:

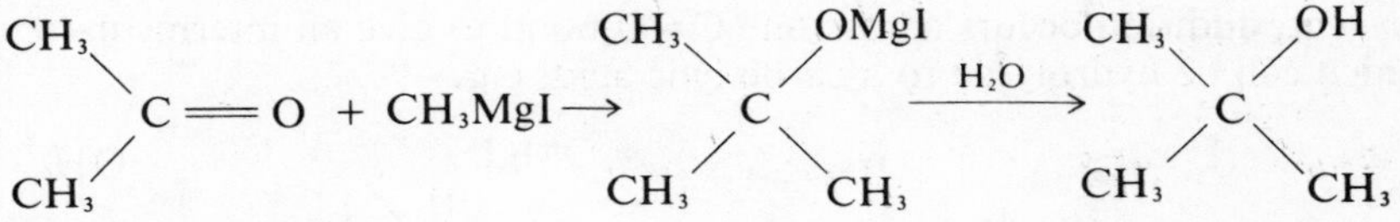

$$CH_3 \diagdown \atop CH_3 \diagup C = O \ + \ CH_3MgI \rightarrow \ CH_3 \diagdown \atop CH_3 \diagup C \diagup^{OMgI} \xrightarrow{H_2O} \ CH_3 \diagdown \atop CH_3 \diagup C \diagup^{OH} \diagdown_{CH_3}$$

Aldehydes

Methanal is converted to primary alcohols, and other aldehydes to secondary alcohols, e.g.

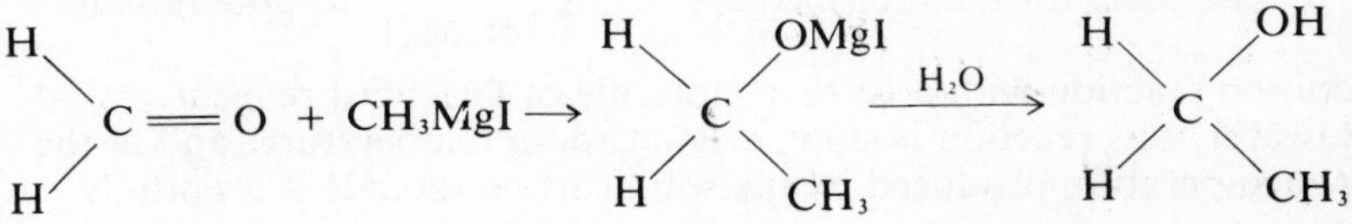

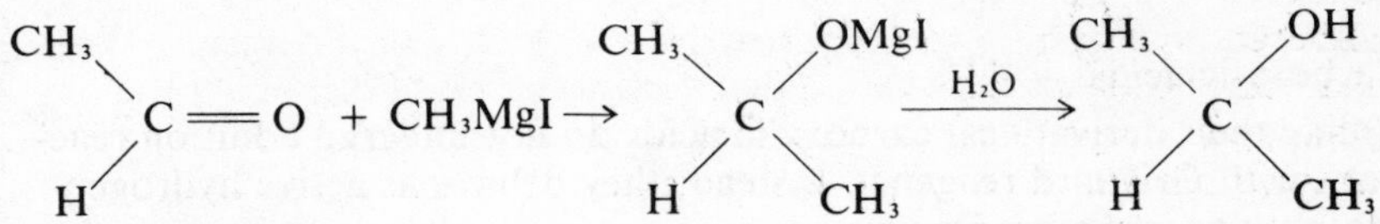

$$H \diagdown \atop H \diagup C = O \ + \ CH_3MgI \rightarrow \ H \diagdown \atop H \diagup C \diagup^{OMgI} \diagdown_{CH_3} \xrightarrow{H_2O} \ H \diagdown \atop H \diagup C \diagup^{OH} \diagdown_{CH_3}$$

$$CH_3 \diagdown \atop H \diagup C = O \ + \ CH_3MgI \rightarrow \ CH_3 \diagdown \atop H \diagup C \diagup^{OMgI} \diagdown_{CH_3} \xrightarrow{H_2O} \ CH_3 \diagdown \atop H \diagup C \diagup^{OH} \diagdown_{CH_3}$$

Acid chlorides

Ketones are formed, in reactions similar to those just described, e.g.

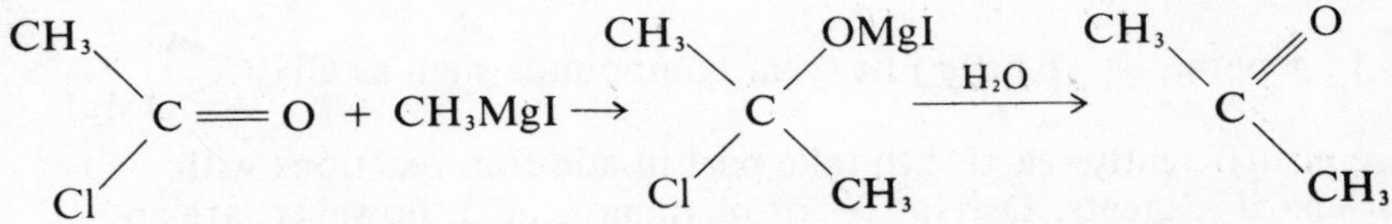

$$CH_3 \diagdown \atop Cl \diagup C = O \ + \ CH_3MgI \rightarrow \ CH_3 \diagdown \atop Cl \diagup C \diagup^{OMgI} \diagdown_{CH_3} \xrightarrow{H_2O} \ CH_3 \diagdown \atop \diagup C = O \diagdown_{CH_3}$$

This is a poor way of preparing ketones, because the product readily reacts with further Grignard reagent to give a tertiary alcohol. Various methods have been proposed for improving the yield of ketone. They include:

(i)   running the Grignard reagent into the acid chloride so that it is never present in excess, and using the theoretical quantity of reagent;

(ii) working at a low temperature;

(iii) using tetrahydrofuran as a solvent, instead of ethoxyethane.

Esters

In contrast to acid chlorides, esters give tertiary alcohols whatever the conditions. This is because ketones are more reactive than esters towards Grignard reagents. However, it has been reported that reasonably good yields of aldehydes can be obtained from esters of methanoic acid.

Carbon dioxide

If small lumps of solid carbon dioxide are stirred into a Grignard reagent, addition occurs across one $C=O$ bond to give an intermediate which can be hydrolysed to a carboxylic acid, e.g.

$$O=C=O \ + \ CH_3MgI \ \longrightarrow \ O=C\begin{smallmatrix}OMgI\\ \\CH_3\end{smallmatrix} \ \xrightarrow{H_2O} \ O=C\begin{smallmatrix}OH\\ \\CH_3\end{smallmatrix}$$

It is possible for the intermediate, $CH_3C\begin{smallmatrix}O\\ \\OMgI\end{smallmatrix}$, to undergo an

addition reaction with a further molecule of Grignard reagent. However, this reaction is slow, even at room temperature, and at the low temperature produced by the solid carbon dioxide it is entirely suppressed.

Carboxylic acids

Unlike their derivatives, carboxylic acids do not undergo addition reactions with Grignard reagents. Instead, they behave as active hydrogen compounds and give rise to alkanes, e.g.

$$CH_3C\begin{smallmatrix}O\\ \\OH\end{smallmatrix} \ + \ CH_3MgI \ = \ CH_3C\begin{smallmatrix}O\\ \\OMgI\end{smallmatrix} \ + \ CH_4$$

In general, as we have just seen, compounds such as $CH_3C\begin{smallmatrix}O\\ \\OMgI\end{smallmatrix}$

are insufficiently reactive to take part in addition reactions with Grignard reagents. Derivatives of methanoic acid, however, are an exception, and it has recently been reported that good yields of aldehydes can be obtained in this way, e.g.

$$HC\begin{smallmatrix}O\\ \\OH\end{smallmatrix} \ + \ C_2H_5MgBr \ = \ HC\begin{smallmatrix}O\\ \\OMgBr\end{smallmatrix} \ + \ C_2H_6$$

$$\begin{smallmatrix}H\\ \\BrMgO\end{smallmatrix}C=O \ + \ C_6H_5MgBr \ \longrightarrow \ \begin{smallmatrix}H\ \ OMgBr\\ \\BrMgO\ \ C_6H_5\end{smallmatrix}C \ \xrightarrow{H_2O} \ \begin{smallmatrix}H\ \ O\\ \\C_6H_5\end{smallmatrix}C$$

Oxygen, sulphur and halogens

With oxygen, Grignard reagents add across the O=O bond to give magnesium salts which can be hydrolysed to alcohols, e.g.

$$O=O + CH_3MgI \longrightarrow CH_3-O-O-MgI$$

$$\xrightarrow{CH_3MgI} 2CH_3-O-MgI \xrightarrow{H_2O} 2CH_3OH$$

Although the yields are good, the method is of limited value as there is a more direct way (hydrolysis) of converting an alkyl halide to an alcohol.

Similar reactions occur with sulphur and the halogens to give thiols, RSH, and alkyl halides respectively.

Oxirane (ethylene oxide)

The three membered oxirane ring is sufficiently strained to be broken by Grignard reagents, with the formation of primary alcohols, e.g.

$$CH_2\overset{O}{-}CH_2 + CH_3MgI \longrightarrow CH_3CH_2CH_2-O-MgI$$

$$\xrightarrow{H_2O} CH_3CH_2CH_2OH$$

The reaction between oxirane and a Grignard reagent thus provides a means of increasing the length of a carbon chain by two atoms. (In contrast, the reaction with methanal increases it by one atom; see above.)

Chapter 12

The C—O bond in alcohols, phenols and ethers

Data on the C—O bond

covalent bond length/nm   0.143
bond dissociation enthalpy/kJ mol^{-1} at 298 K   360

The carbon–oxygen single bond is encountered in alcohols, phenols and ethers. It also occurs, together with the carbon–oxygen double bond, in carboxylic acids, carboxylic acid anhydrides and carboxylic acid esters, but consideration of such compounds will be left until Chapter 13 because their chemistry is governed primarily by the C=O bond.

An *alcohol* is a compound whose molecules possess at least one hydroxyl group, HO. Alcohols may be aliphatic, e.g. ethanol, alicyclic, e.g. cyclohexanol, or aromatic with the HO group in a side chain, e.g. phenylmethanol. Compounds of the last kind are referred to as *alkaryl alcohols*.

C_2H_5OH

ethanol   cyclohexanol 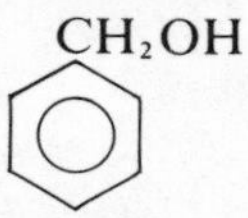   phenylmethanol

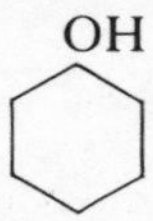

Alcohols are classified as primary, secondary or tertiary, depending upon whether the hydroxyl group is attached to a primary, secondary or tertiary carbon atom. Thus, ethanol and phenylmethanol are primary alcohols, while cyclohexanol is secondary.

An alcohol with one hydroxyl group is described as *monohydric*, one with two hydroxyl groups is *dihydric*, and so on. Dihydric alcohols are called *diols* (commonly 'glycols'), trihydric alcohols are *triols*, etc. Examples are as follows:

CH₂OH
|
CH₂OH

CH₂OH
|
CHOH
|
CH₂OH

1,2-ethanediol (ethylene glycol)   1,2,3-propanetriol (glycerol)

With very few exceptions, alcohols with more than one hydroxyl group are stable only if those groups are attached to different carbon atoms. Thus, 1,2-ethanediol is stable, but 1,1-ethanediol is unstable, except in aqueous solution, and decomposes spontaneously with the elimination of water between adjacent HO groups. This process always leads to the establishment of a C=O bond, i.e. to the formation of an aldehyde or a ketone.

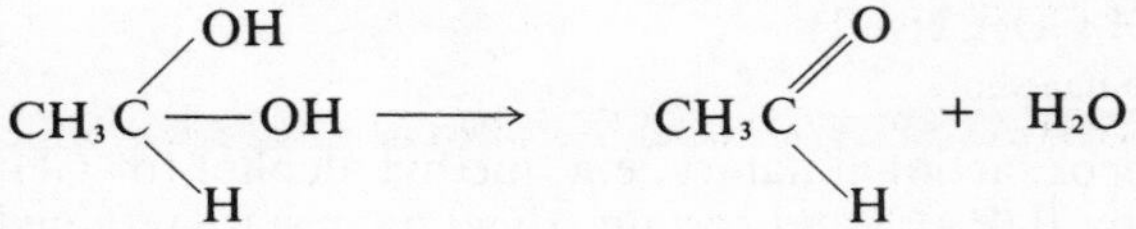

$$CH_3\overset{OH}{\underset{H}{C}}-OH \longrightarrow CH_3\overset{O}{\underset{H}{C}} + H_2O$$

A *phenol* is an aromatic hydroxy compound in which at least one HO group is attached directly to the benzene ring. The simplest compound is phenol itself, C_6H_5OH. Like alcohols, phenols can be monohydric, dihydric or trihydric, according to the number of hydroxyl groups in the molecule.

An *ether* is a compound whose molecules consist of two hydrocarbon radicals joined together via an atom of oxygen. If the two radicals are identical, as in ethoxyethane, C_2H_5—O—C_2H_5, the compound is said to be a *simple ether*; otherwise it is a *mixed ether*. Cyclic ethers, e.g.

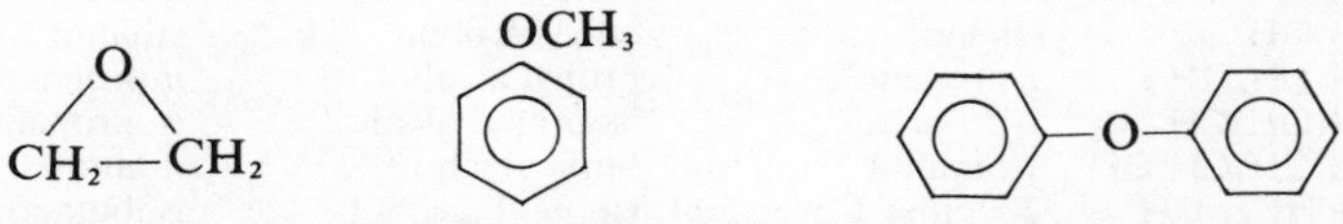

oxirane   methoxybenzene   phenoxybenzene

oxirane, are sometimes referred to as *epoxides*. Diaryl ethers, such as phenoxybenzene (diphenyl ether), are seldom encountered, but phenolic ethers, such as methoxybenzene (anisole), are common.

12.1 Nomenclature of alcohols, phenols and ethers

Alcohols
Alcohols, like alkyl halides, are usually referred to by substitutive names in which the final 'e' of the name of the parent hydrocarbon is

replaced by 'ol'; e.g. methanol for CH_3OH. To distinguish between isomers, the number of the carbon atom bearing the HO group is introduced before the name, e.g. 2-propanol for $CH_3CH(OH)CH_3$.

The HO group should always appear in the carbon chain which has been selected as the basis of the name. Thus, the compound

$$CH_3CH_2CHCH_2CH_3$$
$$|$$
$$CH_2OH$$

is called 2-ethyl-1-butanol: naming it as a pentane derivative is less straightforward.

When HO is not the principal group, it is indicated by the prefix 'hydroxy'. The compound $CH_3CH(OH)CH_2CH_2CHO$ is called 4-hydroxypentanal because, according to the IUPAC rules of priority (see Appendix), the aldehyde group takes precedence over the hydroxyl group and must be cited as a suffix.

Diols and triols are named as such, e.g.

$$HOCH_2CH_2CH_2CH_2OH$$
1,4-butanediol

Older, radicofunctional names, e.g. methyl alcohol for CH_3OH, are still allowed by IUPAC, and certain trivial names are retained, e.g. ethylene glycol for $HOCH_2CH_2OH$. Industrial names, such as isopropanol for $CH_3CH(OH)CH_3$, are not permitted by IUPAC.

Table 12.1 summarises the various names that may be encountered for $C_1 - C_4$ saturated monohydric alcohols.

Table 12.1 Nomenclature of alcohols

Formula	Substitutive name	Radicofunctional name	Industrial name
CH_3OH	methanol	methyl alcohol	methanol
CH_3CH_2OH	ethanol	ethyl alcohol	ethanol
$CH_3CH_2CH_2OH$	1-propanol	propyl alcohol	*n*-propanol
$CH_3CH(OH)CH_3$	2-propanol	isopropyl alcohol	isopropanol
$CH_3CH_2CH_2CH_2OH$	1-butanol	butyl alcohol	*n*-butanol
$(CH_3)_2CHCH_2OH$	2-methyl-1-propanol	isobutyl alcohol	isobutanol
$CH_3CH(OH)CH_2CH_3$	2-butanol	*sec*-butyl alcohol	*sec*-butanol
$(CH_3)_3COH$	2-methyl-2-propanol	*tert*-butyl alcohol	*tert*-butanol

Phenols

Phenols, like alcohols, can be given substitutive names by replacing the last letter 'e' of the name of the parent arene by 'ol', 'diol', etc, but IUPAC recommends the retention of trivial names for simple compounds.

Trivial name:	phenol	m-cresol 3-methylphenol	2-naphthol
Substitutive name:	benzenol	3-methylbenzenol	2-naphthalenol

Trivial name:	pyrocatechol	resorcinol	hydroquinone
Substitutive name:	1,2-benzenediol	1,3-benzenediol	1,4-benzenediol

Ethers

IUPAC allows the use of either substitutive or radicofunctional names, e.g.

	$CH_3-O-CH_2CH_3$	$CH_3CH_2-O-CH_2CH_3$
Substitutive name:	methoxyethane	ethoxyethane
Radicofunctional *name:*	ethyl methyl ether	diethyl ether

Trivial names are retained for simple phenolic ethers, notably the following:

Trivial name:	anisole	phenetole
Substitutive name:	methoxybenzene	ethoxybenzene

Apart from oxirane (see above), the best known cyclic ethers are:

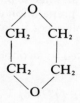

1,4-dioxan   tetrahydrofuran

12.2 Formation of the C—O bond

Both substitution and addition reactions can be employed.

Substitution methods

From C—H; the substitution of H by HO

The oxidation of alkanes to alcohols, RH → ROH, is usually impossible, although alkanes possessing a tertiary hydrogen atom can be attacked by acidified potassium permanganate solution, e.g.

$$(CH_3)_3CH \xrightarrow{\quad MnO_4^- + H^+ \quad} (CH_3)_3COH$$

2-methylpropane   2-methyl-2-propanol

Methane is exceptional, in that it can be oxidised with oxygen under drastic conditions to provide a 30 per cent yield of methanol:

$$2CH_4 + O_2 \xrightarrow[\substack{130—200 \text{ atm } (13\,000—20\,000 \text{ kPa}) \\ \text{metallic catalyst}}]{400 \text{ °C } (673 \text{ K})} 2CH_3OH$$

If other alkanes are treated in this way, they crack to yield products with fewer carbon atoms than the reactants.

Arenes cannot be converted directly into phenols.

From C—X (X = halogen); the substitution of X by HO or RO (§11.5)

Alkyl halides can readily be converted to alcohols or ethers by nucleophilic substitution reactions. Alcohols are formed on refluxing with aqueous alkali, e.g.

$$C_2H_5Br + NaOH = C_2H_5OH + NaBr$$

The method is of limited use because alkyl halides are normally prepared from alcohols in the first place. However, the refluxing of an alkyl halide with a sodium or potassium alkoxide in alcoholic solution provides a valuable route to ethers, especially mixed ethers, e.g.

$$C_2H_5Br + CH_3ONa = C_2H_5—O—CH_3 + NaBr$$

The non-lability of the halogen atom in an aryl halide prevents similar reactions from being used for the laboratory preparation of phenols and phenolic ethers, although substitution will occur if the conditions are exceptionally severe.

Alkyl and aryl halides may also be converted to alcohols and phenols via Grignard reagents.

From C—NH₂; the substitution of NH₂ by HO (§14.6)

Primary amines react with sodium nitrite and dilute hydrochloric acid to give diazonium salts which, on hydrolysis, yield alcohols or phenols, e.g.

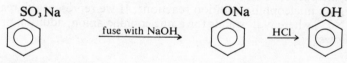

$$C_2H_5NH_2 + HNO_2 = C_2H_5OH + N_2 + H_2O$$

Aliphatic diazonium salts are highly unstable, so that on treatment with sodium nitrite and hydrochloric acid an aliphatic primary amine appears to be converted directly to an alcohol, e.g.

Except in the case of ethylamine, the yield of alcohol is low and the method is of little preparative value.

From C—SO₃H; the substitution of SO₃H by HO (§16.4)

Heating a sodium arylsulphonate with solid sodium hydroxide is one of the main ways of making phenols, e.g.

sodium benzenesulphonate   sodium phenoxide

Alkylsulphonic acids do not undergo this reaction.

Substitution of HO by RO (§12.5)

The etherification of alcohols by dehydration under acidic conditions is one of the most important ways of making simple ethers, e.g.

$$2C_2H_5OH \xrightarrow[\text{130 °C (403 K)}]{\text{conc. } H_2SO_4} C_2H_5—O—C_2H_5 + H_2O$$

This is a nucleophilic substitution reaction of an alcohol, and is often accompanied by the elimination of water to give an alkene. An excess of alcohol is used to lessen alkene formation.

Substitution of RO by HO

In a reaction which is essentially the reverse of etherification, an ether can be cleaved in the presence of a strong acid, usually hydriodic acid, to give an alcohol, e.g.

$$C_2H_5—O—C_2H_5 + HI = C_2H_5OH + C_2H_5I \qquad (§12.5)$$

Alcohols are also formed by the hydrolysis of esters, e.g.

$$CH_3C\underset{OC_2H_5}{\overset{O}{<}} + H_2O \rightleftharpoons CH_3C\underset{OH}{\overset{O}{<}} + C_2H_5OH \qquad (\S13.12)$$

Although these reactions are superficially similar to each other, they proceed by entirely different mechanisms.

Addition methods

From C=C (§7.4)

The hydration of alkenes is the major industrial way of making alcohols and ethers. Water alone is inadequate, and an acid catalyst is always employed (see below).

The oxidation of alkenes by reagents such as potassium permanganate gives a low yield of 1,2-diols, e.g.

$$CH_2{=}CH_2 \xrightarrow{\text{cold alkaline } KMnO_4} HOCH_2CH_2OH$$

Better yields are obtained by oxidising alkenes with osmium(VIII) oxide or peroxoacids.

From C=O (§13.6)

The principal feature of the C=O bond is the readiness with which it undergoes nucleophilic addition reactions. If we represent the reagent by H⁺Nu:⁻, where Nu:⁻ stands for a nucleophilic anion, addition proceeds as follows:

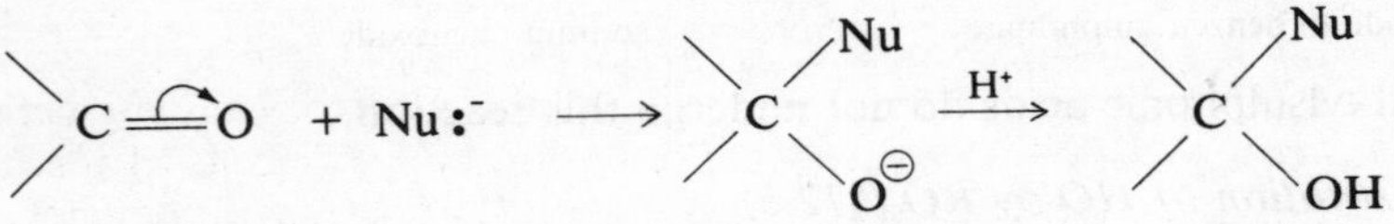

$$\underset{/}{\overset{\backslash}{C}}{=}O \;+\; Nu:^- \;\longrightarrow\; \underset{/}{\overset{\backslash}{C}}\underset{O^{\ominus}}{\overset{Nu}{<}} \;\xrightarrow{H^+}\; \underset{/}{\overset{\backslash}{C}}\underset{OH}{\overset{Nu}{<}}$$

Thus, in all such reactions, the doubly bonded oxygen atom is converted into an HO group.

Of particular interest is the hydrogenation of aldehydes and ketones to primary and secondary alcohols respectively by means of a metal and a protic solvent, e.g.

$$CH_3CHO \xrightarrow{\text{metal + acid}} C_2H_5OH$$

Alternative reducing agents are lithium tetrahydridoaluminate(III) and sodium tetrahydroborate(III).

12.3 Manufacture of alcohols and phenol

Methanol

This compound is manufactured in the UK by the catalytic addition of hydrogen to carbon monoxide:

$$CO + 2H_2 \underset{\substack{50 \text{ atm } (5\,000 \text{ kPa}) \\ \text{copper catalyst}}}{\overset{270\ ^\circ\text{C } (543\ \text{K})}{\rightleftharpoons}} CH_3OH$$

Formerly, the reaction was conducted in the presence of a mixture of zinc oxide and chromium(III) oxide and at a much higher pressure.

Ethanol

Industrial ethanol is made by the acid catalysed hydration of ethene:

$$CH_2{=}CH_2 + H_2O \xrightarrow{\text{H}_3\text{PO}_4 \text{ catalyst}} CH_3CH_2OH$$

First, ethene is heated with steam to a temperature of 300 °C (573 K). The mixture is then compressed to 70 atm (7 000 kPa) and passed into a reactor containing a catalyst of phosphoric acid absorbed on diatomaceous earth. Partial hydration occurs, and dilute aqueous ethanol is condensed from the gases leaving the reactor.

The product is concentrated by fractional distillation to give 'rectified spirit', which is an azeotrope (i.e. a constant boiling mixture) containing approximately 96 per cent ethanol and 4 per cent water. 'Industrial methylated spirit' consists of 95 per cent rectified spirit and 5 per cent methanol, while 'mineralised methylated spirit', sold by pharmacists, contains 90 per cent rectified spirit, 9 per cent methanol, 1 per cent pyridine and a trace of purple dye. 'Absolute alcohol' (99.5 per cent ethanol) is rectified spirit that has been dried by azeotropic distillation with benzene.

Alcoholic drinks are produced by *fermentation*. This term is used to describe various chemical reactions brought about by the catalytic action of *enzymes*, i.e. proteins which are present in the living cells of plants and animals. *Yeasts* and *bacteria* can be used as sources of enzymes. Both are single celled microorganisms, although they differ from each other in that whereas yeast cells have a nucleus, bacteria cells, which are much smaller, do not. Enzymes are produced in small structures (called *ribosomes*) in the *cytoplasm*, i.e. that part of the cell which is not the nucleus.

Fermentation causes carbohydrates, which are essentially polyhydric alcohols, to be broken down into simple alcohols. In wine making, an enzyme called 'zymase', which is present in yeast cells, converts the sugars of a fruit juice into ethanol and carbon dioxide:

$$C_6H_{12}O_6 \xrightarrow{\text{zymase}} 2C_2H_5OH + 2CO_2$$

glucose or fructose

Fermentation proceeds best in the warm (~ 30 °C (303 K)) and ceases when the concentration of ethanol reaches ~ 14 per cent. Alcoholic beverages stronger than this ('spirits') are made by distilling the fermented liquor.

Although most industrial ethanol is manufactured from ethene, some is prepared by the fermentation of molasses, i.e. the brown syrupy residue produced in the refining of cane sugar. After fermentation with yeast, the liquor is distilled to give three fractions: (i) ethanal; (ii) rectified spirit; and (iii) 'fusel oil', consisting of C_3-C_5 alcohols.

2-Propanol

2-Propanol, known commercially as 'isopropanol' or 'IPA', is made from propene by both direct and indirect hydration. In the former process a mixture of propene and steam is passed at a high temperature and pressure over a tungsten oxide catalyst:

$$CH_3CH{=}CH_2 + H_2O = CH_3CH(OH)CH_3$$

In the second and more widely used process, propene is absorbed in 70–93 per cent sulphuric acid at a temperature no greater than 40 °C (313 K) to give isopropyl hydrogen sulphate, which is then hydrolysed by hot water:

$$CH_3CH{=}CH_2 + H_2SO_4 = (CH_3)_2CHO{-}SO_2{-}OH$$
$$(CH_3)_2CHO{-}SO_2{-}OH + H_2O = (CH_3)_2CHOH + H_2SO_4$$

All the lower alcohols are used as solvents for lacquers, varnishes, printing inks, adhesives, etc, and as starting materials in the manufacture of other organic compounds, notably aldehydes, ketones and esters.

Phenol

Phenol is made commercially by the autoxidation of isopropylbenzene, which is obtained from benzene by a Friedel–Crafts reaction with propene:

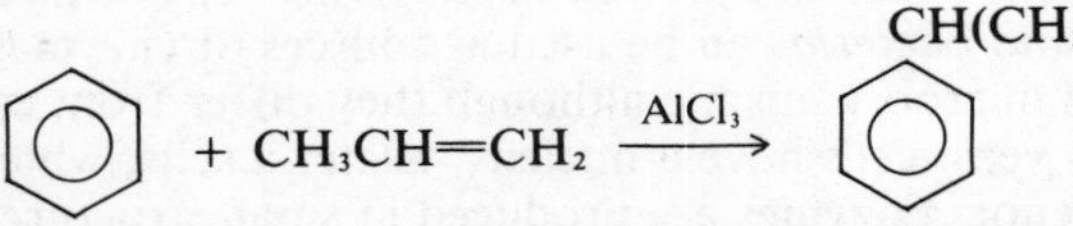

Isopropylbenzene, in the form of a slightly alkaline emulsion in water, is oxidised with air at 110–130 °C (383–403 K) to give the hydroperoxide:

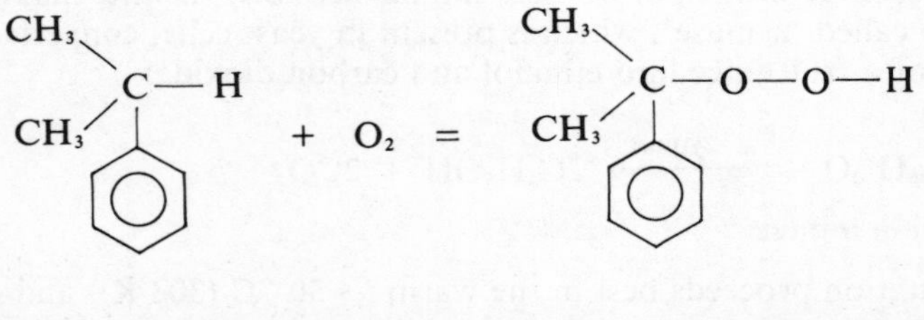

In the final stage, dilute sulphuric acid at 40–60 °C (313–333 K) decomposes the hydroperoxide into a mixture of phenol and propanone:

$$\underset{\underset{\displaystyle \text{(benzene ring)}}{\displaystyle}}{\overset{\displaystyle CH_3}{\underset{\displaystyle CH_3}{}}C-O-O-H} \quad = \quad \underset{\displaystyle \text{(benzene ring)}}{OH} \quad + \quad CH_3COCH_3$$

The manufacture of phenol by the fusion of sodium benzenesulphonate with sodium hydroxide is no longer worked in the UK.

Phenol is used for the manufacture of phenol–formaldehyde plastics (Bakelite), dyes, drugs (especially aspirin) and selective weed killers.

12.4 General properties of alcohols, phenols and ethers

In contrast to hydrocarbons, the simplest of which are gases, all the common alcohols are liquids. Table 12.2 provides an interesting comparison of the boiling temperatures of alcohols and alkanes of comparable relative molecular mass.

Table 12.2 Boiling temperatures of alcohols and alkanes

Alkane	Relative molecular mass	Boiling temperature/ °C (K)	Alcohol	Relative molecular mass	Boiling temperature/ °C (K)
ethane	30	− 88.6 (184.6)	methanol	32	64.5 (337.7)
propane	44	− 42.2 (231.0)	ethanol	46	78.5 (351.7)
butane	58	− 0.5 (272.7)	1-propanol	60	97.2 (370.4)
pentane	72	36.3 (309.5)	1-butanol	74	117 (390)

Because boiling temperature is a function of relative molecular mass, it is reasonable to suppose that alcohols are associated through hydrogen bonding, in exactly the same way as water:

$$\underset{R}{\overset{\delta^-}{O}}-\overset{\delta^+}{H}\cdots\underset{R}{\overset{\delta^-}{O}}-\overset{\delta^+}{H}\cdots\underset{R}{\overset{\delta^-}{O}}-\overset{\delta^+}{H}\cdots\underset{R}{\overset{\delta^-}{O}}-\overset{\delta^+}{H}$$

The dotted lines symbolise hydrogen bonds.

Because of the hydrogen bonding that can exist between alcohol and water molecules, the lower aliphatic alcohols, namely methanol, ethanol, 1-propanol, 2-propanol and 2-methyl-2-propanol, are completely miscible with water in all proportions. (Alkanes, in contrast, are immiscible.)

The remaining C_4 alcohols and the higher members of the homologous series are only partially miscible, with solubility falling to a very low level at about $C_7H_{15}OH$ (1-heptanol = 0.09 g per 100 g of water at 18 °C (291 K)).

Phenols, too, exhibit hydrogen bonding. Phenol itself is a solid with a melting temperature of 40.9 °C (314.1 K), while methylbenzene, with a similar relative molecular mass, is a liquid. Phenol is soluble in water, although aromatic hydrocarbons are not. When phenol is shaken with cold water a two phase liquid system is obtained, the lower of which is a solution of water in phenol and the upper, phenol in water.

Ethers cannot associate through hydrogen bonding. In consequence, the boiling temperatures of ethers are close to those of alkanes of similar relative molecular mass, but very different from those of comparable alcohols; compare Table 12.3 with Table 12.2.

Table 12.3 Boiling temperatures of ethers

Ether	Relative molecular mass	Boiling temperature/ °C (K)
methoxymethane	46	−24.8 (248.4)
methoxyethane	60	7.0 (280.2)
ethoxyethane	74	34.5 (307.7)

Hydrogen bonding is possible, however, between ether and water molecules; thus, the water solubility of ethers is comparable with that of alcohols. For example, the solubilities of ethoxyethane and 1-butanol (isomers of formula $C_4H_{10}O$) are 7.5 and 7.9 g per 100 g of water respectively, at 20 °C (293 K).

12.5 Chemical properties of the C—O bond

The chemistry of alcohols, phenols and ethers can be studied under three main headings: (i) acidic and basic character; (ii) nucleophilic substitution reactions; and (iii) elimination reactions. We shall consider each in turn.

Acidic and basic character

Alcohols and phenols resemble water in that they can donate protons and also accept protons, i.e. they have both acidic and basic character. Ethers, in contrast, are purely basic. We must consider this aspect first, because a knowledge of it is crucial to an understanding of the rest of the chemistry.

Acidity

Alcohols and especially phenols act as weak protic acids, ionising in the presence of water to give a proton and an anion:

$$R\text{—}O\overset{\bullet\bullet}{\underset{\times\times}{}}H \rightleftharpoons RO^- + H^+$$

alkoxide ion

$$Ar\text{—}O\overset{\bullet\bullet}{\underset{\times\times}{}}H \rightleftharpoons ArO^- + H^+$$

phenoxide ion

The acidity of hydroxy compounds varies considerably, depending partly on the ease with which a proton is lost and partly on the stability of the anion which is formed (§3.2). Alcohols are extremely weak acids: the pK_a values of methanol and ethanol are 15.5 and ~16 respectively (cf. 15.7 for water). Aqueous solutions of alcohols are neutral to litmus and other acid—base indicators. Their salts can be made only by adding metal to alcohol; sodium ethoxide, for example, is produced in a vigorous reaction between ethanol and sodium at room temperature:

$$C_2H_5OH + Na = C_2H_5O^-Na^+ + \tfrac{1}{2}H_2$$

Sodium ethoxide remains in solution, but can be isolated as a white solid after evaporation of unreacted ethanol.

Only the alkali metals (group 1A) are sufficiently reactive to dissolve readily in alcohols, although alkoxides can be prepared from group 2A metals (notably magnesium), and also aluminium.

Metallic alkoxides, being salts of strong bases and weak acids, are extensively hydrolysed in solution, which is the reason why they cannot be made by neutralisation methods, e.g.

$$C_2H_5ONa + H_2O \rightleftharpoons C_2H_5OH + NaOH$$

Alkoxides are useful compounds in the laboratory, especially in the Williamson synthesis (§11.5).

Phenols are much more acidic than alcohols (§3.2). Phenol itself — formerly called 'carbolic acid' — has a pK_a value of 10.00. Phenols give acidic solutions in water, which turn litmus red, and their sodium and potassium salts can be made not only from the metals but also from their hydroxides, e.g.

sodium phenoxide

Phenol can be liberated from an alkaline solution by the addition of a strong acid, e.g.

Most phenols are weaker acids than carboxylic acids. (Ethanoic acid has a pK_a value of 4.76.) They are also weaker than carbonic acid, and will therefore not liberate carbon dioxide from carbonates or hydrogen-carbonates. However, the presence of a nitro group on the aromatic ring may increase the acidity of a phenol to such an extent that a reaction with sodium hydrogencarbonate is possible, e.g.

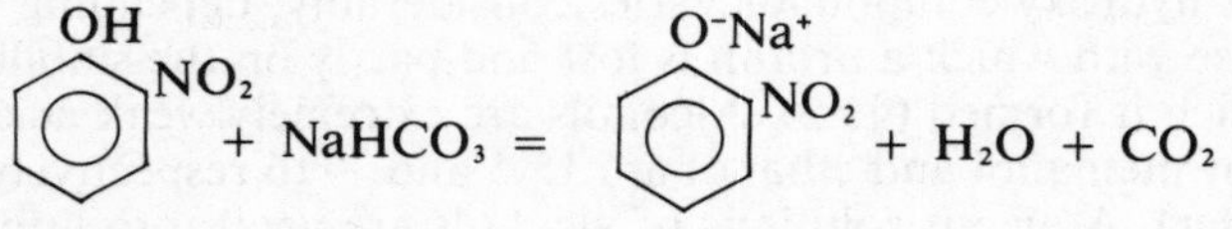

The nitro group has a tendency to withdraw electrons from the aromatic ring. As a result, electrons are drawn into the ring from the HO group. Polarisation of the O—H bond is increased, and loss of a proton is facilitated. 2,4,6-Trinitrophenol (picric acid), with a pK_a value of 0.42, is almost as strong as the common mineral acids and will corrode metals.

Sodium phenoxides, like sodium alkoxides, react readily with alkyl halides in the Williamson synthesis, e.g.

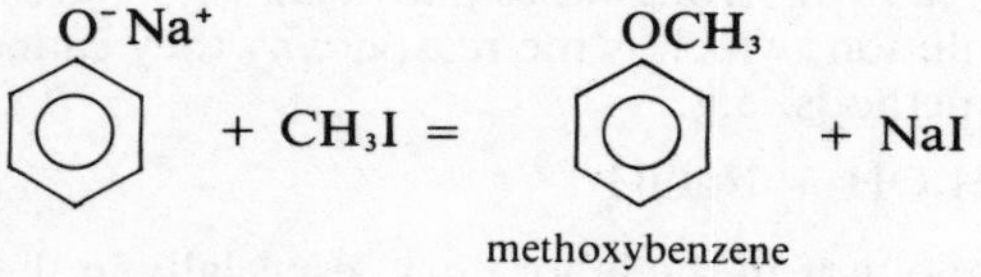

methoxybenzene

Ethers, because they lack an O—H bond, are not acidic.

Basicity

Alcohols and ethers, like water, can function as Lewis bases by donating a lone pair of electrons from the outer shell of the oxygen atom. If the lone pair is donated to a proton the alcohol or ether is said to be a 'base'; if it is donated to some other Lewis acid the alcohol or ether is referred to as a 'nucleophile'. We shall consider each in turn. Phenols seldom behave in this manner because the mesomeric effect (§1.4) reduces the electron availability at the oxygen atom.

Alcohols as bases

In the same way that water molecules are converted into oxonium ions in the presence of protic acids, so alcohols and ethers give substituted oxonium ions, RH_2O^+ and R_2HO^+ respectively:

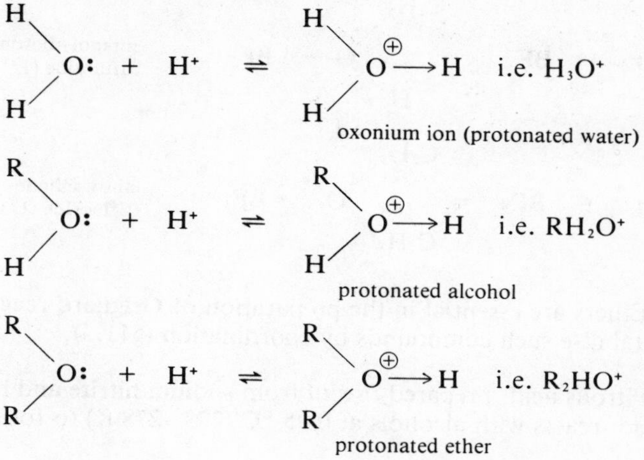

oxonium ion (protonated water)

protonated alcohol

protonated ether

Although it is conventional to write the positive charge on the central oxygen atom, the charge does not reside there but is distributed over the ion as a whole.

The ability to donate a lone pair of electrons increases in the order

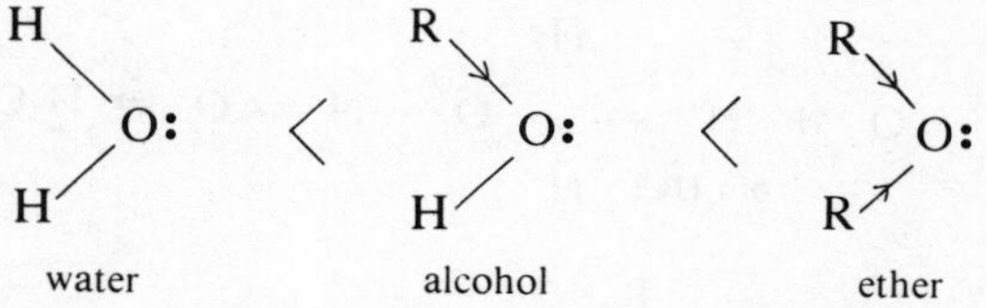

because the $+I$ effect of the alkyl groups increases the electron availability at the oxygen atom. Even so, alcohol and ether salts formed in this way are unstable and exist only in acidic solution.

The reaction between ethers and concentrated sulphuric acid is used in the laboratory for distinguishing between ethers and alkanes, for while ethers dissolve readily in the concentrated acid, alkanes do not. The ether can be recovered by pouring the solution cautiously into ice water.

Alcohols as nucleophiles

Alcohols and, to a lesser extent, ethers can react with a considerable variety of Lewis acids. Coordination of alcohol is often followed by the elimination of another species, as the following examples show.

BF₃ Boron trifluoride coordination compounds are formed with both alcohols and ethers, e.g.

$$\begin{array}{c} C_2H_5 \\ \diagdown \\ \diagup \\ H \end{array} O: \ + \ BF_3 \ = \ \begin{array}{c} C_2H_5 \\ \diagdown \\ \diagup \\ H \end{array} O \longrightarrow BF_3 \qquad \text{ethanol–boron trifluoride (1/1)}$$

$$\begin{array}{c} C_2H_5 \\ \diagdown \\ \diagup \\ C_2H_5 \end{array} O: \ + \ BF_3 \ = \ \begin{array}{c} C_2H_5 \\ \diagdown \\ \diagup \\ C_2H_5 \end{array} O \longrightarrow BF_3 \qquad \text{ethoxyethane–boron trifluoride (1/1)}$$

RMgX Ethers are essential in the preparation of Grignard reagents, for they stabilise such compounds by coordination (§11.5).

HNO₂ Nitrous acid, prepared *in situ* from sodium nitrite and hydrochloric acid, reacts with alcohols at 0–5 °C (273–278 K) to form alkyl nitrites:

$$ROH + HONO = RONO + H_2O$$

The reaction proceeds by coordination of an alcohol molecule to a nitrosyl cation, NO^+, followed by elimination of a proton. The nitrosyl cation is formed by the protonation of nitrous acid:

$$\overset{H}{\underset{H}{\diagdown}}\overset{\bullet\bullet}{O}-N\!=\!O \ + \ H^+ \ \rightleftharpoons \ \overset{H}{\underset{H}{\diagdown}}\overset{\oplus}{O}-N\!=\!O \ \rightleftharpoons \ H_2O \ + \ \overset{\oplus}{N}\!=\!O$$

from HCl

$$\overset{R}{\underset{H}{\diagdown}}O: \ + \ \overset{\oplus}{N}\!=\!O \ \longrightarrow \ \overset{R}{\underset{H}{\diagdown}}\overset{\oplus}{O}\longrightarrow N\!=\!O \ \longrightarrow \ R-O-N\!=\!O \ + \ H^+$$

The reaction is similar to that between nitrous acid and a secondary amine (§14.6).

Alkyl nitrites are isomeric with nitroalkanes, RNO_2. Perhaps the commonest example is 3-methylbutyl nitrite (isoamyl nitrite), $(CH_3)_2CHCH_2CH_2ONO$, a fragrant yellow liquid that is used as an antidote for hydrogen cyanide poisoning.

HNO₃ Nitric acid reacts with alcohols in the cold to give alkyl nitrates:

$$ROH + HONO_2 = RONO_2 + H_2O$$

However, ordinary concentrated nitric acid reacts violently with alcohols by an oxidation–reduction reaction initiated by traces of nitrous acid. The danger can be averted by first boiling the nitric acid with urea:

$$2HNO_2 + NH_2CONH_2 = 2N_2 + CO_2 + 3H_2O$$

Nitrates of certain polyhydric alcohols, notably 1,2,3-propanetriol (glycerol) and cellulose, find application as explosives. 1,2,3-Propane-triyl trinitrate ('nitroglycerine') is manufactured by slowly running 1,2,3-propanetriol into a mixture of concentrated nitric acid and fuming sulphuric acid at a temperature of 10–20 °C (283–293 K):

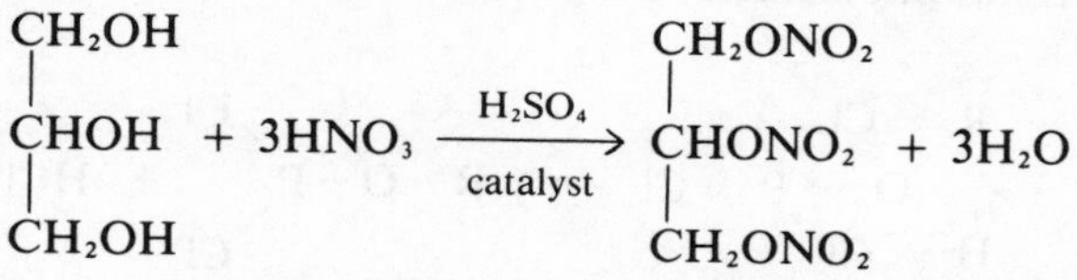

$$
\begin{array}{l}
CH_2OH \\
| \\
CHOH \\
| \\
CH_2OH
\end{array}
+ 3HNO_3 \xrightarrow[\text{catalyst}]{H_2SO_4}
\begin{array}{l}
CH_2ONO_2 \\
| \\
CHONO_2 \\
| \\
CH_2ONO_2
\end{array}
+ 3H_2O
$$

The use of sulphuric acid as a catalyst suggests that nitric acid reacts by a similar mechanism to nitrous acid, for in the presence of sulphuric acid nitric acid yields the nitryl cation, NO_2^+ (§10.2).

Nitroglycerine is an oily liquid which explodes violently on shock. In the pure state it is too dangerous to be useful, but it is much more stable in the form of dynamite, in which it is absorbed on to kieselguhr (silica) or sawdust. 'Gelignite', a powerful explosive used in mining, is a mixture of nitroglycerine, ammonium nitrate and cellulose trinitrate ('gun cotton'). 'Cordite', a smokeless gunpowder, is made from nitroglycerine, cellulose trinitrate and petroleum jelly.

Carbonyl compounds   Alcohols react with a wide range of carbonyl compounds, i.e. substances whose molecules possess the carbonyl group, C=O. Examples are aldehydes, carboxylic acids, carboxylic acid anhydrides and carboxylic acid chlorides. In most cases there occurs an addition–elimination reaction, in which the nucleophilic addition of alcohol across the C=O bond is followed by the loss of a small molecule such as H_2O (Chapter 13).

PCl_3, PBr_3 and PI_3   It is well known that phosphorus trichloride is rapidly attacked by water to give phosphonic acid and hydrogen chloride:

$$3HOH + PCl_3 = H_2PHO_3 + 3HCl$$

The reaction takes place by the coordination of water to PCl_3, followed by the elimination of HCl:

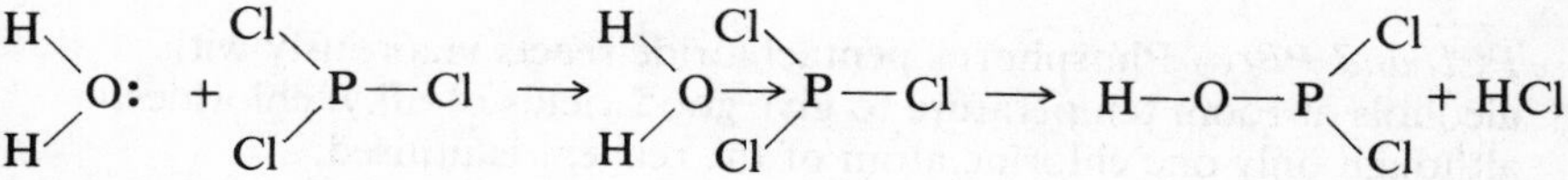

Two repetitions of the process, each with a further molecule of water, give $P(OH)_3$ (which becomes phosphonic acid) and more hydrogen chloride.

In a change which is superficially similar to hydrolysis, phosphorus trichloride reacts readily with an alcohol to give phosphonic acid and an alkyl chloride:

$$3ROH + PCl_3 = H_2PHO_3 + 3RCl$$

The first stage of alcoholysis resembles that of hydrolysis, and gives an ester-acid chloride of phosphonic acid, $ROPCl_2$:

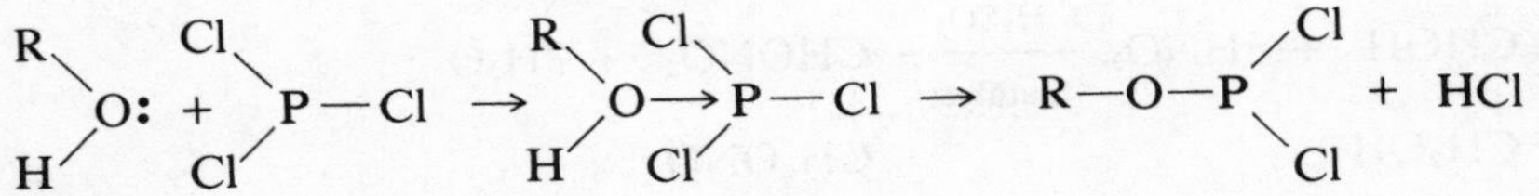

Two repetitions of this process, each with a further molecule of alcohol, would give an ester $P(OR)_3$ of the acid $P(OH)_3$, together with more hydrogen chloride, and indeed this does happen to a considerable extent. However, the intermediate, $ROPCl_2$, can undergo nucleophilic substitution by a chloride ion:

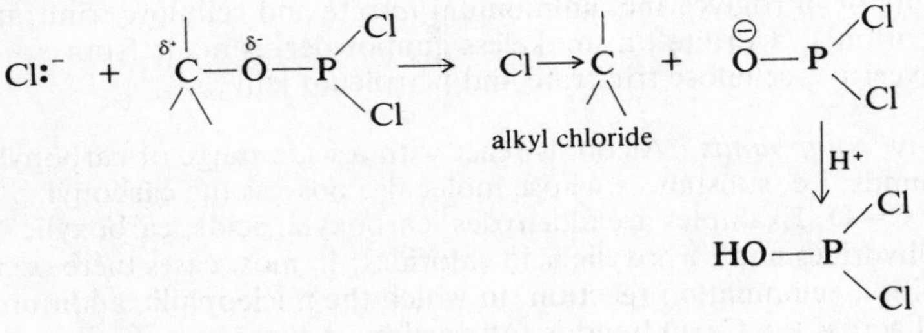

When the process is repeated, twice more, phosphonic acid and more alkyl chloride are obtained.

Phosphorus trichloride is not a highly recommended reagent for the preparation of alkyl chlorides as the yields are low. This is because $ROPCl_2$ reacts with further molecules of alcohol in preference to chloride ions. However, phosphorus tribromide and phosphorus triiodide give good yields of alkyl bromides and iodides respectively. (Br^- and I^- are better nucleophiles than Cl^-.) The reagents are usually made *in situ* from red phosphorus and the appropriate halogen.

PCl₅ and PBr₅ Phosphorus pentachloride reacts vigorously with alcohols at room temperature to give good yields of alkyl chlorides, although only one chlorine atom of the reagent is utilised:

$$ROH + PCl_5 = RCl + POCl_3 + HCl$$

The reaction, which has a mechanism similar to that of the reaction with phosphorus trichloride, serves not only as a means of making alkyl chlorides but also as a test for the presence of a hydroxyl group in an organic compound. (The substance under test must be dried beforehand.) Phosphorus pentabromide, likewise, is excellent for the preparation of alkyl bromides; but it should be noted that phosphorus pentaiodide does not exist.

Phenols, when treated with phosphorus halides, give very low yields of aryl halides. The main products are esters of phosphorus-containing acids.

$SOCl_2$ Thionyl chloride is often suggested as a reagent for the preparation of alkyl chlorides:

$$ROH + SOCl_2 = RCl + SO_2 + HCl$$

It has an advantage in that both the by-products are gaseous, but the reagent is relatively unreactive and is normally used in excess.

Once again the reaction commences by coordination of the alcohol to the reagent, followed by the elimination of HCl:

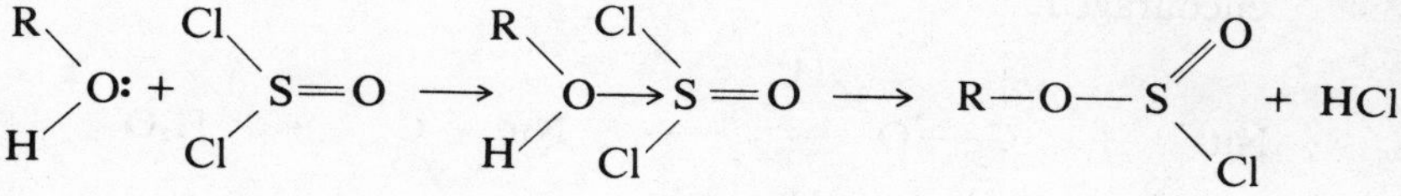

However, the ester-acid chloride, instead of being attacked by Cl^- as in the previous examples, undergoes an *internal* nucleophilic substitution, in which it is attacked by its own chlorine atom:

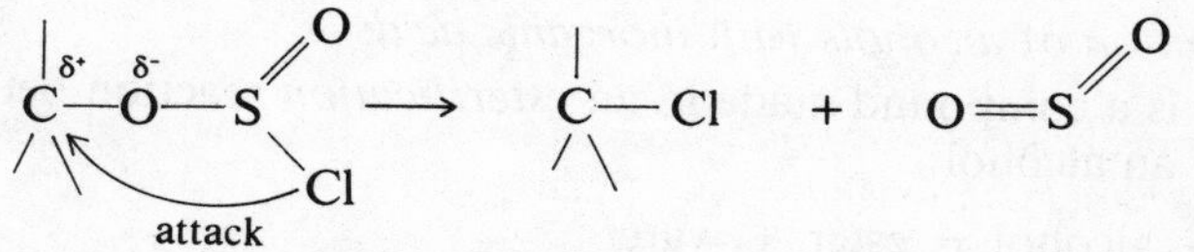

Nucleophilic substitution reactions

Because the C—O bond is polarised in the same way as the C—halogen bond, we might expect alcohols and ethers to take part in nucleophilic substitutions, thus:

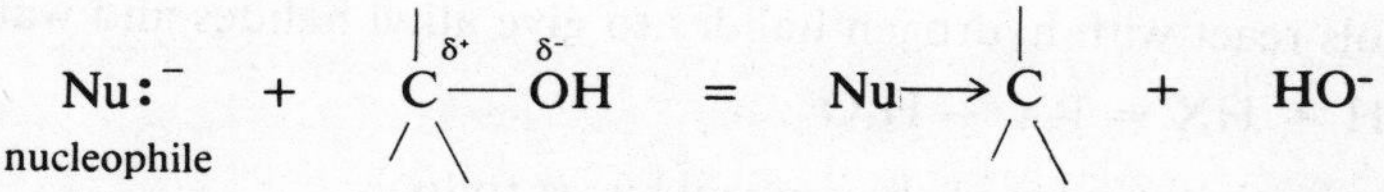

Except under acidic conditions, however, such reactions do not occur, because the C—O bond, with its relatively high bond dissociation enthalpy, is too difficult to break. HO^- and RO^- are 'poor leaving groups'. (The best leaving groups, e.g. I⁻ and Br⁻, form bonds with carbon that are easily broken.)

In the presence of a protic acid the substrate is no longer an alcohol (or ether) but a protonated alcohol (or ether), and the equation must be modified as follows:

$$\text{Nu:}^- \ + \ \overset{\displaystyle |}{\underset{\diagup\ \diagdown}{C}}\!-\!\overset{\oplus}{O}H_2 \ = \ \text{Nu} \longrightarrow \overset{\displaystyle |}{\underset{\diagup\ \diagdown}{C}} \ + \ H_2O$$

Reaction is facilitated for two reasons.

(i)   The positive charge on the protonated alcohol (or ether) increases the attraction for nucleophiles.

(ii) When an alcohol becomes protonated, the proton attracts a lone pair of electrons *away* from the oxygen atom. The oxygen atom therefore becomes electron deficient, and draws towards itself the electrons of the C—O bond. In other words, the C—O bond becomes increasingly polarised, and the loss of H₂O (not HO⁻) is encouraged.

$$\text{Nu:}^- \ + \ \overset{\displaystyle |}{\underset{\diagup\ \diagdown}{C}}\!-\!\overset{\nearrow\ H^+}{O}_{\searrow H} \ \longrightarrow \ \text{Nu} \longrightarrow \overset{\displaystyle |}{\underset{\diagup\ \diagdown}{C}} \ + \ H_2O$$

Lewis acid catalysts, such as aluminium oxide, can also be used.

Phenols rarely participate in nucleophilic substitution reactions because the mesomeric effect strengthens the C—O bond (§1.4).

Esterification of alcohols with inorganic acids

An *ester* is a compound made in an *esterification* reaction between an acid and an alcohol:

acid + alcohol ⇌ ester + water

Perhaps the best known esters are those of carboxylic acids: they will be discussed in Chapter 13. We are concerned here with esters of inorganic acids, notably hydrogen halides and sulphuric acid. (Esters of hydrogen halides are alkyl halides.)

HI, HBr and HCl

Alcohols react with hydrogen halides to give alkyl halides and water:

$$\text{ROH} + \text{HX} \rightleftharpoons \text{RX} + H_2O$$

The reactivity order of the reagents is as follows:

HI > HBr > HCl

Hydriodic acid and hydrobromic acid react readily to give alkyl iodides and bromides respectively. Hydrochloric acid reacts at a reasonable rate only with tertiary alcohols and phenylmethanol. Other alcohols must be saturated with hydrogen chloride gas and then heated in a sealed tube for reaction to occur. Anhydrous zinc chloride, which acts as a Lewis acid catalyst, is usually included.

In the preparation of alkyl bromides, hydrobromic acid is often replaced by a mixture of concentrated sulphuric acid (which supplies protons) and sodium bromide (which supplies bromide ions). 1-Bromobutane, for example, is usually prepared in this manner:

$$CH_3CH_2CH_2\text{—}\underset{\substack{| \\ H}}{\overset{\substack{H \\ |}}{C}}\text{—OH} + H^+ \xrightarrow{\text{from } H_2SO_4} CH_3CH_2CH_2\text{—}\underset{\substack{| \\ H}}{\overset{\substack{H \\ |}}{C}}\text{—}\overset{\oplus}{O}H_2 \xrightarrow[\text{from NaBr}]{Br^-} CH_3CH_2CH_2\text{—}Br\text{→}\underset{\substack{| \\ H}}{\overset{\substack{H \\ |}}{C}} + H_2O$$

1-butanol   protonated 1-butanol   1-bromobutane

The presence of sulphuric acid introduces complications. Because it is a dehydrating agent it partially converts alcohols into ethers and alkenes (see below), and for this reason its use is not recommended with secondary and tertiary alcohols. Also, it is an oxidising agent, and may oxidise primary and secondary alcohols to carboxylic acid esters and ketones respectively. Esters of sulphuric acid may be formed too, but such compounds are relatively involatile and are unlikely to contaminate the final product.

The same technique cannot be used for iodides, owing to the ease with which the iodide ion is oxidised to iodine by concentrated sulphuric acid. Instead, constant boiling hydriodic acid is employed. The yield of alkyl iodide may be poor, because hydriodic acid can reduce such compounds to alkanes.

Alcohols vary considerably in their reactivity towards hydrogen halides. Besides the decline of reactivity that is always associated with the ascent of a homologous series, there is also a considerable difference between the reactivity of primary, secondary and tertiary alcohols:

tertiary > secondary > primary

This is a reflection of differences in reaction mechanism. In general, primary alcohols react by an S_N2 mechanism; cf. primary alkyl halides (§11.5).

$$Br:^- \quad \underset{\substack{H \\ H}}{\overset{CH_3}{C}}\overset{\oplus}{\text{—}O}\underset{H}{\overset{H}{\diagup}} \rightleftharpoons \overset{\delta^-}{Br}\cdots\underset{\substack{H \\ H}}{\overset{CH_3}{C}}\cdots\overset{\delta^+}{O}\underset{H}{\overset{H}{\diagup}} \rightleftharpoons Br\text{→}\underset{\substack{H \\ H}}{\overset{CH_3}{C}}\overset{H}{\diagup} + H_2O$$

approach of Br$^-$ to   transition state   bromoethane
protonated ethanol

Tertiary alcohols react by an S_N1 mechanism, which involves the formation of a carbonium ion; cf. tertiary alkyl halides.

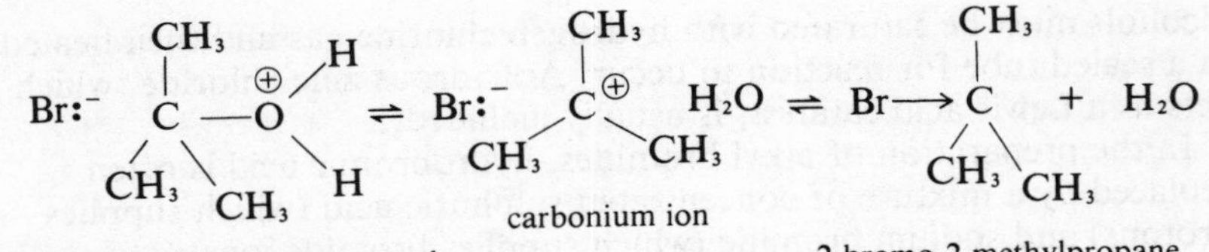

approach of Br⁻ to protonated 2-methyl-2-propanol — carbonium ion — 2-bromo-2-methylpropane

Secondary alcohols react by an intermediate mechanism (cf. secondary alkyl halides). The fact that these reactions are so similar to those of alkyl halides is scarcely surprising because they represent the reverse of alkyl halide hydrolysis.

As we saw above, phosphorus trihalides and pentahalides can be used instead of hydrogen halides for converting alcohols into alkyl halides. The choice of reagent is dictated partly by consideration of yield and partly by ease of purifying the product.

H_2SO_4
Sulphuric acid is diprotic and thus gives rise to two series of esters, e.g.

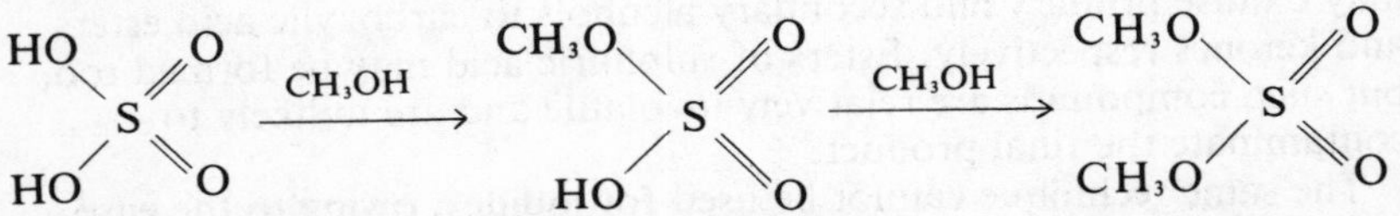

sulphuric acid — methyl hydrogen sulphate — dimethyl sulphate

Although alkyl hydrogen sulphates can be made in fairly good yield by heating the appropriate alcohol with concentrated sulphuric acid on a boiling water bath, attempts to prepare dialkyl sulphates by direct esterification lead, except in the case of methanol and ethanol, to etherification of the alcohol and the formation of alkenes (see below). Dialkyl sulphates are usually made by heating alkyl hydrogen sulphates, e.g.

$$2CH_3OSO_2OH = (CH_3O)_2SO_2 + H_2SO_4$$

Although dimethyl sulphate and diethyl sulphate are effective alkylating agents, they are carcinogenic and their use is now forbidden.
Concentrated sulphuric acid with phenols has the effect of sulphonating the aromatic ring.

Etherification of alcohols
When an alcohol is heated in the presence of a protic acid (usually concentrated sulphuric acid) or a Lewis acid (usually aluminium oxide) it undergoes etherification through the loss of a molecule of water between two molecules of the alcohol, e.g.

$$2C_2H_5OH \xrightarrow[\text{or } Al_2O_3 \text{ at } 250\,°C\,(523\,K)]{\text{conc. } H_2SO_4 \text{ at } 130\,°C\,(403\,K)} C_2H_5-O-C_2H_5 + H_2O$$

The temperature is somewhat critical, for at lower temperatures esters are formed (e.g. ethyl hydrogen sulphate), and at higher temperatures the principal product is an alkene (e.g. ethene).

Etherification should be regarded as a nucleophilic substitution reaction involving two molecules of alcohol, one of which, in the protonated state, acts as the substrate, while the other behaves as the nucleophilic reagent. The reaction may have either an S_N1 or S_N2 mechanism, depending on the type of alcohol.

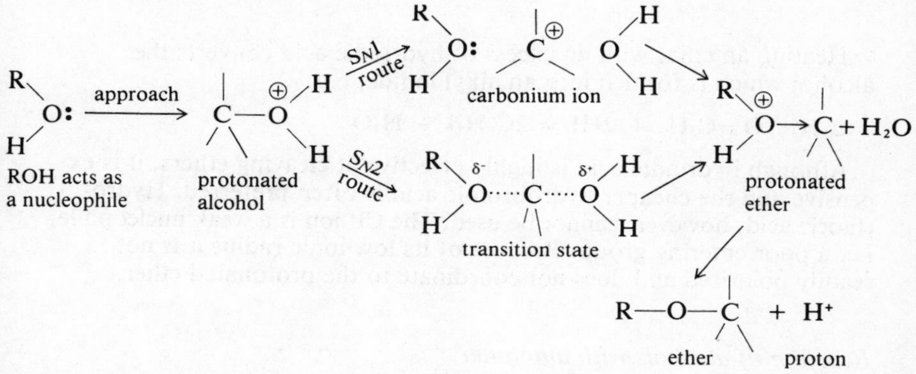

Secondary and, in particular, tertiary alcohols follow the S_N1 route and are dehydrated so easily that dilute sulphuric acid is adequate.

Cleavage of ethers

Etherification is a reversible reaction, and in the presence of a strong acid catalyst an ether may revert to the alcohol or alcohols from which it is derived, e.g.

$$C_2H_5-O-C_2H_5 + H_2O \xrightarrow[\text{heat under pressure}]{\text{dilute } H_2SO_4} 2C_2H_5OH$$

When concentrated hydriodic acid — the most usual reagent — is used, the product is a mixture of an alcohol and an alkyl iodide, e.g.

$$C_2H_5-O-C_2H_5 + HI(aq) \xrightarrow{\text{reflux}} C_2H_5OH + C_2H_5I$$

The I^- ion acts as a nucleophile in attacking the protonated ether:

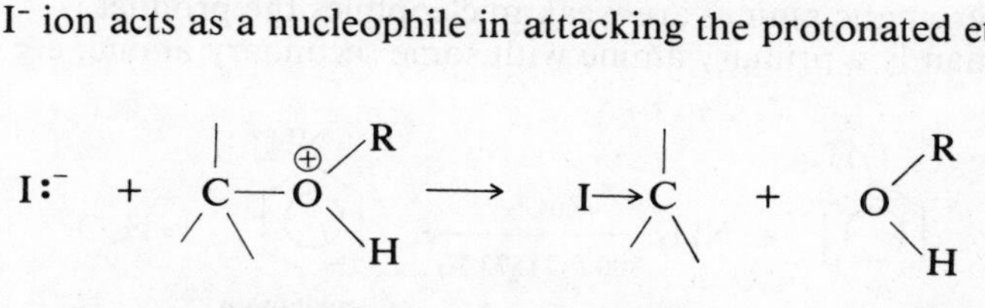

Within this overall framework, an S_N1 or S_N2 route is possible.

When a mixed ether is refluxed with hydriodic acid, the iodine atom will attach itself to the *smaller* group, e.g.

$$CH_3—O—C_2H_5 + HI = CH_3I + C_2H_5OH$$

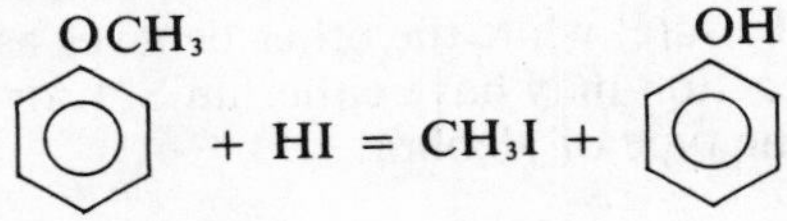

Heating an ether with an excess of hydriodic acid converts the alcohol which is formed into an alkyl iodide, e.g.

$$C_2H_5—O—C_2H_5 + 2HI = 2C_2H_5I + H_2O$$

Athough hydriodic acid is highly effective at cleaving ethers, it is expensive and the cheaper hydrobromic acid is often preferred. Hydrochloric acid, however, cannot be used. The Cl⁻ ion is a weak nucleophile, i.e. a poor entering group. Because of its low ionic radius it is not readily polarised and does not coordinate to the protonated ether.

Reaction of alcohols with ammonia

Alcohols (but not ethers) react with ammonia at high temperatures in the presence of a Lewis acid catalyst, usually zinc chloride, to give primary amines. The primary amines subsequently react with more alcohol to give secondary amines, and these in turn give rise to tertiary amines, e.g.

$$CH_3OH + NH_3 \xrightarrow[\text{300 °C (573 K) under pressure}]{\text{ZnCl}_2 \text{ catalyst}} CH_3NH_2 + H_2O$$
methylamine

$$CH_3NH_2 + CH_3OH = (CH_3)_2NH + H_2O$$
dimethylamine

$$(CH_3)_2NH + CH_3OH = (CH_3)_3N + H_2O$$
trimethylamine

Quaternary ammonium hydroxides, $[R_4N]^+ HO^-$, are not formed, because they are unstable at the temperature of this reaction. (Cf. the reaction between alkyl halides and ammonia, §11.5.)

Phenols, too, react with ammonia under these conditions, but because aromatic amines are weak nucleophiles the product is predominantly a primary amine with some secondary amine, e.g.

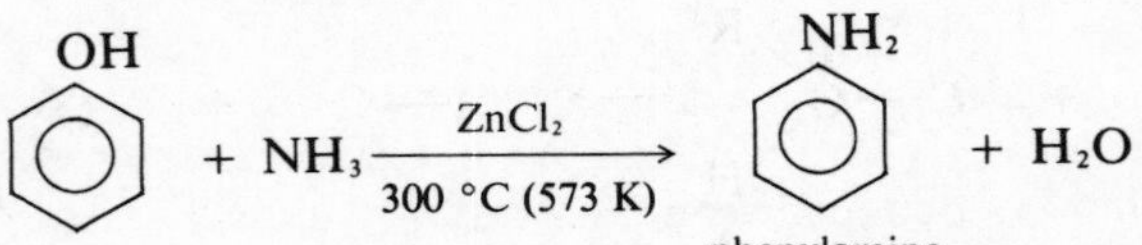

$$NH_2 + OH = \text{(ring)}NH\text{(ring)} + H_2O$$

diphenylamine

Elimination reactions of alcohols

Elimination of water (dehydration)

When an alcohol is heated with an acid catalyst, etherification is always accompanied by a competing reaction leading to the formation of an alkene. One molecule of water is lost from one molecule of the alcohol in a 1,2-elimination; cf. 1,2-elimination of HX from an alkyl halide. The term 'β-elimination' is also used, because the hydrogen atom is lost from the β-carbon atom. (The HO group is said to reside on the α-carbon atom.) For example

$$\underset{\beta \quad \alpha}{CH_3CH_2OH} \xrightarrow[\substack{\text{or conc. } H_3PO_4 \text{ at } 210 \text{ °C (483 K)} \\ \text{or } Al_2O_3 \text{ at } 350 \text{ °C (623 K)}}]{\text{conc. } H_2SO_4 \text{ at } 180 \text{ °C (453 K)}} CH_2{=}CH_2 + H_2O$$

An excess of acid is necessary to ensure an adequate rate of reaction.

The substitution and elimination reactions brought about by sulphuric acid have basically similar mechanisms. The main difference is that in substitution (i.e. the formation of an alkyl hydrogen sulphate) the HSO_4^- ion behaves as a nucleophile, donating a lone pair of electrons to the α-carbon atom, while in elimination the HSO_4^- ion behaves as a base and coordinates to a proton from the β-position:

HSO$_4^-$ ion approaches
a protonated alcohol

HSO$_4^-$ ion reaches a
β-hydrogen atom.
C—O and C—H bonds
are strongly polarised

HSO$_4^-$ coordinates to
H$^+$. H$_2$O molecule leaves

When studying alkyl halides (§11.5) we saw that tertiary halides, which underwent substitution reactions by an S_N1 mechanism, could also take part in elimination reactions by an E1 mechanism, while primary halides, which took part in S_N2 reactions, also entered into E2 reactions. We might expect, therefore, tertiary alcohols to eliminate water by an E1 mechanism, via a carbonium ion, and primary alcohols to do so by the E2 route via a transition state. It is interesting, however, that *all* alcohols eliminate water by a mechanism which is essentially E1:

$$HSO_4^- \quad H^{\delta+} \qquad\qquad HSO_4^- \quad H^{\delta+} \qquad\qquad H_2SO_4$$

| protonated alcohol | carbonium ion | alkene |

An E2 mechanism is impossible because HSO_4^- ions do not allow the necessary transition state to be achieved. HSO_4^-, a relatively weak base, is unable to coordinate to a β-hydrogen atom with its partial positive charge. It can coordinate only to a fully charged proton, which it abstracts from a carbonium ion.

To summarise, the dehydration of any alcohol proceeds in three stages:

(i) protonation of the alcohol;
(ii) decomposition of the protonated alcohol to give a carbonium ion;
(iii) loss of a proton from the carbonium ion to give an alkene.

Consider, for example, the conversion of 1-butanol to 1-butene:

$$CH_3CH_2CH_2CH_2OH + H^+ \rightleftharpoons CH_3CH_2CH_2CH_2\overset{\oplus}{O}H_2$$

$$CH_3CH_2CH_2CH_2\overset{\oplus}{O}H_2 \rightleftharpoons CH_3CH_2CH_2\overset{\oplus}{C}H_2 + H_2O$$

$$CH_3CH_2CH_2\overset{\oplus}{C}H_2 \rightleftharpoons CH_3CH_2CH{=}CH_2 + H^+$$

It will be seen that all stages of the reaction are reversible. As we have seen, the reverse change, i.e. the acid catalysed hydration of an alkene to an alcohol, is important in the chemical industry as a means of manufacturing alcohols.

In the above example the principal product is 2-butene; not 1-butene. This is because the primary carbonium ion, $CH_3CH_2CH_2\overset{\oplus}{C}H_2$, tends to rearrange to a more stable secondary carbonium ion:

$$CH_3CH_2CH_2\overset{\oplus}{C}H_2 \rightleftharpoons CH_3CH_2\overset{\oplus}{C}HCH_3$$

$$CH_3CH_2\overset{\oplus}{C}HCH_3 \rightleftharpoons CH_3CH{=}CHCH_3 + H^+$$

Secondary and, in particular, tertiary alcohols form carbonium ions more readily than primary alcohols and can therefore be dehydrated under relatively mild conditions. Secondary alcohols respond to orthophosphoric acid at about 140 °C (413 K), and tertiary alcohols at about 100 °C (373 K).

In the elimination of water from secondary or tertiary alcohols Saytzeff's rule is followed, i.e. the thermodynamically more stable alkene is formed through the loss of a hydrogen atom from the β-

carbon atom which possesses the *lower* number of hydrogen atoms, e.g.

$$CH_3CH(OH)CH_2CH_3 \xrightarrow{\text{conc. } H_2SO_4}$$

→ $CH_3CH{=}CHCH_3$ main product

→ $CH_2{=}CHCH_2CH_3$ minor product

The dehydration of alcohols by Lewis acid catalysts takes place by an adsorption mechanism. According to the *multiple adsorption theory* of Burk and Balandin, a strain is produced in certain bonds of a molecule when it is adsorbed at two or more 'active centres' on the surface of a catalyst. An alcohol molecule becomes adsorbed at two sites on the surface of alumina, the hydroxyl oxygen atom being attracted to an Al^{3+} ion and a β-hydrogen atom to an oxide ion. This weakens the C—O and C—H bonds to such an extent that they can break, with the formation of water and an alkene:

ions at the catalyst surface

Substitution versus elimination

When studying the reactions between alkyl halides and bases (§11.5), we saw that high temperatures bring about elimination rather than substitution. In the same way, when an alcohol is treated with an acid catalyst, high temperatures favour elimination (i.e. alkene formation) in preference to substitution (i.e. etherification). The difference in temperatures when concentrated sulphuric acid is used is shown in Table 12.4; with alumina as the catalyst a temperature of 250 °C (523 K) is recommended for etherification, and 350 °C (623 K) for alkene formation.

Table 12.4 The action of concentrated sulphuric acid on ethanol at various temperatures

Temperature/ °C (K)	Type of reaction	Mechanism of reaction	Product
20 (293)	neutralisation	protonation of the alcohol	unstable oxonium salt, $C_2H_5OH_2^{\oplus}HSO_4^-$
100 (373)	esterification	nucleophilic substitution (HSO_4^- as nucleophile)	ethyl hydrogen sulphate, $C_2H_5OSO_2OH$
130 (403)	etherification	nucleophilic substitution (C_2H_5OH as nucleophile)	ethoxyethane, $C_2H_5{-}O{-}C_2H_5$
180 (453)	alkene formation	elimination of water	ethene, $CH_2{=}CH_2$

Alkene formation has a high activation energy and cannot occur at low temperatures. Etherification, in contrast, has a relatively low activation energy and is able to take place. When the temperature is raised the rates of both reactions increase, but alkene formation is promoted to a greater extent. This is because the proportion of molecules with the necessary activation energy for alkene formation increases more than the proportion with the activation energy needed for etherification.

Elimination of hydrogen (oxidation)

The oxidation of alcohols is the most important way of preparing aldehydes and ketones. The reaction should be regarded essentially as a dehydrogenation, in which two hydrogen atoms are lost from the same position on the chain. One is an α-hydrogen atom, while the other is the hydroxyl hydrogen atom. Thus, primary alcohols are oxidised to aldehydes, and secondary alcohols to ketones.

Tertiary alcohols cannot be oxidised in this way because they lack an α-hydrogen atom. Nevertheless, in the presence of acidic oxidising agents, they readily undergo oxidative cleavage, probably via alkenes.

$$CH_3 - \overset{\overset{\displaystyle O - H}{|}}{\underset{\underset{\displaystyle H}{|}}{C}} - H \xrightarrow[-H_2]{\text{oxidation}} CH_3 \, C\!\!\underset{H}{\overset{O}{\diagup\!\!\!\diagdown}}$$

ethanol   ethanal

primary alcohol

$$CH_3 - \overset{\overset{\displaystyle O - H}{|}}{\underset{\underset{\displaystyle CH_3}{|}}{C}} - H \xrightarrow[-H_2]{\text{oxidation}} \overset{CH_3}{\underset{CH_3}{\diagdown\!\!\diagup}}C\!=\!O$$

2-propanol   propanone

secondary alcohol

$$CH_3 - \overset{\overset{\displaystyle CH_3}{|}}{\underset{\underset{\displaystyle CH_3}{|}}{C}} - OH \xrightarrow[- H_2O \, (H^+ \text{ catalyst})]{\text{acidic oxidising agent}} \overset{H}{\underset{H}{\diagdown\!\!\diagup}}C\!=\!C\overset{CH_3}{\underset{CH_3}{\diagup\!\!\diagdown}}$$

2-methyl-2-propanol   2-methylpropene

tertiary alcohol

$$\xrightarrow[\text{cleavage}]{\text{oxidative}} \overset{H}{\underset{H}{\diagdown\!\!\diagup}}C\!=\!O + O\!=\!C\overset{CH_3}{\underset{CH_3}{\diagup\!\!\diagdown}}$$

propanone

The elimination of hydrogen can be achieved catalytically, by passing the alcohol vapour over any metal which can act as a hydrogen acceptor. While palladium has been recommended for the purpose, copper or silver, at a temperature of 400−450 °C (673−723 K), is cheaper and highly effective.

Alternatively, an oxidising agent such as acidified sodium dichromate solution or alkaline potassium permangante solution may be employed to eliminate hydrogen in the form of water. Dichromate is usually preferred because, unlike permanganate, it will not normally attack any C=C bonds that there may be in the molecule. The reaction commences by an esterification reaction between the alcohol and chromic acid, represented as H_2CrO_4, e.g.

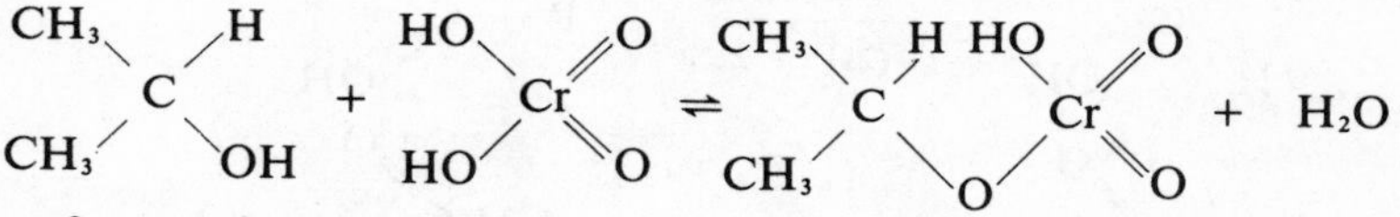

| 2-propanol | chromic acid | isopropyl hydrogen chromate |

The electron attracting hydrogen chromate group causes the α-hydrogen atom to acquire a partial positive charge so that it is open to attack by H_2O in an elimination reaction which is very similar to that of alkene formation (see above).

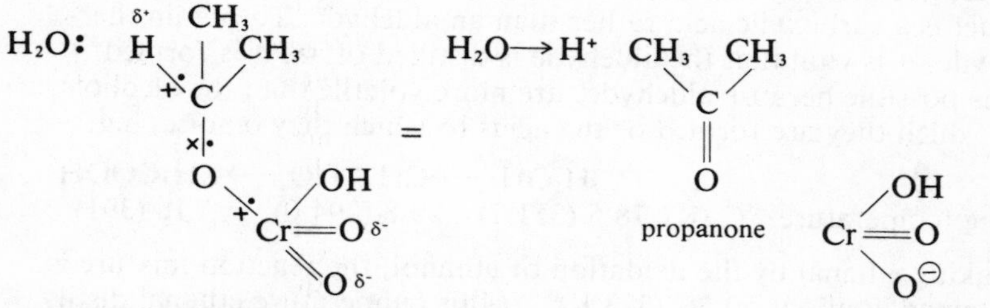

H_2O approaches isopropyl hydrogen chromate   oxonium ion   chromate(IV) ion

In this way, the oxidation state of the carbon is increased from 0 to +2, while that of the chromium is reduced from +6 to +4. Chromic(IV) acid, H_2CrO_3, is unstable, and rapidly disproportionates to give the chromium(III) ion and chromic acid.

In aqueous solution, aldehydes and ketones form hydrates by the nucleophilic addition of water across the C=O bond (§13.7):

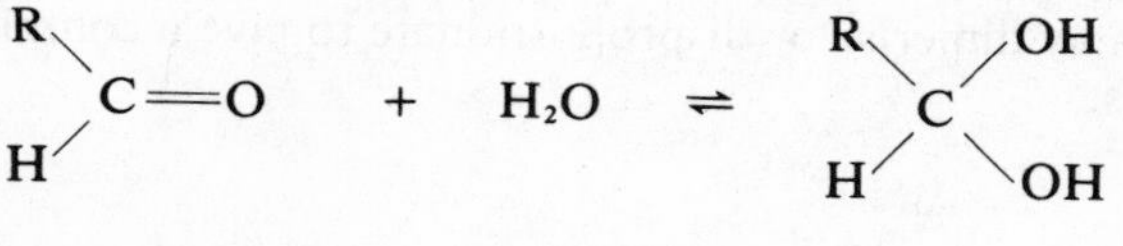

aldehyde hydrate

$$\underset{R}{\overset{R}{\diagdown}} C = O \; + \; H_2O \; \rightleftharpoons \; \underset{R}{\overset{R}{\diagdown}} C \underset{\diagdown OH}{\overset{\diagup OH}{}} \qquad \text{ketone hydrate}$$

These compounds are 1,1-diols. Aldehyde hydrates bear a structural resemblance to primary and secondary alcohols in that they possess the $\diagup\overset{H}{\underset{OH}{C}}$ grouping. They can therefore lose two atoms of hydrogen from the same position on the chain and become oxidised to carboxylic acids:

$$\underset{H}{\overset{R}{\diagdown}} C \underset{\diagdown OH}{\overset{\diagup OH}{}} \quad \xrightarrow[-(2H^+ + 2e^-)]{\text{oxidation}} \quad R - C \underset{\diagdown OH}{\overset{\diagup O}{}}$$

E.g. $\qquad CH_3C\overset{O}{\underset{H}{\diagdown}} \qquad \longrightarrow \qquad CH_3C\underset{\diagdown OH}{\overset{\diagup O}{}}$

Ketone hydrates, on the other hand, do not possess the $\diagup\overset{H}{\underset{OH}{C}}$ grouping and are not susceptible to oxidation.

Thus, if a primary alcohol is refluxed with an oxidising agent, the product is a carboxylic acid rather than an aldehyde. To obtain the aldehyde, it is vital that the aldehyde is distilled off as it is formed. This is possible because aldehydes are more volatile than the alcohols from which they are formed or the acids to which they oxidise, e.g.

$$C_2H_5OH \longrightarrow CH_3CHO \longrightarrow CH_3COOH$$
boiling temperature/°C (K)   78.5 (351.7)   20.8 (294.0)   118 (391)

In making ethanal by the oxidation of ethanol, the reaction mixture is maintained at about 50 °C (323 K). At this temperature ethanal distils off rapidly, while most of the other liquids remain in the distillation flask. In the preparation of ketones such precautions are unnecessary, and the secondary alcohol is merely refluxed with the oxidant until reaction is complete.

Phenols resemble tertiary alcohols in yielding no simple oxidation products, but oxidation of the aromatic ring occurs very easily. Phenol itself gives a tarry mass often containing 1,4-benzoquinone, especially if chromic acid is used as the oxidant.

The first stage in the oxidation of a phenol is believed to be the loss of the hydroxyl hydrogen atom to give a phenoxy free radical, ArO·. Such free radicals may dimerise or disproportionate to give a complex mixture of products.

Chapter 13

The C=O bond in organic carbonyl compounds

Data on the C=O bond

covalent bond length/nm	0.122
bond dissociation enthalpy/kJ mol^{-1} at 298 K	743

The simplest compounds to contain the carbon−oxygen double bond (often called the 'carbonyl group') are aldehydes, $RC{\overset{O}{\underset{H}{\diagdown}}}$, and ketones, $RC{\overset{O}{\underset{R}{\diagdown}}}$. Other carbonyl compounds possess additional bonds which considerably affect their chemistry. Chief among them are carboxylic acids, $RC{\overset{O}{\underset{OH}{\diagdown}}}$, and their derivatives. These derivatives include salts, $RC{\overset{O}{\underset{O^-M^+}{\diagdown}}}$, acid anhydrides, $(RCO)_2O$, acid chlorides, $RC{\overset{O}{\underset{Cl}{\diagdown}}}$, esters, $RC{\overset{O}{\underset{OR'}{\diagdown}}}$, and amides, $RC{\overset{O}{\underset{NH_2}{\diagdown}}}$.

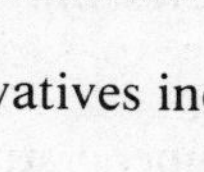

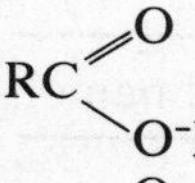

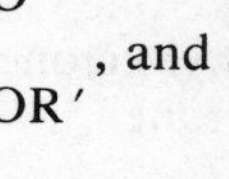

13.1 Nomenclature of carbonyl compounds

Aldehydes

Systematic names of aliphatic aldehydes are derived from those of the corresponding hydrocarbons by replacing the terminal 'e' by 'al'. Dialdehydes are named as 'dials'. Aromatic aldehydes are named by adding the suffix '-carbaldehyde' to the name of the ring system (Table 13.1).

Trivial names, which relate to the names of the carboxylic acids to which the aldehydes are oxidised, are still approved by IUPAC.

Table 13.1 Nomenclature of aldehydes

Formula	Substitutive name	Trivial name
HCHO	methanal	formaldehyde*
CH_3CHO	ethanal	acetaldehyde
CH_3CH_2CHO	propanal	propionaldehyde
$CH_3CH_2CH_2CHO$	butanal	butyraldehyde
$(CH_3)_2CHCHO$	2-methylpropanal	isobutyraldehyde
OHCCHO	ethanedial	glyoxal
C_6H_5CHO	benzenecarbaldehyde	benzaldehyde

*An aqueous solution of formaldehyde is often called 'formalin'.

Ketones

Aliphatic ketones are usually given substitutive names formed by adding '-one', '-dione', etc to the name of the corresponding hydrocarbon, with elision of the final 'e'. An older system of radicofunctional names, in which a ketone $R'COR''$ is named by citing the radicals R' and R'' in alphabetical order, followed by the word 'ketone', is also approved by IUPAC (Table 13.2).

Table 13.2 Nomenclature of aliphatic ketones

Formula	Substitutive name	Radicofunctional name
CH_3COCH_3	propanone	dimethyl ketone
$CH_3COCH_2CH_3$	2-butanone	ethyl methyl ketone
$CH_3CH_2COCH_2CH_3$	3-pentanone	diethyl ketone
$CH_3COCOCH_3$	2,3-butanedione	——

The only trivial names which are recognised by IUPAC are as follows:

CH_3COCH_3 acetone
$CH_3COCOCH_3$ biacetyl

To name an aromatic ketone on the IUPAC system, we look for the acid corresponding to the acyl group ($RC\overset{O}{\diagdown}$ or $ArC\overset{O}{\diagdown}$) which is present in the compound, and change its name ending from '-ic acid' to '-ophenone'. For example, the acyl group in $C_6H_5COCH_3$ is $CH_3C\overset{O}{\diagdown}$. This is derived from the acid $CH_3C\overset{O}{\underset{OH}{\diagdown}}$, which has the trivial name of acetic acid; hence the compound $C_6H_5COCH_3$ is called acetophenone. $C_6H_5COCH_2CH_3$ is exceptional, in that it is called propiophenone; not propionophenone.

Carboxylic acids

Although IUPAC recommends the retention of trivial names for simple aliphatic acids, substitutive names derived from those of the corresponding hydrocarbons by the change of ending from '-e' to '-oic acid' are gaining wide acceptance. The same principles apply to dicarboxylic acids as to monocarboxylic acids (Table 13.3).

Table 13.3 Nomenclature of aliphatic carboxylic acids

Formula	Substitutive name	Trivial name
HCOOH	methanoic acid	formic acid
CH_3COOH	ethanoic acid	acetic acid
CH_3CH_2COOH	propanoic acid	propionic acid
$CH_3CH_2CH_2COOH$	butanoic acid	butyric acid
$(CH_3)_2CHCOOH$	2-methylpropanoic acid	isobutyric acid
$CH_3CH_2CH_2CH_2COOH$	pentanoic acid	valeric acid
HOOC—COOH	ethanedioic acid	oxalic acid
HOOC—CH_2—COOH	propanedioic acid	malonic acid
HOOC—$(CH_2)_2$—COOH	butanedioic acid	succinic acid
HOOC—$(CH_2)_3$—COOH	pentanedioic acid	glutaric acid
HOOC—$(CH_2)_4$—COOH	hexanedioic acid	adipic acid

Trivial names are also retained for aromatic acids, but these compounds can be given substitutive names by means of the suffix '-carboxylic acid' (Table 13.4).

Table 13.4 Nomenclature of aromatic acids

Formula	Substitutive name	Trivial name
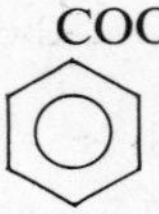 COOH (benzene ring)	benzenecarboxylic acid	benzoic acid
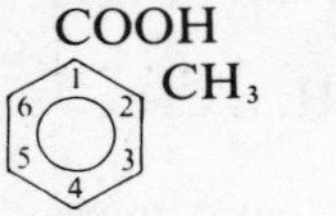 COOH CH₃ (benzene ring, positions 1-6)	2-methylbenzenecarboxylic acid	*o*-toluic acid
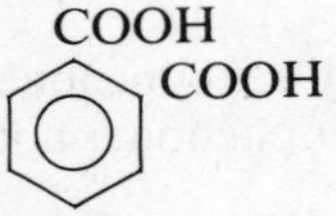 COOH COOH (benzene ring)	1,2-benzenedicarboxylic acid	phthalic acid
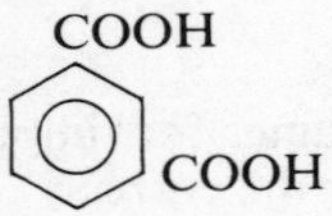 COOH COOH (benzene ring)	1,3-benzenedicarboxylic acid	isophthalic acid
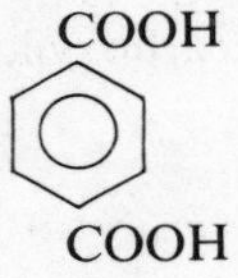 COOH COOH (benzene ring)	1,4-benzenedicarboxylic acid	terephthalic acid
 COOH OH (benzene ring, positions 1-6)	2-hydroxybenzenecarboxylic acid	salicylic acid

Acid anhydrides

Such compounds are named after their parent acids by replacing the word 'acid' by 'anhydride'. Thus, the compound $(CH_3CO)_2O$ is called ethanoic anhydride or acetic anhydride.

Acid chlorides (acyl chlorides)

Acid chlorides are derived from acids, in theory and in practice, by the substitution of the HO group by Cl. They are often referred to as 'acyl chlorides', and are given radicofunctional names in which the name of the acyl radical is followed by the word 'chloride'. Acyl radicals have both trivial and systematic names taken from those of the corresponding carboxylic acids. Consequently, acid chlorides (like acids) have two series of names (Table 13.5).

Table 13.5 The naming of acid chlorides

Radical	Trivial name	Systematic name
CH_3CO	acetyl	ethanoyl
CH_3CH_2CO	propionyl	propanoyl
C_6H_5CO	benzoyl	benzenecarbonyl

Acid chloride	Trivial name	Systematic name
CH_3COCl	acetyl chloride	ethanoyl chloride
CH_3CH_2COCl	propionyl chloride	propanoyl chloride
C_6H_5COCl	benzoyl chloride	benzenecarbonyl chloride

Salts and esters

These compounds, also, are named after the parent acids, e.g.

CH_3COONa	sodium ethanoate or sodium acetate
$CH_3COOC_2H_5$	ethyl ethanoate or ethyl acetate

Amides

Amides are derived from acids by the substitution of the hydroxyl group, HO, by the amino group, NH_2. Their names are derived from those of the parent acids by replacement of the ending '-oic acid' or '-ic acid' by '-amide', e.g.

CH_3CONH_2
$\left\{ \begin{array}{l} \text{ethanamide (from 'ethanoic acid')} \\ \text{acetamide (from 'acetic acid')} \end{array} \right.$

An ending of the type '-carboxylic acid' must be replaced by '-carboxamide', e.g.

$C_6H_5CONH_2$
$\left\{ \begin{array}{l} \text{benzamide (from 'benzoic acid')} \\ \text{benzenecarboxamide (from 'benzenecarboxylic acid')} \end{array} \right.$

N-Substituted amides are named in accordance with the same rules, e.g.

$CH_3CONHC_2H_5$ *N*-ethylethanamide or *N*-ethylacetamide

However, *N*-phenyl substituted amides, because they are derived from $C_6H_5NH_2$, commonly called aniline, may be named in trivial fashion as 'anilides', e.g.

$CH_3CONHC_6H_5$ acetanilide or *N*-phenylethanamide or *N*-phenylacetamide

N-Substituted amides which are derived from aromatic amines other than aniline (phenylamine) must be named systematically.

13.2 Formation of the C=O bond

Like the C=C bond, the C=O bond is established principally by elimination reactions. The addition of ozone across a C=C bond, or of sulphuric acid and water across a C≡C bond, can also be employed.

Elimination methods

From $\overset{\diagdown}{\underset{\diagup}{C}}-\overset{H}{\underset{\diagup}{C}}-OH$ *by the elimination of* H_2

Both in the laboratory and the chemical industry the oxidation of alcohols is the most important way of making aldehydes and ketones. In the laboratory an oxidising agent is used (§12.5), while in industry the favoured technique is dehydrogenation (see below).

From $\overset{\diagdown}{\underset{\diagup}{C}}-\overset{Cl}{\underset{Cl}{C}}$ *by the action of alkali (§11.5)*

On treatment with aqueous alkali, 1,1-dihalides give rise to aldehydes or ketones, e.g.

$$CH_3CHCl_2 + 2NaOH = CH_3CHO + 2NaCl + H_2O$$

The method is important in the manufacture of benzaldehyde (see below), but is of limited use for the preparation of other aldehydes and ketones because most 1,1-dihalides have to be made from aldehydes and ketones by the action of phosphorus pentachloride or pentabromide.

Addition methods

From C=C (§7.4)

The ozonolysis of alkenes, i.e. ozonisation followed by hydrolysis, results in the cleavage of the C=C bond and the formation of aldehydes and ketones:

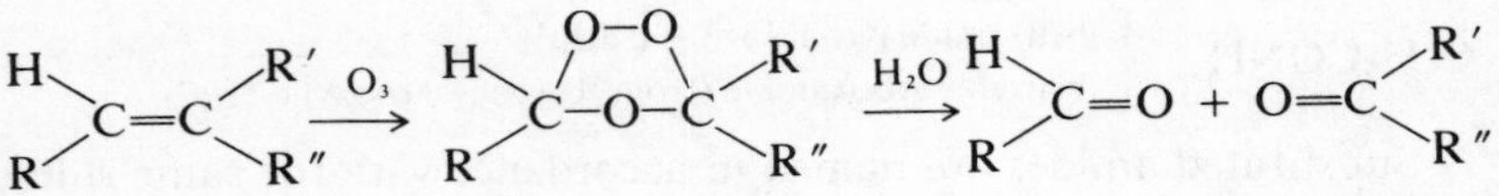

The reaction is important, not so much as a method of preparing aldehydes and ketones but as a means of locating the position of a double bond.

From C≡C (§8.4)

Alkynes are readily hydrated to aldehydes and ketones by the action of warm dilute sulphuric acid in the presence of mercury(II) sulphate, e.g.

$$CH{\equiv}CH \xrightarrow[\text{HgSO}_4]{\text{dilute H}_2\text{SO}_4} CH_2{=}CHOH \xrightarrow{\text{rearrangement}} CH_3CHO$$

Until 1940 ethanal was manufactured in this way in the UK.

13.3 Interconversion of carbonyl compounds

Many types of carbonyl compounds can be interconverted, and carboxylic acids, acid anhydrides, acid chlorides, esters and amides are usually prepared in this manner. Such reactions do not appear to concern the C=O bond, although, as we shall see later, the bond is very much involved in the reaction mechanisms.

Carboxylic acids

The two main routes to carboxylic acids are (i) the oxidation of aldehydes (§13.7), and (ii) the hydrolysis of acid derivatives, including acid anhydrides, acid chlorides, esters and amides. We shall study all these hydrolysis reactions later in this chapter.

Acid anhydrides

Although certain dicarboxylic acids can be converted to their anhydrides on treatment with a dehydrating agent, most acid anhydrides must be made by heating the acid chloride with the sodium salt of the same acid:

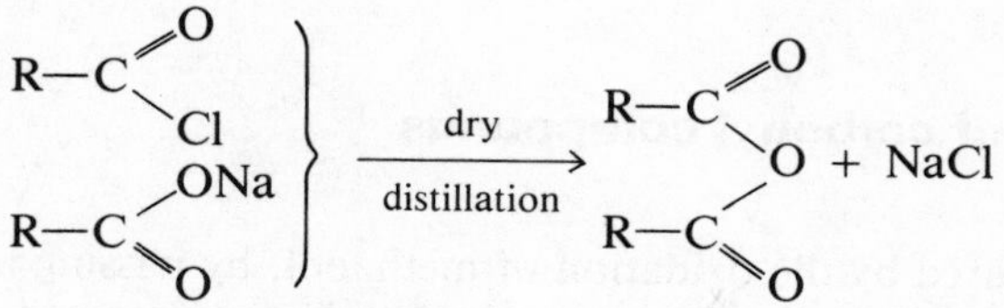

The reaction should be regarded as a nucleophilic substitution at the C—Cl bond, with RCOO⁻ acting as the nucleophile.

If an acid chloride is heated with the sodium salt of a different acid, the product is a *mixed anhydride*.

Acid chlorides

In the same way that an alcohol can readily be converted to an alkyl chloride by means of phosphorus trichloride, phosphorus pentachloride or thionyl chloride, so the same reagents can be employed for converting a carboxylic acid into its acid chloride (§12.5).

Esters

Although direct esterification between carboxylic acids and alcohols is usually employed, it is also possible to make esters from acid anhydrides, acid chlorides or carboxylates. All these methods will be discussed later.

Amides

In a study of the chemistry of amides, the following sequence of reactions should be borne in mind:

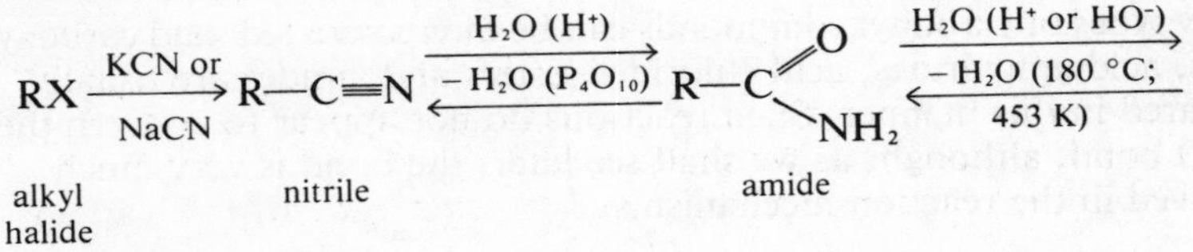

$$RX \xrightarrow[\text{NaCN}]{\text{KCN or}} R-C\equiv N \underset{-H_2O (P_4O_{10})}{\overset{H_2O (H^+)}{\rightleftharpoons}} R-C\overset{O}{\underset{NH_2}{\big<}} \xrightarrow[-H_2O (180 \,^\circ C; \; 453 \, K)]{H_2O (H^+ \text{ or } HO^-)}$$

alkyl halide   nitrile   amide

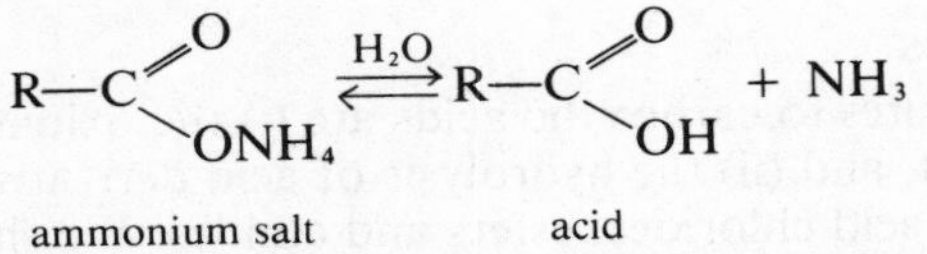

$$R-C\overset{O}{\underset{ONH_4}{\big<}} \overset{H_2O}{\rightleftharpoons} R-C\overset{O}{\underset{OH}{\big<}} + NH_3$$

ammonium salt   acid

There are thus two principal routes to an amide. Either a nitrile can be partially hydrolysed — although it is difficult to stop at the amide stage — or a carboxylic acid can be treated with ammonia and the resulting ammonium salt heated at 180 °C (453 K) to bring about partial dehydration.

To improve the yield it is better to use an acid derivative rather than the carboxylic acid itself. Acid anhydrides, acid chlorides and esters can all be employed.

13.4 Manufacture of carbonyl compounds

Aldehydes

Methanal is manufactured by the oxidation of methanol, by passing a mixture of methanol vapour and air over a silver catalyst maintained at 560–650 °C (833–923 K).

$$CH_3OH \xrightarrow{Ag} HCHO + H_2$$

Also

$$2CH_3OH + O_2 = 2HCHO + 2H_2O$$

To avoid continued oxidation to methanoic acid, the reaction is not taken to completion. The product is cooled, and methanal obtained from it by distillation. Most methanal is immediately dissolved in water to give 'formalin', which should contain about 40 per cent HCHO and 8 per cent CH_3OH.

The evaporation of an aqueous solution of methanal gives the polymer 'paraformaldehyde' as a white powder, and some methanal is sold in this form.

Large quantities of methanal are used in making phenol–formaldehyde, urea–formaldehyde and melamine–formaldehyde plastics. (Systematic names, e.g. phenol–methanal, are seldom used.) Small amounts of formalin are employed for the preservation of anatomical specimens.

Ethanal is made in a similar way to methanal, by the oxidation of ethanol in the presence of a silver catalyst. Ethanal is used for the manufacture of pentaerythritol, $C(CH_2OH)_4$ (required for the manufacture of alkyd resins), lactic acid, chloral, metaldehyde and other synthetic products.

Benzaldehyde is obtained from methylbenzene, either by direct oxidation or via (dichloromethyl)benzene. Oxidation may be carried out in the liquid or vapour phase:

$$CH_3\text{-}\underset{}{\bigcirc} \xrightarrow[\substack{\textit{or} \text{ vapour phase oxidation with air} + N_2 \text{ at } 500\ ^\circ C\ (773\ K) \\ V_2O_5 \text{ catalyst}}]{\text{liquid phase oxidation with } MnO_2 + \text{ dil. } H_2SO_4 \text{ at } 40\ ^\circ C\ (313\ K)} CHO\text{-}\underset{}{\bigcirc}$$

Alternatively:

$$CH_3\text{-}\underset{}{\bigcirc} \xrightarrow[\text{(no catalyst)}]{Cl_2,\ \text{heat}} CHCl_2\text{-}\underset{}{\bigcirc} \xrightarrow[\substack{Na_2CO_3(aq) \text{ or} \\ 70\% \ H_2SO_4}]{\text{hydrolysis with}} CHO\text{-}\underset{}{\bigcirc}$$

Benzaldehyde, which has the flavour of almonds, is used in the food industry. Other applications include the making of dyes and perfumes.

Ketones

Propanone is a by-product in the manufacture of phenol by the autoxidation of isopropylbenzene (§12.3). Additional amounts are made by the dehydrogenation of 2-propanol:

$$CH_3CH(OH)CH_3 \xrightarrow[350-380\ ^\circ C\ (623-653\ K)]{Cu,\ \text{brass or ZnO catalyst}} CH_3COCH_3 + H_2$$

Propanone is a most useful solvent, as it will dissolve a great many organic compounds and is miscible with water. It is also used in the manufacture of other solvents and poly(methyl 2-methylpropenoate), i.e. polymethyl methacrylate or 'Perspex' (Table 7.2).

Acetophenone is obtained partly by the catalytic oxidation of ethylbenzene and partly as a by-product from the manufacture of phenol by the autoxidation of isopropylbenzene. It is used in perfumery.

Carboxylic acids

Methanoic acid is obtained:

(i) as a by-product in the manufacture of ethanoic acid by the oxidation of C_5–C_7 alkanes (see below);

(ii) from calcium methanoate, which is a by-product in the manufacture of pentaerythritol;

(iii) from sodium methanoate, which is obtained by a reaction between carbon monoxide and pulverised sodium hydroxide at a temperature of 200 °C (473 K) and a pressure of 6–10 atm (600 – 1 000 kPa).

Methanoic acid is used for the preparation of its esters, and in the rubber, textile and electroplating industries.

Ethanoic acid and *propanoic acid* are obtained by the aerial oxidation of C_5–C_7 alkanes in the liquid phase at a high temperature and pressure. Ethanoic acid is required largely for the manufacture of ethanoic anhydride and ethanoate esters. It is also used as a solvent and in the food industry.

Benzoic acid is made by the catalytic aerial oxidation of methylbenzene in the liquid phase:

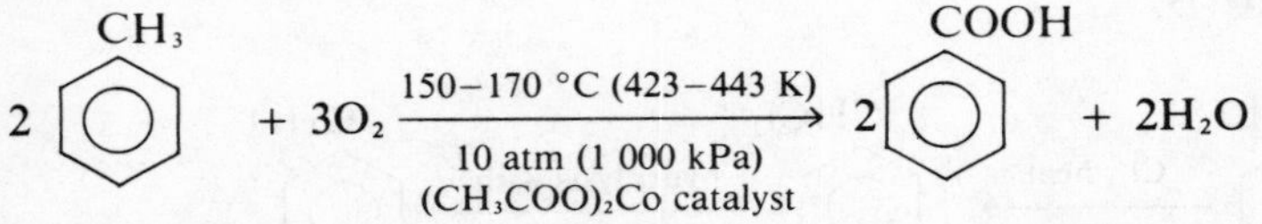

$$2 \underset{}{\bigcirc}\!\!-CH_3 + 3O_2 \xrightarrow[\substack{10 \text{ atm (1 000 kPa)} \\ (CH_3COO)_2Co \text{ catalyst}}]{150-170 \text{ °C } (423-443 \text{ K})} 2 \underset{}{\bigcirc}\!\!-COOH + 2H_2O$$

Most benzoic acid is converted into sodium benzoate, which is used as a corrosion inhibitor and a preservative for food and pharmaceutical products.

Acid anhydrides

Ethanoic anhydride cannot be prepared by the direct dehydration of ethanoic acid. It is mostly manufactured by passing ethenone (ketene) vapour into glacial ethanoic acid:

$$CH_2{=}C{=}O + CH_3COOH = (CH_3CO)_2O$$

The ethenone is obtained either by the catalytic dehydration of ethanoic acid or by the cracking of propanone:

$$CH_3COOH \xrightarrow[\text{trimethyl phosphate catalyst}]{650-700 \text{ °C } (923-973 \text{ K})} CH_2{=}C{=}O + H_2O$$

or

$$CH_3COCH_3 \xrightarrow[(1\,023-1\,123 \text{ K})]{750-850 \text{ °C}} CH_2{=}C{=}O + CH_4$$

Ethanoic anhydride is used for the production of cellulose diethanoate and triethanoate, which are used, in the form of fibre, by the textile industry. Cellulose triethanoate is also used in making photographic film. A further use for ethanoic anhydride lies in the manufacture of drugs, notably aspirin.

1,2-Benzenedicarboxylic anhydride (phthalic anhydride) is manufactured on a large scale by the catalytic aerial oxidation of naphthalene or 1,2-dimethylbenzene:

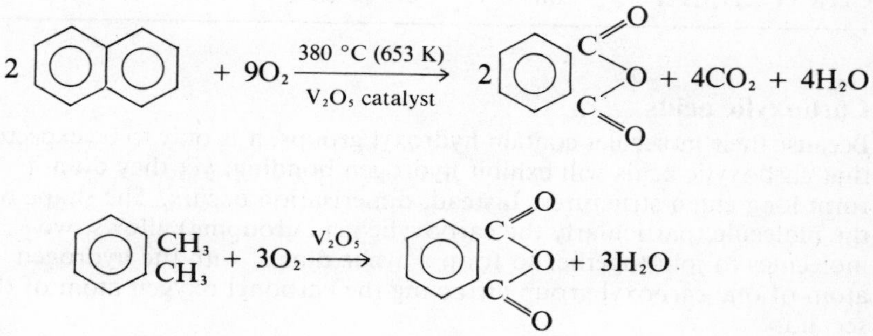

This compound has two main uses: (i) in the production of polyester resins, especially alkyd resins for the paint industry; and (ii) for making dialkyl 1,2-benzenedicarboxylates, which are used as plasticisers for the softening of high polymers. Smaller amounts are converted into dyes, e.g. phenolphthalein, and pigments, e.g. phthalocyanine blue.

13.5 General properties of carbonyl compounds

Aldehydes and ketones

Aliphatic aldehydes, such as methanal and ethanal, have pungent, acrid odours, whereas ketones, such as propanone, are pleasant smelling. All the common aldehydes and ketones are liquids, except for methanal which is a gas. Their boiling temperatures are somewhat higher than those of comparable alkanes, because of the association that results from charge separation within the carbonyl group: $\overset{\delta^+}{C}=\overset{\delta^-}{O}$. (Oxygen is much more electronegative than carbon.)

Aldehyde and ketone molecules form hydrogen bonds with water molecules; thus the lower aldehydes and ketones are soluble in water. As in other homologous series, solubility decreases with increasing chain length, so that whereas ethanal is completely miscible with water, propanal is soluble only to the extent of 20 g per 100 g of water at 20 °C (293 K) (Table 13.6).

Table 13.6 Comparison of the boiling temperatures and water solubilities of aldehydes, ketones and alkanes

Compound	Relative molecular mass	Boiling temperature/ °C (K)	Solubility/g per 100 g of water at 20 °C (293 K)
CH_3CH_2CHO	58	48.8 (322.0)	20.0
CH_3COCH_3	58	56.2 (329.4)	∞
$CH_3CH_2CH_2CH_3$	58	−0.5 (272.7)	0.037

Carboxylic acids

Because their molecules contain hydroxyl groups, it is only to be expected that carboxylic acids will exhibit hydrogen bonding, yet they do not form long chain structures. Instead, dimerisation occurs. The shape of the molecule (particularly the carboxylic acid grouping) allows two molecules to join together to form a cyclic dimer, with the hydrogen atom of one carboxyl group attracting the carbonyl oxygen atom of the second:

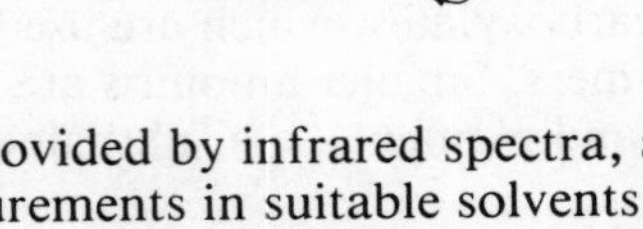

Evidence for this is provided by infrared spectra, and by relative molecular mass measurements in suitable solvents, such as benzene.

The lower carboxylic acids — even methanoic acid — are all pungent smelling liquids, with boiling temperatures much higher than would be expected from their relative molecular masses. Up to and including C_2H_5COOH they are completely miscible with water. Thereafter they are only partially miscible, and the solubility decreases sharply with increasing chain length. Relative molecular mass measurements indicate that in aqueous solution carboxylic acids are monomeric. This suggests that hydrogen bonding between acid and water molecules is sufficiently strong to overcome the hydrogen bonding between acid molecules themselves.

Salts

All common salts of carboxylic acids are crystalline solids with a high solubility in water. (Lead(II) ethanoate is one of the few soluble salts of lead.) Sodium and potassium salts are alkaline in aqueous solution, owing to hydrolysis. The sodium salts of long chain acids, e.g. sodium octadecanoate (sodium stearate), $C_{17}H_{35}COONa$, are soaps.

Acid anhydrides

Methanoic anhydride, $(HCO)_2O$, does not exist. (The anhydride of methanoic acid is carbon monoxide.) Anhydrides of other carboxylic acids look and smell very much like their parent acids, but owing to the greater number of carbon atoms their solubility in water is considerably lower. Ethanoic anhydride, for instance, has a solubility in the cold of only 13.6 g per 100 g of water; an excess of anhydride will form a separate *lower* phase. (Density = 1.087 g cm^{-3} (1.087 kg dm^{-3}) at 20 °C (293 K).) Ethanoic acid, in contrast, is infinitely soluble in water.

Acid chlorides

The acid chloride of methanoic acid, HCOCl, is very unstable and decomposes spontaneously into carbon monoxide and hydrogen chloride. The other carboxylic acids, however, give stable acid chlorides which are colourless liquids with unpleasant acrid smells. They fume in moist air, because of hydrolysis to the parent acids and hydrogen chloride. Acid chlorides are immiscible with water — they form a separate lower phase — but aliphatic ones are rapidly hydrolysed when an aqueous mixture is shaken. Aromatic acid chlorides, such as benzoyl chloride, are attacked by water much more slowly.

Esters

Hydrogen bonding is impossible between the molecules of an ester, and the lower esters are a family of volatile liquids with boiling temperatures close to those of comparable hydrocarbons but far removed from those of carboxylic acids. They have fruity odours, and are in fact responsible for the natural flavours of many fruits. Butyl ethanoate is commonly known as 'pear drops'. Esters of low relative molecular mass are fairly soluble in water because hydrogen bonding can occur between molecules of ester and those of water, cf. ethers, aldehydes and ketones.

Esters of 1,2,3-propanetriol (glycerol) and the long chain fatty acids constitute animal and vegetable oils and fats. They have a special importance in that on hydrolysis with aqueous sodium hydroxide they form soaps.

Amides

The simplest amide, that of methanoic acid, is a liquid with a boiling temperature of 193 °C (466 K). Other amides are solids. (Ethanamide has a melting temperature of 82.3 °C (355.5 K).) The high melting and boiling temperatures of amides, compared with those of carboxylic acids, suggest the existence of association due to hydrogen bonding:

Hydrogen bonding with water molecules leads to the lower members of the homologous series having a high solubility in water. Methanamide is miscible in all proportions, and ethanamide dissolves to the extent of 97.5 g per 100 g of water at 20 °C (293 K).

13.6 Chemical properties of the C=O bond

The C=O bond can take part in a great many chemical reactions, although, as we shall see, its reactivity depends very much on the type of molecule in which it occurs. All its reactions are essentially addition reactions. If we represent the reagent by HA, where H stands for hydrogen and A for a suitable group of atoms, we can write a general equation as follows:

$$\text{\Large >}C{=}O + H{-}A = \text{\Large >}C\!\!\begin{array}{c}OH\\ \diagdown A\end{array}$$

The adduct thus contains two functional groups, one of which is always HO. In other words the product is always, in part, an alcohol.

Superficially, addition to the C=O bond resembles addition to the C=C bond. Both reactions are accompanied by loss of the double bond. However, the C=O bond is quite different from the C=C bond in that there is considerable polarisation. Oxygen, because it is more electronegative than carbon, exerts a negative inductive effect, and the π electrons of the double bond are displaced towards the oxygen atom, thus:

$$\text{\Large >}C{=}\!O \quad \text{or} \quad \text{\Large >}C^{\delta^+}{=}O^{\delta^-}$$

Consequently, reagents which add to the C=O bond are totally different from those which add to the C=C bond. Chlorine, bromine, hydrogen bromide, hydrogen iodide and sulphuric acid, well known for their reactions with alkenes, do not take part in addition reactions with carbonyl compounds. Any reagent that attacks the carbonyl carbon atom must possess a negative charge ($-$ or δ^-), i.e. it must be a nucleophile. **Thus, there is nucleophilic addition to the C=O bond, in contrast to electrophilic addition to the C=C bond.**

The nucleophile may be an anion or a molecule, and the reaction may be uncatalysed or acid catalysed. There are thus four basic reaction mechanisms to consider.

I Uncatalysed addition of a nucleophilic anion

Coordination of a nucleophilic anion, Nu:$^-$, to the carbonyl carbon atom is accompanied by coordination of the carbonyl oxygen atom to a proton. (In theory, other positively charged species could join the oxygen atom, but only combination with a proton gives stable products.) The question immediately arises, 'Which species attacks first?' Do we get nucleophilic attack on carbon, followed by electrophilic attack on oxygen (route 1) or vice versa (route 2)?

Route 1 $\overset{\delta^+}{C}{=}\overset{\delta^-}{O}$ + Nu:⁻ → C(Nu)(O⁻) $\xrightarrow{H^+}$ C(Nu)(OH)

alkoxide ion

Route 2 $\overset{\delta^+}{C}{=}\overset{\delta^-}{O}$ + H⁺ → C—OH $\xrightarrow{Nu:^-}$ C(Nu)(OH)

carbonium ion

In uncatalysed reactions the first route is considered to be the more likely. Alkoxide ions, because they contain electronegative oxygen bearing a negative charge, are more stable and are thus more readily formed than carbonium ions.

II Acid catalysed addition of a nucleophilic anion

In highly acidic conditions route 2 is the more probable. It will be seen that protons catalyse the reaction by converting the partial positive charge on the carbonyl carbon atom into a full positive charge, thereby increasing the attraction for nucleophiles.

III Uncatalysed addition of a nucleophilic molecule

In many reactions a nucleophilic molecule, which we can represent by H—Nu:, takes the place of the anion Nu:⁻ and the proton H⁺. (Examples are H_2O and NH_3.) In such reactions, the initial attack of the nucleophile leads to the formation of a transition state which, because it possesses both positive and negative charges, is referred to as a *zwitterion*. This is converted into the final adduct by the transfer of a proton.

Nu—H
$\overset{\delta^+}{C}{=}\overset{\delta^-}{O}$ → C(⊕Nu—H)(O⊖) $\xrightarrow{-H^+}$ C(Nu)(O⊖) $\xrightarrow{H^+}$ C(Nu)(OH)

coordination of   zwitterion   alkoxide ion
the nucleophile

It will be seen that during the transfer of the proton an alkoxide ion is momentarily formed. **Uncatalysed addition across the C=O bond always proceeds via an alkoxide ion.**

IV Acid catalysed addition of a nucleophilic molecule

Molecular nucleophiles are relatively weak, because they possess only a partial negative charge, and may require the assistance of an acid catalyst. In such circumstances a proton attacks first, at the oxygen atom, followed by the molecule H—Nu: to the carbon atom. The reaction is completed by regeneration of the proton catalyst:

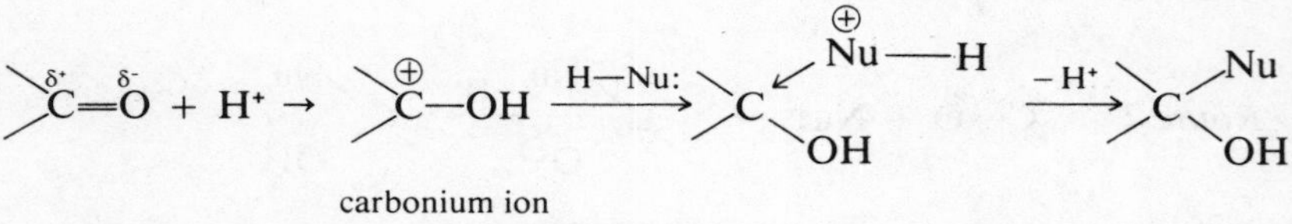

$$\overset{\delta^+}{\diagup}\overset{\delta^-}{C=O} + H^+ \rightarrow \diagup\overset{\oplus}{C}-OH \xrightarrow{H-Nu:} \diagup C\overset{\overset{\oplus}{Nu}-H}{\diagdown OH} \xrightarrow{-H^+} \diagup C\overset{Nu}{\diagdown OH}$$

carbonium ion

Acid catalysed addition across the C=O bond always proceeds via a carbonium ion.

Addition—elimination reactions

In order to produce stable end-products, addition reactions across the C=O bond are frequently followed by the loss of water or other small molecule, such as HCl or C_2H_5OH (Table 13.7). Such reactions are known as *addition—elimination* or *condensation reactions*. Almost all the reactions of carboxylic acids and their derivatives are of this type.

Table 13.7 Addition—elimination reactions of carbonyl compounds

Carbonyl compound	Molecule eliminated
aldehyde, RCHO	H_2O
ketone, R_2CO	H_2O
carboxylic acid, RCOOH	H_2O
acid anhydride, $(RCO)_2O$	RCOOH
acid chloride, RCOCl	HCl
ester, RCOOR'	R'OH
amide, $RCONH_2$	NH_3

Reactivity of carbonyl compounds

The rates of carbonyl addition reactions vary tremendously, depending on the following factors.

(i)   The nature of the reagent. In general, ionic nucleophiles react more rapidly than molecular nucleophiles.

(ii) Reaction conditions, especially temperature and the presence or absence of a catalyst.

(iii) The nature of the carbonyl compound. This factor is particularly important, for carbonyl compounds range in reactivity from acid chlorides, which are highly reactive, to carboxylic acid salts, which undergo no addition reactions at all. The order of reactivity is as follows:

acid chloride > acid anhydride > aldehyde > ketone ≫ ester > amide > acid ⋙ salt

There are three effects which have a bearing on the reactivity of carbonyl compounds and we shall now discuss them.

Inductive effect

All alkyl groups exert an electron releasing or $+I$ effect (§1.4). In an aldehyde other than methanal, this effect reduces the positive charge on the carbonyl carbon atom and renders the atom less susceptible to nucleophilic attack. Ethanal is therefore less reactive than methanal. The relatively low reactivity of ketones, compared with aldehydes, is accounted for very largely by the existence of *two* inductive effects, one from each alkyl group:

$$
\underset{H}{\overset{H}{>}}C{=}O \qquad \underset{H}{\overset{CH_3}{>}}C{=}O \qquad \underset{CH_3}{\overset{CH_3}{>}}C{=}O
$$

decreasing order of reactivity $\longrightarrow$

The inductive effect of an ethyl group is slightly greater than that of a methyl group; thus propanal is slightly less reactive than ethanal, and 3-pentanone is less reactive than propanone.

The inductive effect also explains the high reactivity of acid chlorides:

$$
R{-}C\underset{Cl}{\overset{O}{<}}
$$

The chlorine atom has a $-I$ effect, i.e it withdraws electrons from the carbonyl carbon atom and so increases its positive charge.

Mesomeric effect

The generally low reactivity of carboxylic acids, in comparison with aldehydes and ketones, is accounted for by the mesomeric effect (§1.4). Electrons in the delocalised π orbital (Fig. 1.17) tend to shield the carbonyl carbon atom from attack by nucleophilic reagents. Carboxylic acids are therefore relatively unreactive in addition–elimination reactions, and we say that these compounds have a low *carbonyl activity*.

The mesomeric effect is also present in carboxylic acid derivatives. In consequence, esters and amides, like carboxylic acids, have a relatively low carbonyl activity. Acid chlorides, however, are highly reactive because the negative inductive effect of the chlorine atom more than compensates for the mesomeric effect.

Steric hindrance

The inductive effect is not the only reason for the reactivity difference between aldehydes and ketones. A contributory factor is that ketones suffer from *steric hindrance*. In an aldehyde molecule the functional group (C=O) is joined on one side to a hydrogen atom which, being very small, permits ready access of reagents. A ketone molecule, however, has its carbonyl group shielded by hydrocarbon groups on both sides. A reagent, especially if it is bulky, is hindered in its approach to the carbonyl carbon atom and can join it only with difficulty.

Steric hindrance is encountered also in esterification reactions between alcohols and organic acids.

13.7 Aldehydes and ketones

The nucleophilic reagents which react with aldehydes and ketones are conveniently classified as (i) Lewis bases, (ii) C—H acidic compounds, and (iii) cryptobases. Lewis bases have been discussed in Chapter 3. C—H acidic compounds include hydrogen cyanide and many carbonyl compounds (notably aldehydes, ketones and esters) with an α-hydrogen atom. Cryptobases are reagents with hidden basic characteristics. The best known examples are certain organometallic compounds (e.g. Grignard reagents) and complex aluminium hydrides and borohydrides.

Reactions with Lewis bases

The reagents that concern us here include water, alcohols, sodium hydrogensulphite, ammonia, and derivatives of ammonia, both organic and inorganic. While many of these compounds react at a reasonable rate in neutral or weakly alkaline conditions, there are some — notably alcohols and 2,4-dinitrophenylhydrazine — which are too feebly basic to do so. In such cases a strong acid (HCl or H_2SO_4) is employed as a catalyst. The concentration of acid catalyst is critical, for the acid affects not only the carbonyl compound but also the reagent. Too high a concentration converts the reagent into an unreactive form, e.g.

$$:NH_2R + H^+ \rightleftharpoons \overset{\oplus}{N}H_3R \qquad \textit{no lone pair of electrons on N}$$

ammonia   substituted
derivative   ammonium ion

Certain amphoteric reagents, notably hydroxylamine, function best in strongly alkaline solution.

H_2O

Aldehydes in aqueous solution exist in equilibrium with 1,1-diols, commonly called 'aldehyde hydrates', e.g.

$$CH_3C{\overset{O}{\underset{H}{\diagup\diagdown}}} + H_2O \rightleftharpoons CH_3C{\overset{OH}{\underset{H}{\diagup\diagdown}}}OH \qquad \text{mechanism III; §13.6}$$

1,1-ethanediol (acetaldehyde hydrate)

In the case of methanal, equilibrium lies almost entirely to the right-hand side; for ethanal, the extent of hydrate formation at room temperature has been estimated at 58 per cent. Ketones do not undergo extensive hydration into $R_2C(OH)_2$ because the carbonyl carbon atom is insufficiently positively charged to attract H_2O — a weak nucleophile.

Most 1,1-diols are unstable and cannot be isolated because H_2O, which is a very stable molecule, can readily be formed from the diol by the reverse reaction:

$$R-C\overset{OH}{\underset{H}{-OH}} \rightleftharpoons R-C\overset{OH}{\underset{H}{-O^{\ominus}}} + H^+ \rightleftharpoons R-C\overset{O}{\underset{H}{\diagdown}} + H_2O$$

alkoxide ion

In contrast, monohydric alcohols and 1,2-diols are stable, for they can only be dehydrated via a carbonium ion (§12.5), a much more difficult route.

Halogenation at the α-position, i.e. at the carbon atom next to the carbonyl group, enhances the susceptibility of the carbonyl carbon atom to nucleophilic attack, and at the same time increases the stability of the aldehyde hydrate. Thus, 2,2,2-trichloro-1,1-ethanediol is sufficiently stable to be isolated as a crystalline solid. It can be dehydrated only by concentrated sulphuric acid, because the alkoxide ion is stabilised by delocalisation of the negative charge. (The charge is shared among the oxygen atom and the three chlorine atoms.)

trichloroethanal
(chloral)

alkoxide ion

2,2,2-trichloro-1,1-ethanediol
(chloral hydrate)

H_2O + oxidant

Aldehydes are readily oxidised in aqueous solution to give carboxylic acids. The reactions involve the elimination of hydrogen from aldehyde hydrates, and are thus similar to the oxidation of alcohols by means of a dichromate solution (§12.5).

$$R-C\overset{O}{\underset{H}{\diagdown}} \xrightarrow{H_2O} R-C\overset{OH}{\underset{H}{-OH}} \xrightarrow{-(2H^+ + 2e^-)} R-C\overset{O}{\underset{OH}{\diagdown}}$$

Ketones are unable to take part in this reaction, for they do not possess a hydrogen atom on the carbonyl carbon atom. (With a powerful oxidant, such as HNO_3, ketones undergo oxidation with cleavage of C—C bonds; §6.4.)

Mild oxidants, such as Fehling's solution or Tollens' reagent, which do not attack ketones, are useful in the laboratory for distinguishing between aldehydes and ketones. Fehling's solution is a mixture of copper(II) sulphate and an alkaline solution of potassium sodium 2,3-dihydroxybutanedioate (potassium sodium tartrate, or 'Rochelle salt'). It functions as an oxidant because it contains copper(II) as a complex 2,3-dihydroxybutanedioate, which can be reduced to copper(I) in the form of copper(I) oxide:

$$RCHO + 2Cu^{2+} + 5HO^- = RCOO^- + Cu_2O + 3H_2O$$

Fehling's solutions I and II (the $CuSO_4$ and the alkaline solution of Rochelle salt) are mixed together in roughly equal proportions, the aldehyde is added and the mixture boiled. The purple colour of the solution gradually fades and a reddish brown precipitate of copper(I) oxide appears as redox occurs.

Tollens' reagent is an ammoniacal solution of silver nitrate, prepared by adding one or two drops of sodium hydroxide solution to silver nitrate solution and then adding aqueous ammonia until the precipitate of silver oxide dissolves. The reagent contains the diamminesilver(I) ion, and can oxidise an aldehyde because silver(I) in this form can readily be reduced to silver(0), i.e. metallic silver.

$$RCHO + 2[Ag(NH_3)_2]^+ + 3HO^- = RCOO^- + 2Ag + 4NH_3 + 2H_2O$$

The reaction is usually conducted in such a way that silver is deposited on the walls of a test tube. For this purpose, the test tube must be chemically clean, and the silver must be liberated slowly. This entails the use of a very dilute solution and a moderate temperature — no more than 50 °C (323 K).

Unlike Fehling's solution, which can be stored without deterioration, Tollens' reagent must be prepared at the time it is required. On standing, silver nitride, Ag_3N, is formed, which is highly explosive.

H_2O + acid (polymerisation)

Aldehyde hydrates readily undergo etherification to give polyethers in the presence of an acid catalyst; cf. the etherification of alcohols in the presence of acids (§12.5). For the simplest possible case,

Many molecules of aldehyde hydrate may be involved, and a variety of chain and ring structures can be formed.

The best known polymer of methanal is a long chain compound commonly called paraformaldehyde. It is made by the evaporation to dryness of an acidic solution of methanal:

HOCH$_2$ $\boxed{\text{OH} + n\text{H}}$ OCH$_2$ $\boxed{\text{OH} + \text{H}}$ OCH$_2$OH

$$= \text{HOCH}_2(\text{OCH}_2)_n\text{OCH}_2\text{OH} + (n + 1)\text{H}_2\text{O}$$

paraformaldehyde

Paraformaldehyde is a white amorphous powder, which can readily be depolymerised on heating to 180−200 °C (453−473 K). For this reason it is often used as a convenient source of methanal.

Ethanal, on treatment with acid, gives a cyclic trimer or tetramer according to the conditions.

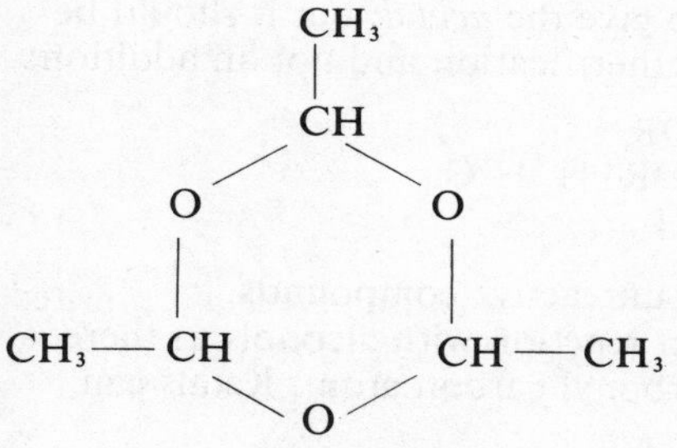

2,4,6-trimethyl-1,3,5-trioxane
(paraldehyde)

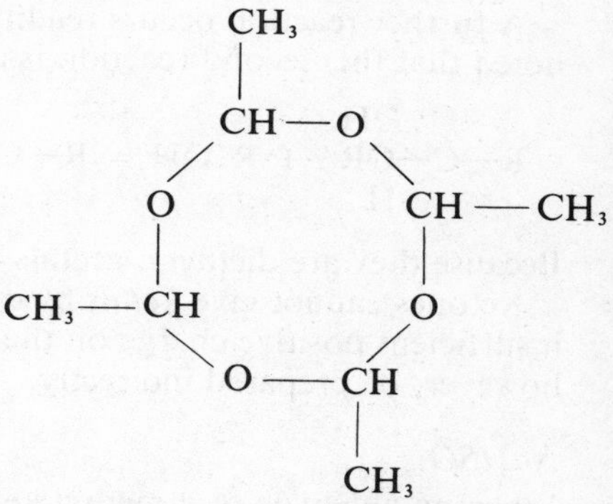

2,4,6,8-tetramethyl-1,3,5,7-tetraoxocane
(metaldehyde)

The trimer, usually called paraldehyde, is a liquid with a boiling temperature of 124 °C (397 K). It is made by treating ethanal with a little sulphuric acid or hydrochloric acid at room temperature. (Very often the change is spontaneous, owing to the presence of a trace of sulphuric acid remaining from the preparation.) An equilibrium mixture is established, which contains about 95 per cent of the trimer:

$$3\text{CH}_3\text{CHO} \rightleftharpoons (\text{CH}_3\text{CHO})_3$$

Paraldehyde can be depolymerised by adding a small amount of sulphuric acid to maintain the equilibrium, and then distilling off the more volatile monomer.

The tetramer, metaldehyde, is a white solid with a melting temperature of 246 °C (519 K). It is made by treating ethanal with a little concentrated sulphuric acid at 0 °C (273 K), and is used in firelighters and for killing slugs and snails. Both the trimer and the tetramer are inert, with no addition or reducing properties.

Ketones, because they do not form hydrates, do not polymerise in this way.

ROH

If an aldehyde is allowed to react with an alcohol in the presence of anhydrous mineral acid (usually HCl), nucleophilic addition occurs to give a *hemiacetal*:

$$R-C\overset{O}{\underset{H}{\diagdown}} + R'OH \rightleftharpoons R-\overset{OH}{\underset{H}{\overset{|}{C}}}-OR' \qquad \text{mechanism IV; §13.6}$$

$$\text{e.g. } CH_3C\overset{O}{\underset{H}{\diagdown}} + CH_3OH \rightleftharpoons CH_3\overset{OH}{\underset{H}{\overset{|}{C}}}-OCH_3 \qquad \text{1-methoxyethanol}$$

A further reaction occurs readily to give the *acetal*; but it should be noted that this second reaction is an etherification and not an addition:

$$R-\overset{OH}{\underset{H}{\overset{|}{C}}}-OR' + R'OH = R-\overset{OR'}{\underset{H}{\overset{|}{C}}}-OR' + H_2O$$

Because they are diethers, acetals are unreactive compounds.

Ketones cannot give *ketals* by direct reaction with alcohols as there is insufficient positive charge on the carbonyl carbon atom. Ketals can, however, be prepared indirectly.

NaHSO₃

When an aldehyde or a methyl ketone is shaken at room temperature with a saturated aqueous solution of sodium hydrogensulphite, a white crystalline precipitate is obtained of a so-called 'bisulphite addition compound' — in reality, the sodium salt of a hydroxysulphonic acid. The reaction is uncatalysed; HSO_3^- adds first, to carbon, followed by the transfer of H^+ to oxygen (mechanism I, §13.6), e.g.

$$\underset{CH_3}{\overset{CH_3}{\diagup}}C=O + Na^+HSO_3^- \rightleftharpoons \underset{CH_3}{\overset{CH_3}{\diagup}}C\underset{SO_3H}{\overset{O^\ominus Na^+}{\diagup}} \rightleftharpoons \underset{CH_3}{\overset{CH_3}{\diagup}}C\underset{SO_3^\ominus Na^+}{\overset{OH}{\diagup}}$$

sodium 2-hydroxy-2-propanesulphonate
(acetone sodium bisulphite)

A proton is transferred in the final stage of the mechanism because the SO_3H group is more acidic than the HO group and has a greater tendency to ionise.

Hydrogensulphite addition compounds can be hydrolysed back to the compounds from which they are derived by treatment with aqeuous sodium carbonate or hydrochloric acid. Because of this, and because they are formed only by aldehydes and ketones, hydrogensulphite addition compounds can be used in the purification of aldehydes and ketones. The impure compound is mixed with sodium hydrogensulphite solution, and the resulting precipitate is filtered off and washed. Subsequent treatment with sodium carbonate or hydrochloric acid liberates the aldehyde or ketone, which can be recovered by distillation or ether extraction.

Aldehydes undergo the reaction more easily than ketones, especially the higher ketones, a fact which is utilised in *Schiff's reagent* for distinguishing between aldehydes and ketones. This reagent is a solution of the dyestuff rosaniline ('magenta'), which has been decolorised by the passage of sulphur dioxide. Aldehydes will restore the magenta colour to Schiff's reagent at room temperature, as they can abstract sulphur dioxide by an addition reaction, but ketones react only very slowly.

RNH_2 and R_2NH

Primary and secondary amines react readily with aldehydes and ketones by uncatalysed nucleophilic addition followed by the elimination of water. (Tertiary amines cannot add to the $C=O$ bond because they do not possess an active hydrogen atom on N.) The elimination of water proceeds differently in the two cases so that different products result.

Primary amines

The addition of primary amines to aldehydes and ketones is represented by the equation:

$$\ce{>C=O} + RNH_2 \rightleftharpoons \ce{>C<^{OH}_{NHR}} \qquad \text{mechanism III; §13.6}$$

The adducts are unstable — so much so that they cannot be isolated — and undergo spontaneous loss of water to give more stable products. There are two possible ways in which water could be eliminated:

$$\ce{-CH<-C<^{OH}_{NHR}} = \ce{-C=C<^{|}_{NHR}} + H_2O \qquad (1)$$

$$\ce{>C<^{OH}_{NHR}} = \ce{>C=NR} + H_2O \qquad (2)$$

Schiff's base

From our knowledge of the dehydration of alcohols to alkenes we might expect reaction (1), but in practice reaction (2) occurs because it is easier for a proton to be lost from a nitrogen atom than a carbon atom. (The reason is that the N—H bond is more polarised and hence more easily broken than the C—H bond.) **Wherever possible, in the elimination of water, a hydrogen atom leaves a nitrogen atom rather than a carbon atom.**

The products of these reactions are therefore substituted imines, usually known as *Schiff's bases* or *azomethines*. A well known example is *N*-benzylidenephenylamine, made by warming together benzaldehyde and phenylamine:

$$C_6H_5C\overset{O}{\underset{H}{\diagdown}} + C_6H_5NH_2 \rightleftharpoons C_6H_5C\overset{OH}{\underset{H}{\diagup}}NHC_6H_5 \rightleftharpoons C_6H_5CH{=}NC_6H_5 + H_2O$$

Schiff's bases can be reduced to secondary amines by the addition of hydrogen across the double bond (§14.2).

Secondary amines

The adducts from secondary amines cannot lose hydrogen from nitrogen. Instead, a β-hydrogen atom is lost and the product is an *enamine*:

$$-\overset{H}{\underset{|}{\underset{|}{C}}}-\overset{H}{\underset{|}{C}}{=}O + R_2NH \rightleftharpoons -\overset{H}{\underset{|}{C}}-\overset{\underset{|}{NR_2}}{\underset{|}{C}}-OH \rightleftharpoons {>}C{=}C\overset{\diagup}{\underset{NR_2}{\diagdown}} + H_2O$$

enamine

Note the reversible nature of these reactions. Enamines, like Schiff's bases and many other derivatives of aldehydes and ketones, can be hydrolysed back to the parent compounds.

When a secondary amine reacts with an aldehyde which does not possess an α-hydrogen atom (e.g. methanal or benzaldehyde), neither of these ways of eliminating water is possible and a more complex reaction occurs.

NH₃

Aldehydes react with ammonia in essentially the same way that they react with primary amines, but with two important differences:

(i) the adduct, a so-called 'aldehyde ammonia', can be isolated if the temperature is sufficiently low;

(ii) the condensation product, an aldimine, undergoes further changes.

For example, at 0 °C (273 K) gaseous ammonia will react with ethanal dissolved in ether to give a white crystalline precipitate of 1-aminoethanol (acetaldehyde ammonia):

$$CH_3C\overset{\displaystyle O}{\underset{\displaystyle H}{\big\backslash}} + NH_3 \rightleftharpoons CH_3C\overset{\displaystyle OH}{\underset{\displaystyle H}{\big|}}-NH_2 \qquad \text{mechanism III; §13.6}$$

Treatment of an aldehyde ammonia with dilute acid brings about hydrolysis to the original aldehyde.

Aldehyde ammonias are unstable — so much so that they cannot be isolated in the pure state — and readily undergo the elimination of water to give aldimines, e.g.

$$CH_3C\overset{\displaystyle OH}{\underset{\displaystyle H}{\big|}}-NH_2 = CH_3C\overset{\displaystyle NH}{\underset{\displaystyle H}{\big\backslash\!\!\!/}} + H_2O$$

This imine immediately polymerises to a cyclic trimer:

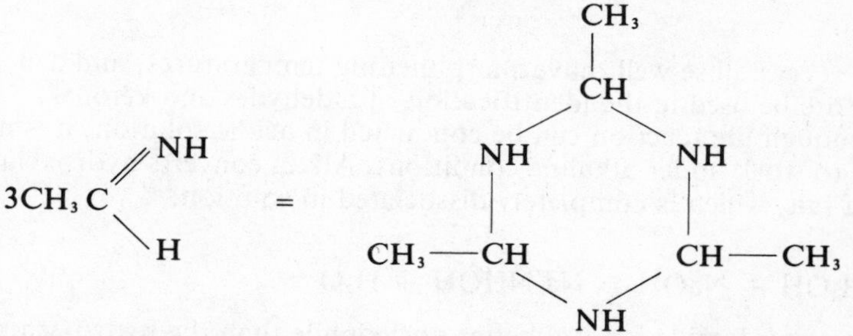

The situation with methanal is more complicated, for the cyclic triimine reacts with more methanal and ammonia to give a white solid called hexamethylenetetramine:

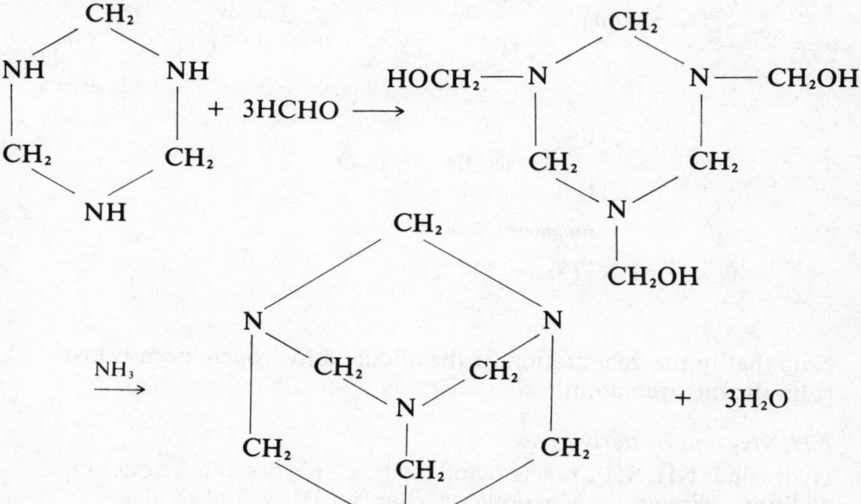

Hexamethylenetetramine, commonly known as hexamine or urotropine, is used for curing Bakelite plastics, and in medicine as a urinary antiseptic.

Ketones, unlike aldehydes, do not undergo addition reactions with ammonia. An entirely different reaction occurs, in which ammonia causes the ketone to undergo an aldol condensation (see below), and then reacts with the product.

NH_2OH

Hydroxylamine, NH_2OH, undergoes addition–elimination reactions with aldehydes and ketones to give stable solids called *oximes*:

$$\text{C=O} + H_2NOH = \text{C=NOH} + H_2O$$

oxime

Oximes crystallise well, have sharp melting temperatures, and can therefore be used in the identification of aldehydes and ketones.

Although the reaction can be conducted in acidic solution, it is more usual to work under alkaline conditions. Alkali converts hydroxylamine into a salt, which is completely dissociated in solution:

$$NH_2OH + NaOH = Na^+\overset{\ominus}{N}HOH + H_2O$$

The hydroxylamide ion is a better nucleophile than the hydroxylamine molecule, because of the full negative charge on the nitrogen atom, and attacks the aldehyde or ketone by mechanism I (§13.6), e.g.

$$\begin{array}{c}CH_3 \\ \diagdown \\ C=O \\ \diagup \\ CH_3\end{array} + \overset{\ominus}{N}HOH \rightleftharpoons \begin{array}{c}CH_3 \diagdown \nearrow NHOH \\ C \\ CH_3 \diagup \searrow O^{\ominus}\end{array} \underset{\text{(from } H_2O)}{\overset{H^+}{\rightleftharpoons}} \begin{array}{c}CH_3 \diagdown \nearrow NHOH \\ C \\ CH_3 \diagup \searrow OH\end{array}$$

adduct

$$\rightleftharpoons \begin{array}{c}CH_3 \\ \diagdown \\ C=NOH \\ \diagup \\ CH_3\end{array} + H_2O$$

propanone oxime

$$\theta_{C,m} = 61\ °C\ (T_m = 334\ K)$$

Note that in the dehydration of the adduct a hydrogen atom is lost from the nitrogen atom.

NH_2NH_2 *and its derivatives*

Hydrazine, NH_2NH_2, reacts readily with aldehydes and ketones in addition–elimination reactions (mechanism III, §13.6) to give *hydrazones*:

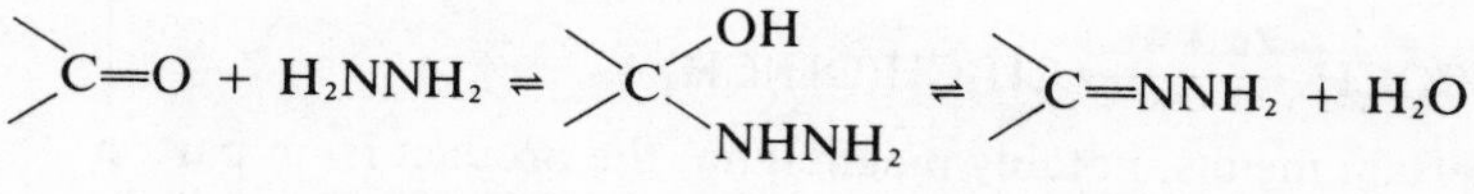

$$\text{C=O} + H_2NNH_2 \rightleftharpoons \underset{NHNH_2}{\overset{OH}{\text{C}}} \rightleftharpoons \text{C=NNH}_2 + H_2O$$

hydrazone

Propanone, for example, gives propanone hydrazone, $(CH_3)_2C{=}NNH_2$.

The use of hydrazine in characterising aldehydes and ketones is limited by the fact that most hydrazones are liquids. In addition, a further reaction with the aldehyde or ketone can occur to give an *azine*:

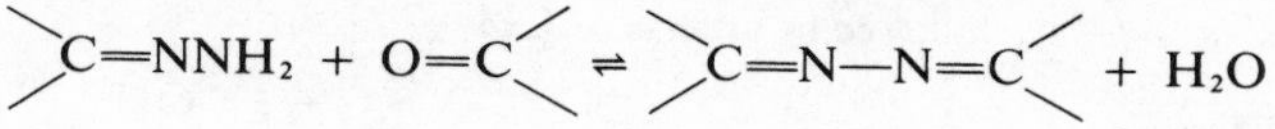

$$\text{C=NNH}_2 + \text{O=C} \rightleftharpoons \text{C=N—N=C} + H_2O$$

azine

To overcome this difficulty, it is necessary to use a substituted hydrazine, such as phenylhydrazine or 2,4-dinitrophenylhydrazine, e.g.

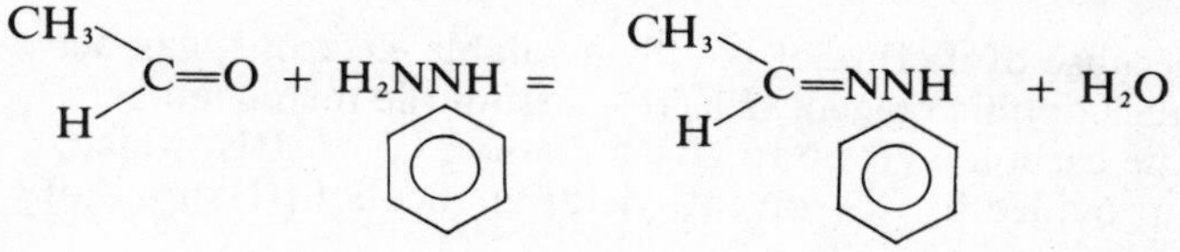

$$\underset{H}{\overset{CH_3}{\text{C}}}{=}\text{O} + H_2NNH\bigcirc = \underset{H}{\overset{CH_3}{\text{C}}}{=}\text{NNH}\bigcirc + H_2O$$

ethanal phenylhydrazone

$$\theta_{C,m} = 101 \text{ °C } (T_m = 374 \text{ K})$$

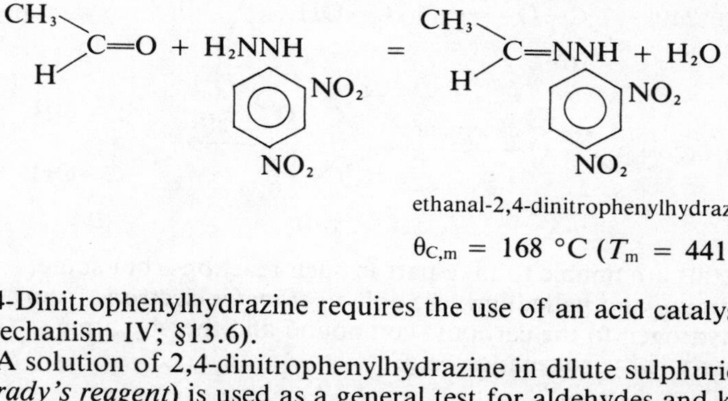

$$\underset{H}{\overset{CH_3}{\text{C}}}{=}\text{O} + H_2NNH\bigcirc^{NO_2}_{NO_2} = \underset{H}{\overset{CH_3}{\text{C}}}{=}\text{NNH}\bigcirc^{NO_2}_{NO_2} + H_2O$$

ethanal-2,4-dinitrophenylhydrazone

$$\theta_{C,m} = 168 \text{ °C } (T_m = 441 \text{ K})$$

2,4-Dinitrophenylhydrazine requires the use of an acid catalyst (mechanism IV; §13.6).

A solution of 2,4-dinitrophenylhydrazine in dilute sulphuric acid (*Brady's reagent*) is used as a general test for aldehydes and ketones. Nearly all such compounds, at room temperature, give an immediate orange precipitate of the 2,4-dinitrophenylhydrazone. The derivative can be filtered off, recrystallised from ethanol, dried, and its melting temperature determined. The result, when compared with tabulated values in reference books, provides a guide to the identity of the original aldehyde or ketone.

Metals

Aldehydes and ketones can be reduced to their corresponding primary or secondary alcohols by the use of a metal and a protic solvent, e.g.

$$CH_3COCH_3 \xrightarrow{\text{Zn + HCl}} CH_3CH(OH)CH_3$$

With certain metals, notably magnesium, the product is, in part, a 1,2-diol, e.g.

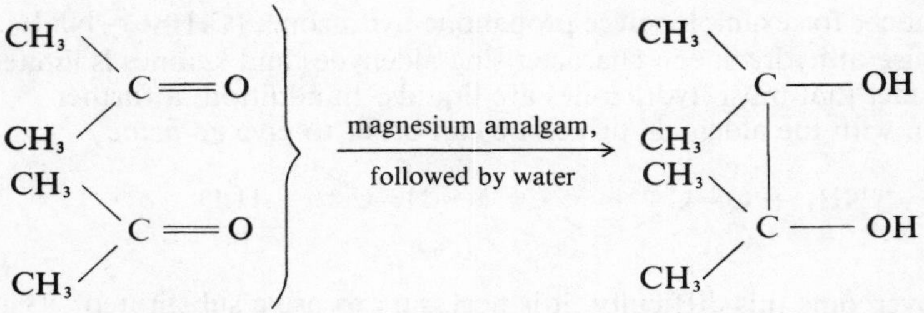

2,3-dimethyl-2,3-butanediol (pinacol)

A metal, because of its flux of readily available electrons, can act almost as a nucleophilic reagent. Electrons from the metal can be accepted by the carbonyl group to give a dianion (I) or (III), which, with protons provided by the solvent, yields an alcohol (II) or a diol (IV) respectively.

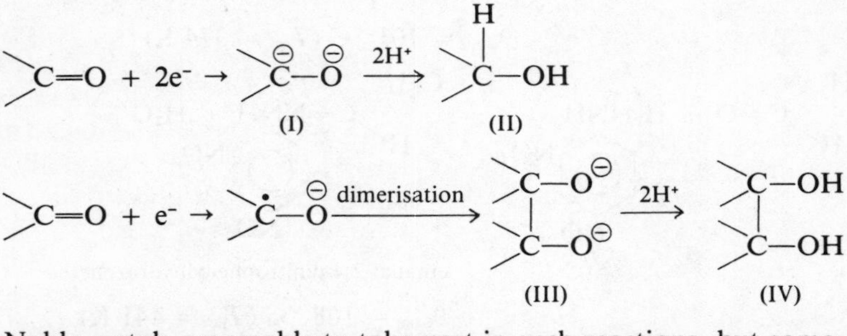

Noble metals are unable to take part in such reactions, but some, notably platinum and palladium, are able to *transfer* electrons from molecular hydrogen to the carbonyl compound and thereby bring about its catalytic hydrogenation, e.g.

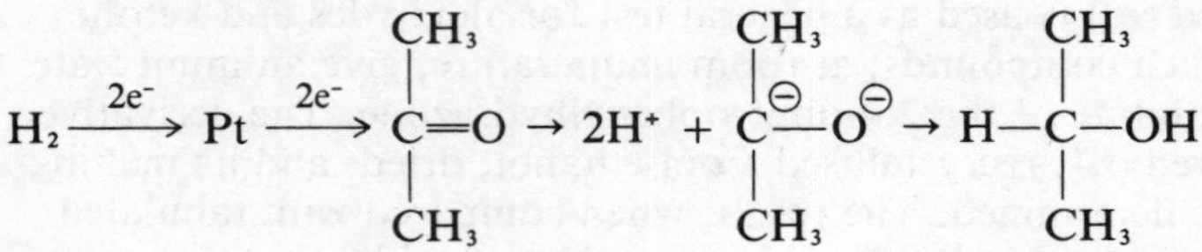

Reactions with C—H acidic compounds

Although relatively few C—H compounds are acidic (§10.2), there are some which, because of their structures, have a slight tendency to ionise in water. An example is hydrogen cyanide, where polarisation of the

C≡N bond induces a corresponding electronic displacement in the
C—H bond:

$$\overset{\delta+}{H}\frown\overset{\frown}{C}\overset{\delta-}{\equiv}\overset{}{N}$$

In such cases, neutralisation with a base (e.g. NaOH) occurs to an
appreciable extent:

$$HCN + NaOH \rightleftharpoons Na^+CN^- + H_2O$$

The anion, :CN^-, is a nucleophile and can react with aldehydes and
ketones. The hydrogen cyanide molecule, in contrast, cannot behave as
a nucleophile because its carbon atom does not possess a lone pair of
electrons.

Other C—H acidic compounds include aldehydes, ketones and esters
with an α-hydrogen atom. These compounds *by themselves* are not
nucleophilic reagents, but in the presence of bases they give rise to
carbanions which, with a negative charge and a lone pair of electrons
on a carbon atom, are excellent nucleophiles. Many of them attack
aldehydes and ketones at room temperature, and very often the
nucleophilic addition is followed by the elimination of water.

Most of these reactions lie outside the scope of this book, and we
shall discuss only the addition of hydrogen cyanide and the aldol
condensation.

HCN

A mixture of sodium cyanide and ethanoic acid reacts with aldehydes
and ketones at room temperature to give hydroxynitriles, commonly
known as 'cyanohydrins'. The reaction is reversible, and yields at
equilibrium are better with aldehydes than with ketones, e.g.

$$CH_3C\overset{O}{\underset{H}{<}} + \;:CN^- \rightleftharpoons CH_3C\overset{O^\ominus}{\underset{H}{\overset{|}{-}CN}} \xrightarrow{H^+ \text{ (from } CH_3COOH)} CH_3C\overset{OH}{\underset{H}{\overset{|}{-}CN}}$$

from NaCN

2-hydroxypropanenitrile
(acetaldehyde cyanohydrin

Provided that a basic catalyst is included to form CN^- ions, the
reaction can also be accomplished by means of liquid or aqueous
hydrogen cyanide.

Cyanohydrins are useful intermediates in the preparation of
α-hydroxyacids, e.g.

$$CH_3CH(OH)CN \xrightarrow{\text{acidic hydrolysis}} CH_3CH(OH)COOH$$

2-hydroxypropanoic acid (lactic acid)

Aldehydes or ketones with an α-hydrogen atom (aldol addition and aldol condensation)

When ethanal is treated with calcium hydroxide solution at room temperature, it dimerises to give a syrupy liquid commonly called aldol:

$$2CH_3CHO \xrightarrow{\text{Ca(OH)}_2} CH_3CH(OH)CH_2CHO$$

3-hydroxybutanal (aldol)

Any reaction of this type, in which an aldehyde or a ketone (or a mixture of such compounds) possessing an α-hydrogen atom dimerises in the presence of a base, is known as an *aldol addition*.

Ethanal has acidic hydrogen atoms at the α-position (§10.2), and when treated with a base (usually the hydroxide of a metal in group 1A or 2A) is partially converted into a carbanion:

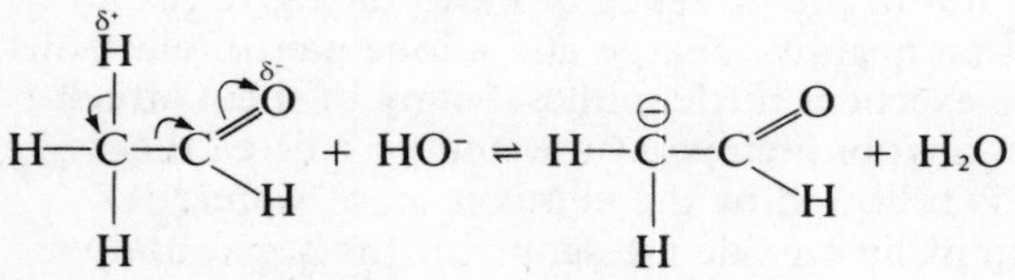

The nucleophilic carbanion attacks a second molecule of ethanal, and the addition is completed by a proton from the water (mechanism I; §13.6).

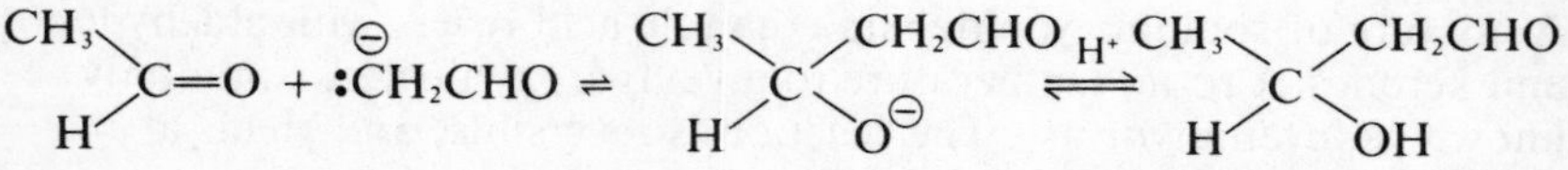

With calcium hydroxide as the catalyst, the reaction stops mainly at this stage, but with sodium hydroxide, which is a stronger base, repeated aldol additions take place to give a long chain polymer which appears as a brown resin.

In most aldol additions it is possible to isolate the product provided that the temperature is kept at about 20 °C (293 K). However, on warming, or in the presence of an acid, the adduct undergoes the elimination of water to give an unsaturated aldehyde or ketone, e.g.

$$CH_3CH(OH)CH_2CHO \xrightarrow{\text{warm}} CH_3CH{=}CHCHO + H_2O$$

2-butenal (crotonaldehyde)

An aldol addition followed by the elimination of water is known as an *aldol condensation*.

Certain aldol additions are of commercial importance. One example is the mixed aldol addition between methanal and ethanal, in the presence of calcium hydroxide, to give pentaerythritol, $C(CH_2OH)_4$, widely used in the manufacture of alkyd resins for the paint industry.

Reactions with cryptobases

The word 'cryptic' means 'hidden'; thus, a *cryptobase* is a chemical species with hidden basic characteristics. A simple example is the tetrahydridoaluminate(III) ion, $[AlH_4]^-$. This ion is *not* a Lewis base, for it does not possess a lone pair of electrons and cannot form coordinate bonds. But it can serve as a source of the hydride ion, H^-, which *is* a base. The species $[AlH_4]^-$ is therefore referred to as a cryptobase.

Cryptobases can be very useful for adding certain anions, notably H^- and R^- (an alkyl anion) to the $C{=}O$ bond. Suppose, for example, that we wished to introduce H^- into an aldehyde or a ketone. What reagent could we use? Not sodium hydride or calcium hydride, because these compounds are insoluble in organic solvents and are decomposed by water. But we could use lithium tetrahydridoaluminate(III), $Li[AlH_4]$, which can serve as a source of hydride ions and which is soluble in ether.

$Li[AlH_4]$ and $Na[BH_4]$

Both these reagents readily reduce virtually all aldehydes and ketones to their respective primary and secondary alcohols in high yield. Sodium tetrahydroborate(III) is often preferred to lithium tetrahydrido-aluminate(III) because it can be used in the presence of ethanol or water.

The reaction mechanism involves the transfer of the true base, H^-, from the cryptobase to the carbonyl compound via a transition state, not unlike that involved in an S_N2 reaction.

$[AlH_4]^-$ ion   aldehyde or ketone   transition state

$$\rightarrow AlH_3 + H{-}\overset{|}{\underset{|}{C}}{-}O^{\ominus}$$

alkoxide ion

The alkoxide ion coordinates to the aluminium hydride molecule:

$$H{-}\overset{|}{\underset{|}{C}}{-}O^{\ominus} + AlH_3 \rightarrow [H{-}\overset{|}{\underset{|}{C}}{-}O{-}AlH_3]^-$$

alkoxytrihydridoaluminate(III) ion

All four hydrogen atoms of the $[AlH_4]^-$ ion are utilised, because the $[H{-}\overset{|}{\underset{|}{C}}{-}O{-}AlH_3]^-$ ion, like $[AlH_4]^-$, is a cryptobase and reacts with another molecule of aldehyde or ketone. The process is repeated twice more, so that ultimately the lithium tetrahydridoaluminate(III) is converted into a lithium tetraalkoxyaluminate(III):

$$Li[AlH_4] \rightarrow Li[Al(O-\overset{|}{\underset{|}{C}}-H)_4]$$

A primary or secondary alcohol is obtained when this complex salt is hydrolysed with dilute sulphuric acid.

RMgX

Grignard reagents may be used for converting aldehydes and ketones to secondary and tertiary alcohols respectively (§11.5). Essentially, the reaction can be represented thus:

$$X-Mg-R + \overset{}{\underset{}{>}}C\overset{=}{=}O \rightarrow X-\overset{\delta^-}{Mg}\cdots R\cdots C\overset{\delta^-}{=}O \rightarrow X^- Mg^{2+} R-\overset{|}{\underset{|}{C}}-O^{\ominus}$$

| Grignard reagent | aldehyde or ketone | transition state | alkoxide ion |

In reality the transition state is more complex than that shown. It is cyclic, and is composed of one molecule of the carbonyl compound and two of the Grignard reagent.

Aldehydes without an α-hydrogen atom (Cannizzaro reaction)

We have seen that if an aldehyde possesses an α-hydrogen atom it undergoes the aldol addition when it is treated with alkali. If it does not possess an α-hydrogen atom it may instead take part in a *Cannizzaro reaction*, which is a disproportionation involving two molecules of the aldehyde. One molecule is oxidised to the corresponding carboxylic acid anion, while the other is reduced to the corresponding primary alcohol, e.g.

$$2HCHO + NaOH \xrightarrow[\text{(303 K)}]{30\,°C} HCOONa + CH_3OH$$

methanal   50% solution   sodium methanoate   methanol

$$2 \underset{\text{benzaldehyde}}{\overset{CHO}{\bigcirc}} + \underset{\substack{60\% \\ \text{solution}}}{KOH} \xrightarrow[\text{(333 K)}]{60\,°C} \underset{\substack{\text{potassium} \\ \text{benzoate}}}{\overset{COOK}{\bigcirc}} + \underset{\text{phenylmethanol}}{\overset{CH_2OH}{\bigcirc}}$$

It is a feature of such reactions that a strong base is required. Under such conditions, the aldehyde hydrate is partially converted into an alkoxide ion which is a cryptobase, capable of transferring H^- to a molecule of aldehyde:

$$R-\overset{\displaystyle OH}{\underset{\displaystyle OH}{C}}-H \ + \ KOH \ \rightleftharpoons \ R-\overset{\displaystyle O^{\ominus}K^{+}}{\underset{\displaystyle OH}{C}}-H \ + \ H_2O$$

aldehyde hydrate   aldehyde hydrate
alkoxide

$$R-\overset{\displaystyle O^{\ominus}}{\underset{\displaystyle OH}{C}}-H \ + \ \overset{\displaystyle R}{\underset{\displaystyle H}{C}}{=}O \ \rightarrow \ R-\overset{\displaystyle \overset{\delta^-}{O}}{\underset{\displaystyle OH}{C}}\cdots H\cdots \overset{\displaystyle R}{\underset{\displaystyle H}{C}}{=}\overset{\delta^-}{O} \ \rightarrow \ R-C\overset{\displaystyle O}{\underset{\displaystyle OH}{}} \ + \ H-\overset{\displaystyle R}{\underset{\displaystyle H}{C}}-O^{\ominus}$$

transition state

$$\rightarrow \ R-C\overset{\displaystyle O}{\underset{\displaystyle O^{\ominus}}{\Big\langle}} \ + \ RCH_2OH$$

13.8 Carboxylic acids

Because of the structure of the carboxyl group, $-C\overset{\displaystyle O}{\underset{\displaystyle OH}{\Big\langle}}$, carboxylic
acids display two sets of reactions, namely those of the C—OH bond
(cf. alcohols), and those of the C=O bond (cf. aldehydes and ketones).

Reactions of the C—OH bond (including the O—H bond)

When we discussed the chemistry of alcohols in Chapter 12, we did so
under three main headings: (i) acidic and basic character; (ii) nucleophilic
substitution reactions; and (iii) elimination reactions. We shall now
consider carboxylic acids under the same three headings.

Acidic and basic character
Acidity
The acidic character of carboxylic acids is due to the presence of the
hydroxyl group, which allows ionisation to occur in the presence of
water:

$$R-C\overset{\displaystyle O}{\underset{\displaystyle O\overset{\bullet}{\underset{\times}{}}H}{\Big\langle}} \ \rightleftharpoons \ R-C\overset{\displaystyle O}{\underset{\displaystyle O^{\ominus}}{\Big\langle}} \ + \ H^+$$

carboxylate ion

Methanoic acid and ethanoic acid have pK_a values of 3.75 and 4.76
respectively, showing that carboxylic acids are more strongly acidic
than alcohols or phenols. (The corresponding values for methanol and
phenol are 15.5 and 10.00.) The increased acidity is due to the negative
inductive effect of the carbonyl oxygen atom, which, transmitted
through the molecule, increases the polarisation of the O—H bond:

$$R-C\overset{\displaystyle O}{\underset{\displaystyle O-H}{}}$$

In this way the hydrogen atom acquires an increased positive charge and can more easily break away as a proton in the presence of water.

Carboxylic acids can be neutralised by bases to give salts, e.g.

$$CH_3COOH + NaOH \rightleftharpoons CH_3COO^-Na^+ + H_2O$$

sodium ethanoate

They react with strongly electropositive metals with the evolution of hydrogen, e.g.

$$2CH_3COOH + 2Na = 2CH_3COONa + H_2$$

Because they are stronger acids than carbonic acid, they will liberate carbon dioxide from carbonates and hydrogencarbonates, e.g.

$$2CH_3COOH + Na_2CO_3 = 2CH_3COONa + H_2O + CO_2$$

They are, however, much weaker than the principal mineral acids. Ethanoic acid, for instance, is only about 0.4 per cent ionised in molar aqueous solution, whereas hydrochloric acid is almost totally ionised. Halogenated acids are stronger than the unsubstituted acids, due to the negative inductive effect of a halogen atom (§3.2).

Basicity

Carboxylic acids can act as bases by donating a lone pair of electrons from either the carbonyl oxygen atom or the hydroxyl oxygen atom:

$$R-C\overset{\displaystyle O}{\underset{\displaystyle OH}{}} + H^+ \rightleftharpoons R-\overset{\oplus}{C}\overset{\displaystyle OH}{\underset{\displaystyle OH}{}} \qquad I$$

carbonium ion

$$R-C\overset{\displaystyle O}{\underset{\displaystyle OH}{}} + H^+ \rightleftharpoons R-C\overset{\displaystyle O}{\underset{\displaystyle \overset{\oplus}{O}H_2}{}} \qquad II$$

substituted oxonium ion

In general, the former is energetically favoured, and when we speak of a 'protonated carboxylic acid' we normally mean a carbonium ion, I. The latter type of acid–base reaction is favoured only when the resulting complex can undergo further change to form stable products.

Carboxylic acids act as nucleophiles in their reactions with phosphorus trichloride:

$$3RCOOH + PCl_3 = 3RCOCl + H_2PHO_3$$

The reaction is exactly analogous to that of phosphorus trichloride with alcohols. Thus, the first step involves coordination to the reagent from the hydroxyl oxygen atom of the carboxylic acid:

$$R\!-\!\overset{\displaystyle O}{\underset{\displaystyle \underset{H}{|}}{\underset{\displaystyle O:}{C}}} \;+\; PCl_3 \;\longrightarrow\; R\!-\!\overset{\displaystyle O}{\underset{\displaystyle \underset{H}{|}}{\underset{\displaystyle \overset{\oplus}{O}\rightarrow \overset{\ominus}{P}}{C}}}\!\!\overset{\displaystyle Cl}{\underset{\displaystyle Cl}{-\,Cl}}$$

This is followed by the elimination of HCl and, finally, nucleophilic substitution by a chloride ion (§12.5).

Phosphorus trichloride converts carboxylic acids to acid chlorides in reasonably good yields, and is a satisfactory reagent for the purpose provided that the acid chloride has a boiling temperature well below 200 °C (473 K) — the decomposition temperature of phosphonic acid — so that the two compounds can be separated by distillation.

Phosphorus pentachloride reacts in a very similar fashion to the trichloride:

$$RCOOH + PCl_5 = RCOCl + POCl_3 + HCl \qquad \text{(cf. §12.5)}$$

Reactions are vigorous, yields of acid chlorides are good, and the products can readily be separated by fractional distillation provided that their boiling temperatures are not too close together. (The boiling temperature of $POCl_3$ is 107 °C (380 K).)

Another common reagent is thionyl chloride:

$$RCOOH + SOCl_2 = RCOCl + SO_2 + HCl \qquad \text{(cf. §12.5)}$$

Thionyl chloride is the least reactive of these three reagents and must be used in excess in order to obtain a good yield. The acid chloride and excess reagent are subsequently separated by fractional distillation. The use of thionyl chloride is thus restricted to the preparation of those acid chlorides which boil well above its boiling temperature of 77 °C (350 K).

Nucleophilic substitution reactions

Difficult to achieve with alcohols, nucleophilic substitution reactions are virtually impossible with carboxylic acids. There is, for example, no reaction with HCl or HBr, although under drastic conditions HI can react as a reducing agent. Acid catalysts, which facilitate such reactions with alcohols, are ineffective with carboxylic acids because, as we have seen, the proton attaches itself to the carbonyl oxygen atom rather than the HO group.

Ammonia and the amines, which react with alcohols by nucleophilic substitution, do so with carboxylic acids by nucleophilic addition across the C=O bond (see below).

Elimination reactions

Elimination of water

Carboxylic acids, like alcohols, form unsaturated compounds by the loss of water in 1,2-elimination reactions. Acid catalysts are ineffective (see above), and very high temperatures must be employed, e.g.

$$CH_3C\begin{smallmatrix}O\\\\OH\end{smallmatrix} \xrightarrow[\text{(973 – 1 173 K)}]{700-900\ °C} CH_2{=}C{=}O\ +\ H_2O$$

ethenone (ketene)

Ethenone is manufactured for use as an intermediate in the production of ethanoic anhydride (§13.4).

Methanoic acid is unique in that it cannot undergo 1,2-elimination of water because the molecule possesses only one carbon atom. However, that carbon atom carries a hydrogen atom in addition to the hydroxyl group, and with concentrated sulphuric acid methanoic acid undergoes 1,1-elimination of water to give carbon monoxide. (The term 'α-elimination' can also be used, to indicate that both groups are lost from the same carbon atom.)

$$HCOOH \xrightarrow[\text{moderate heat}]{\text{conc. } H_2SO_4} CO + H_2O$$

Salts of methanoic acid behave in the same way. This is one of the commonest methods of preparing carbon monoxide in the laboratory.

Elimination of hydrogen

Methanoic acid bears a structural resemblance to primary and secondary alcohols in that the molecule possesses a carbon atom bearing both a hydrogen atom and an HO group. Dehydrogenation can therefore occur on heating:

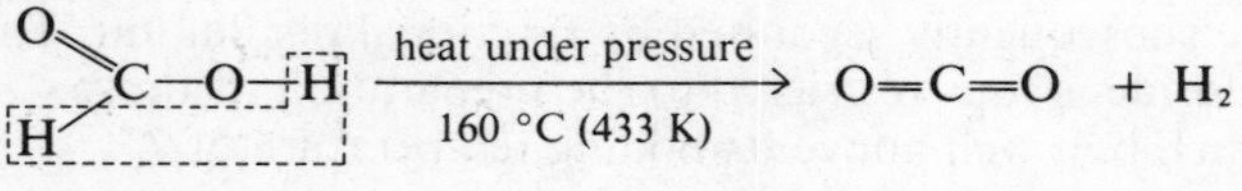

$$\begin{smallmatrix}O\\\\H\end{smallmatrix}{C}{-}O{-}H \xrightarrow[\text{160 °C (433 K)}]{\text{heat under pressure}} O{=}C{=}O\ +\ H_2$$

methanoic acid   carbon dioxide

$$cf.\quad R{-}\underset{\underset{H}{|}}{\overset{\overset{H}{|}}{C}}{-}O{-}H \xrightarrow[\text{Cu catalyst}]{450\ °C\ (723\ K)} R{-}C\begin{smallmatrix}O\\\\H\end{smallmatrix}\ +\ H_2$$

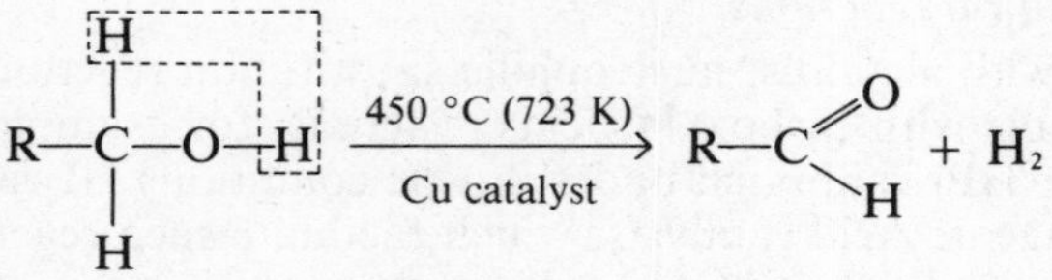

primary alcohol   aldehyde

Hydrogen is also eliminated from sodium methanoate (or potassium methanoate) when it is heated in the dry state:

$$2HCOONa = \begin{array}{c} COONa \\ | \\ COONa \end{array} \qquad + H_2$$

sodium ethanedioate (sodium oxalate)

A test for the residual ethanedioate is provided by the addition of calcium chloride solution, which gives a white precipitate of calcium ethanedioate.

Alternatively, methanoic acid (like primary and secondary alcohols) may be treated with an oxidising agent so that hydrogen is eliminated in the form of water. Such reactions occur readily, and both methanoic acid and the methanoates are powerful reducing agents. They will decolorise acidified potassium permanganate solution in the cold, and reduce silver nitrate solution and Tollens' reagent to silver. Mercury(II) chloride solution is reduced to give a white precipitate of mercury(I) chloride and, eventually, a grey precipitate of metallic mercury.

Acids other than methanoic acid, because they lack the $\begin{array}{c} {}^H \\ C {}^{\diagdown} \\ {}^{OH} \end{array}$ grouping, do not give off hydrogen on heating and do not function as reducing agents.

Reactions of the C=O bond

Carboxylic acids take part in nucleophilic addition reactions across the C=O bond. As with aldehydes and ketones, some of the reactions do not need a catalyst whereas others require the presence of acid.

Carboxylic acids have a relatively low carbonyl activity, due mainly to the mesomeric effect (§13.6). The inductive effect of the alkyl group also plays a part, and methanoic acid, in which this effect is absent, is somewhat more reactive than the other acids. Many reagents which react with aldehydes and ketones, e.g. hydrogen cyanide, hydroxylamine and 2,4-dinitrophenylhydrazine, do not attack carboxylic acids. The principal reagents that do attack carboxylic acids are alcohols, ammonia, primary and secondary amines, and lithium tetrahydridoaluminate(III).

The products of nucleophilic addition can never be isolated, for they are unstable 1,1-diols. As we should expect, they immediately stabilise themselves by the elimination of a molecule of water between the two hydroxyl groups:

$$R-C{\diagdown}^O_{OH} + HNu \rightarrow R-C{\diagdown}^{OH}_{OH}-Nu \rightarrow R-C{\diagdown}^O_{Nu} + H_2O$$

Derivatives of carboxylic acids behave in a similar manner. Always, nucleophilic addition across the C=O bond is followed, in order to form a stable product, by the elimination of a small molecule, such as HCl or C_2H_5OH (Table 13.7). The effect of such elimination is to re-form the C=O bond: in this way, acids and their derivatives are converted into one another, and the C=O bond *appears* to play no part in the reaction.

Many of the addition–elimination reactions of carboxylic acids and their derivatives are reversible, so that equilibrium mixtures may be established. A well known example concerns esterification:

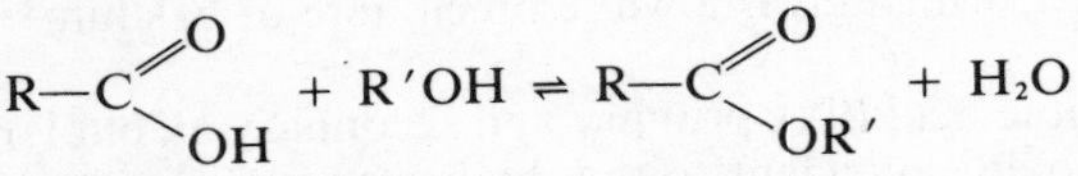

$$R-C{\overset{O}{\underset{OH}{\Big<}}} + R'OH \rightleftharpoons R-C{\overset{O}{\underset{OR'}{\Big<}}} + H_2O$$

In contrast, if an acid chloride is allowed to react with an alcohol, the reaction proceeds almost entirely to completion, even when the hydrogen chloride that is produced is retained in the system:

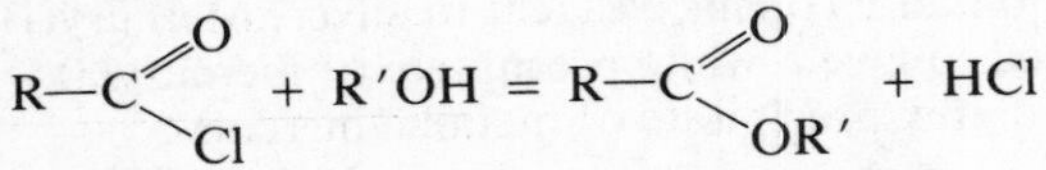

$$R-C{\overset{O}{\underset{Cl}{\Big<}}} + R'OH = R-C{\overset{O}{\underset{OR'}{\Big<}}} + HCl$$

The position of equilibrium depends on the difference between the reactivities of the carbonyl compounds involved (see order of reactivities, §13.6). If the compounds are of comparable reactivity (e.g. acid and ester), reactions proceed with roughly equal ease in both directions, but if the compounds have very different carbonyl activities (e.g. acid chloride ≫ ester), this is not so and the position of equilibrium lies far to one side.

ROH

Carboxylic acids react with alcohols in the presence of acid catalysts to give esters. Nucleophilic addition of alcohol across the C=O bond (mechanism IV; §13.6) is at once followed by the loss of water:

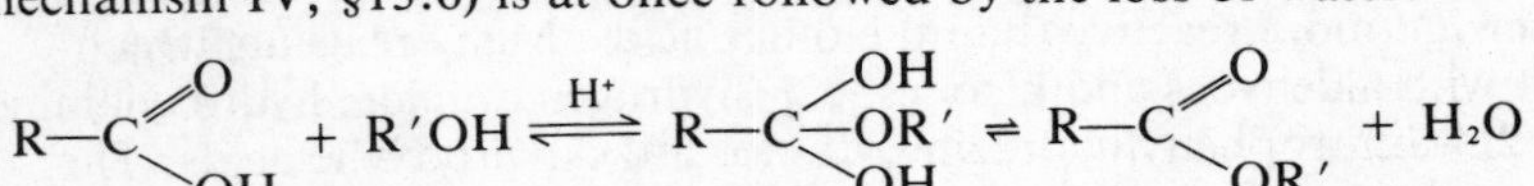

$$R-C{\overset{O}{\underset{OH}{\Big<}}} + R'OH \underset{H^+}{\rightleftharpoons} R-C{\overset{OH}{\underset{OH}{\overset{|}{\underset{|}{-}}}}OR'} \rightleftharpoons R-C{\overset{O}{\underset{OR'}{\Big<}}} + H_2O$$

Elimination of water from the adduct occurs essentially in three stages, as for the dehydration of alcohols to alkenes:

(i)   protonation of an HO group;
(ii) loss of H_2O to form a carbonium ion;
(iii) loss of a proton to form the final product.

Almost certainly, the addition and elimination stages of esterification overlap each other to some extent, so that the proton which is lost in the final stage of the addition joins an HO group to give the protonated adduct. In this way, the unstable adduct, $R-C{\overset{OH}{\underset{OH}{\overset{|}{\underset{|}{-}}}}OR'}$, is never actually formed. The complete mechanism is thus as follows.

$$R-C \overset{O}{\underset{OH}{\Big\langle}} + H^+ \rightarrow R-\overset{\oplus}{\underset{OH}{C}} \overset{OH}{\underset{}{}} \xrightarrow{R'OH} R-\overset{\overset{\overset{R'}{\underset{O}{\oplus}}H}{\Big\downarrow}}{\underset{OH}{C}} \overset{OH}{\underset{}{}} \xrightarrow[\text{transfer}]{H^+} R-\overset{OR'}{\underset{OH}{\overset{\oplus}{C}}-OH_2}$$

$$\xrightarrow{-H_2O} R-\overset{\oplus}{C} \overset{OR'}{\underset{OH}{\Big\langle}} \xrightarrow{-H^+} R-C \overset{OR'}{\underset{O}{\Big\langle}}$$

Esterification is a reversible reaction, for the reason mentioned above. If, for example, ethanoic acid and ethanol are refluxed together in equimolar proportions, equilibrium is established when approximately two-thirds of the acid and alcohol have been transformed. If an involatile acid is reacted with an involatile alcohol, equilibrium can be destroyed by distilling off the water as it is formed, but this technique cannot be used when the lower acids and alcohols are involved. In such cases the yield of ester may be increased by using an excess of alcohol: in this way as much as possible of the more expensive acid is esterified.

The acids commonly employed as catalysts are concentrated sulphuric acid and anhydrous hydrogen chloride. The technique consists of refluxing the mixture of carboxylic acid, alcohol and acid catalyst until equilibrium has been established — usually about half an hour — and then distilling off the ester. The concentration of catalyst is critical, for too much acid protonates not only the carboxylic acid but also the alcohol, to give RH_2O^+, which, because of its positive charge, will not attack a carbonyl carbon atom.

The use of a base catalyst may appear tempting, for this would convert an alcohol into an alkoxide ion, RO^-; an excellent nucleophile. Unfortunately, a base would also convert the carboxylic acid into a carboxylate ion, which undergoes no nucleophilic addition reactions.

The esterification of carboxylic acids by secondary alcohols is slower than by primary alcohols due to steric hindrance. Because of its relatively large bulk, a molecule of a secondary alcohol can approach a protonated acid only with difficulty. If an alcohol contains both primary and secondary hydroxyl groups, e.g. 1,2,3-propanetriol, $CH_2OH \cdot CHOH \cdot CH_2OH$, the primary HO groups always esterify more readily than the secondary ones.

Curiously, tertiary alcohols, which would be expected to present the most steric hindrance, react at much the same rate as primary alcohols. Isotopic tracer experiments have shown that in such cases esterification proceeds by an entirely different mechanism. A tertiary carbonium ion is formed, which joins the *hydroxyl* oxygen atom of the carboxylic acid to give a protonated ester:

$$R_3COH + H^+ \rightleftharpoons R_3\overset{\oplus}{C}OH_2 \rightleftharpoons R_3C^{\oplus} + H_2O$$

$$R-C\overset{O}{\underset{OH}{\diagup}} + R_3C^{\oplus} \rightleftharpoons R-C\overset{O}{\underset{\underset{CR_3}{\overset{\oplus}{OH}}}{\diagup}} \rightleftharpoons R-C\overset{O}{\underset{OCR_3}{\diagup}} + H^+$$

NH₃

At room temperature, ammonia behaves as a base and neutralises carboxylic acids to give ammonium salts:

$$R-C\overset{O}{\underset{OH}{\diagup}} + NH_3 \rightleftharpoons R-C\overset{O}{\underset{O^{\ominus}NH_4^+}{\diagup}}$$

At 150–200 °C (423–473 K) ammonia behaves as a nucleophile, and amides are formed in an addition–elimination reaction:

$$R-C\overset{O}{\underset{OH}{\diagup}} + NH_3 \rightleftharpoons R-\underset{OH}{\overset{OH}{C}}-NH_2 \rightleftharpoons R-C\overset{O}{\underset{NH_2}{\diagup}} + H_2O$$

Amides of high boiling acids can therefore be prepared by passing a current of gaseous ammonia through the carboxylic acid maintained at 150–200 °C (423–473 K). Lower amides, notably ethanamide, are made by distilling the corresponding ammonium salt in the dry state at about 180 °C (453 K). Dissociation occurs into ammonia and the carboxylic acid, which recombine, at this temperature, to give the amide. Yields by this latter method are poor.

RNH₂ and R₂NH

Primary and secondary amines, like ammonia, react with carboxylic acids to give mono- and disubstituted amides respectively:

$$R-C\overset{O}{\underset{OH}{\diagup}} + R'NH_2 = R-C\overset{O}{\underset{NHR'}{\diagup}} + H_2O$$

$$R-C\overset{O}{\underset{OH}{\diagup}} + R'_2NH = R-C\overset{O}{\underset{NR'_2}{\diagup}} + H_2O$$

The reactions are of little importance, for substituted amides are best prepared from acid chlorides or esters.

Li[AlH₄]

Carboxylic acids are readily reduced to primary alcohols by means of lithium tetrahydridoaluminate(III). The technique consists of running a solution of the acid, in dry ether, into one of lithium tetrahydridoaluminate(III), also in dry ether, at such a rate that the mixture does not boil uncontrollably. For the sake of safety, it is advisable to carry out the reaction in an atmosphere of nitrogen. When the reaction is complete, excess lithium tetrahydridoaluminate(III) is destroyed by means of ethyl ethanoate, followed by water, after which the alcohol is liberated by the addition of dilute sulphuric acid.

This is the only way there is of reducing carboxylic acids. Sodium tetrahydroborate(III) is ineffective.

RMgX

Carboxylic acids do not undergo nucleophilic attack by Grignard reagents, but behave as active hydrogen compounds in converting Grignard reagents into alkanes (§11.5).

13.9 Salts of carboxylic acids

Carboxylate ions, because of their delocalised negative charge (§1.4), have an extremely low carbonyl activity and do not participate in addition–elimination reactions.

Their principal property is their ability to donate a lone pair of electrons, so that they can act as nucleophiles and ligands. They behave as nucleophiles in substitution reactions with alkyl halides, e.g.

$$CH_3COOAg + C_2H_5I = CH_3COOC_2H_5 + AgI \qquad (§11.5)$$

Similar reactions with acyl halides give rise to acid anhydrides, e.g.

$$CH_3COONa + CH_3COCl = (CH_3CO)_2O + NaCl$$

Carboxylate ions act as ligands in coordinating to certain transition metal cations, notably Fe^{3+} and Cr^{3+}, to give complex ions of the type $[M_3^{III}(RCOO)_6O]^+$. The colour of the complex depends upon the particular carboxylate ion, a feature that may be employed in identifying such ions. Neutral solutions of ethanoates, in the cold, when treated with neutral iron(III) chloride solution, give a blood-red coloration; when boiled, this changes to a brown precipitate of a basic iron(III) ethanoate. Methanoates behave in a similar fashion.

13.10 Acid anhydrides

Carboxylic acid anhydrides are characterised by addition–elimination reactions, i.e. nucleophilic addition across a $C=O$ bond, followed by the elimination of a molecule of carboxylic acid. The $C-O$ bond, in contrast, is unreactive; cf. ethers.

The mechanism of addition is a little different from usual, in that the proton attaches itself to the ether oxygen atom rather than the carbonyl oxygen atom. This gives an adduct which, on cleavage, gives two molecules of carbonyl compounds. The energy required to break a C—O bond in the final stage is available because a comparable bond is being formed (C—O → C=O).

Acid anhydrides have a high carbonyl activity, and their chemistry is much more extensive than that of carboxylic acids. They will react with hydroxylamine (to give hydroxamic acids), with hydrazine (to give hydrazides) and with C—H acidic compounds. In the scope of their reactions, they compare with aldehydes and ketones rather than carboxylic acids.

H_2O

On hydrolysis, acid anhydrides give the parent acids, e.g.

$$(CH_3CO)_2O + H_2O = 2CH_3COOH$$

The reaction is slow to start at room temperature because of the immiscibility of acid anhydrides with water, but occurs rapidly on warming or shaking **and can be violent.**

ROH and ArOH

Unlike carboxylic acids, acid anhydrides will form esters with phenols as well as with alcohols. Because a molecule of carboxylic acid is eliminated, only half the molecule of acid anhydride is effectively utilised, e.g.

$$(CH_3CO)_2O + C_2H_5OH = CH_3COOC_2H_5 + CH_3COOH$$
$$(CH_3CO)_2O + C_6H_5OH = CH_3COOC_6H_5 + CH_3COOH$$

Most of these esterifications occur readily when the reactants are warmed together. In difficult cases an acid or base catalyst can be used.

NH_3, RNH_2 and R_2NH

The reactions between acid anhydrides and ammonia or amines occur very readily and are frequently used for the preparation of amides and substituted amides. Again, only half the acid anhydride is utilised, e.g.

$$(CH_3CO)_2O + NH_3 = CH_3CONH_2 + CH_3COOH$$
$$\downarrow NH_3$$
$$CH_3COONH_4$$

$Li[AlH_4]$

Like carboxylic acids, acid anhydrides are reduced to primary alcohols by means of lithium tetrahydridoaluminate(III) in dry ether. Ethanoic anhydride, for example, gives ethanol.

RMgX

Acid anhydrides react with Grignard reagents in the same manner as aldehydes and ketones. The products are ketones, which, on reaction with a second molecule of Grignard reagent, give tertiary alcohols. Alkanes cannot be formed, because acid anhydrides, in contrast to carboxylic acids, do not possess active hydrogen atoms.

13.11 Acid chlorides

Acid chlorides possess the $-C\diagup_{\diagdown Cl}^{O}$ grouping and may therefore display nucleophilic substitution reactions typical of the C—Cl bond and nucleophilic addition—elimination reactions typical of the C=O bond.

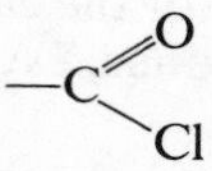

Certain nucleophilic reagents, e.g. KCN, can attack only the C—Cl bond:

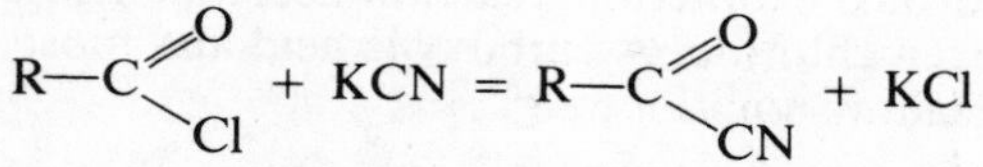

$$R-C\diagup_{\diagdown Cl}^{O} + KCN = R-C\diagup_{\diagdown CN}^{O} + KCl \qquad (cf. \ §11.5)$$

Others, e.g. C—H acidic compounds, react only with the C=O bond.

However, most common nucleophiles, including H_2O, ROH, NH_3, RNH_2 and R_2NH, are able to attack both C—Cl and C=O bonds, and on reacting with acid chlorides would give identical products by either mode of attack. For example, the reaction between an acid chloride and an alcohol, which yields an ester and hydrogen chloride, could be formulated as a nucleophilic substitution:

protonated ester

Cl⁻ abstracts H⁺

ester

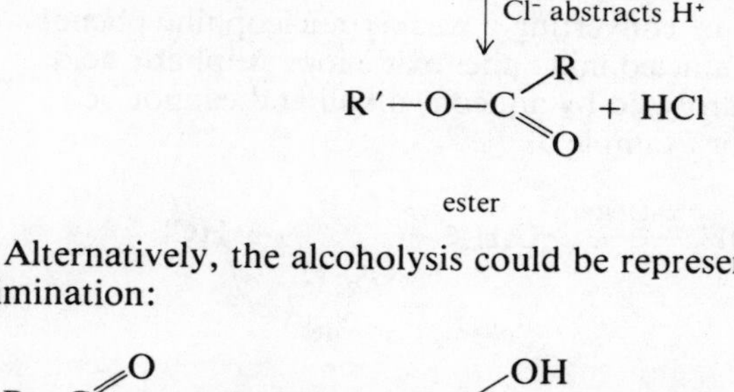

Alternatively, the alcoholysis could be represented as addition—elimination:

$$R-C\diagup_{\diagdown Cl}^{O} + R'OH \rightleftharpoons R-\underset{\underset{Cl}{|}}{\overset{\overset{OH}{|}}{C}}-OR' \rightarrow R-C\diagup_{\diagdown OR'}^{O} + HCl$$

Which mechanism is correct? Although we cannot be certain, we think that it is probably the latter. Nucleophilic substitution is unlikely because the chloride ion is not a good base for abstracting a proton in the final stage.

Hydrolysis, ammonolysis and aminolysis are likely to have a similar mechanism to alcoholysis. **All occur by nucleophilic addition across the C=O bond, followed by the elimination of HCl.**

Because of the inductive effect of the chlorine atom, which reinforces that of the oxygen atom, acid chlorides have an exceptionally high reactivity. They are the most reactive of the carbonyl compounds, and will take part in a wide range of reactions, even with reagents whose nucleophilicity is low.

H_2O

When water is cautiously added to an aliphatic acid chloride at room temperature, two liquid phases can be observed for a short time, but almost at once there is a vigorous, exothermic reaction accompanied by the evolution of some hydrogen chloride. A carboxylic acid and most of the hydrogen chloride remain in solution, e.g.

$$CH_3COCl + H_2O = CH_3COOH + HCl$$

This reaction can be violent.

Aromatic acid chlorides, such as benzoyl chloride, are more resistant to hydrolysis and may require a strong acid to act as catalyst. Alternatively, hydrolysis can be conducted in the presence of a base, which provides the powerful nucleophile HO⁻ (cf. hydrolysis of esters, §13.12).

ROH and ArOH

Aliphatic acid chlorides, like acid anhydrides, readily undergo esterification with both alcohols and phenols. Aromatic acid chlorides are less reactive, but respond to the use of acid catalysts. The esterification of an aromatic acid chloride by a phenol is best performed in the presence of aqueous alkali, a reaction known as the *Schotten–Baumann reaction*. The alkali has the effect of converting a weakly nucleophilic phenol molecule into a strongly nucleophilic phenoxide ion. Aliphatic acid chlorides are rapidly hydrolysed by aqueous alkali and cannot be esterified in this way. For example

$$C_6H_5C\overset{O}{\underset{Cl}{\big<}} + C_6H_5OH \xrightarrow{NaOH(aq)} C_6H_5C\overset{O}{\underset{OC_6H_5}{\big<}} + HCl$$

phenyl benzoate

NH₃, RNH₂ and R₂NH

Concentrated aqueous ammonia readily converts acid chlorides to amides at room temperature, e.g.

$$CH_3C{\overset{O}{\underset{Cl}{}}} + 2NH_3 = CH_3C{\overset{O}{\underset{NH_2}{}}} + NH_4Cl$$

Primary and secondary amines, likewise, react with acid chlorides to give substituted amides, e.g.

$$CH_3C{\overset{O}{\underset{Cl}{}}} + 2C_2H_5NH_2 = CH_3C{\overset{O}{\underset{NHC_2H_5}{}}} + C_2H_5\overset{\oplus}{N}H_3Cl^-$$

         N-ethylethanamide ethylammonium chloride

To avoid the loss of 1 mol of amine as the amine salt, reactions with aromatic acid chlorides (which are relatively resistant to hydrolysis) may be conducted in the presence of dilute aqueous sodium hydroxide, e.g.

$$C_6H_5C{\overset{O}{\underset{Cl}{}}} + C_6H_5NH_2 \xrightarrow{NaOH(aq)} C_6H_5C{\overset{O}{\underset{NHC_6H_5}{}}} + HCl$$

           N-phenylbenzamide (as NaCl)

This is another type of Schotten–Baumann reaction.

The ethanoylation (acetylation) and benzoylation of amines are of great importance in their identification. In the same way that aldehydes and ketones are characterised by the melting temperatures of their 2,4-dinitrophenylhydrazones, so primary and secondary amines are characterised by the melting temperatures of their ethanoyl and benzoyl derivatives.

H_2

As we might expect from their high reactivity, acid chlorides can readily be reduced to aldehydes and thence to primary alcohols. Hydrogen is used as the reducing agent, in the presence of palladium supported on barium sulphate:

$$R{-}C{\overset{O}{\underset{Cl}{}}} \xrightarrow[\text{Pd catalyst}]{H_2} R{-}C{\overset{O}{\underset{H}{}}} \xrightarrow[\text{Pd catalyst}]{H_2} RCH_2OH$$

By utilising the fact that aldehydes are less reactive than acid chlorides we can, if we so wish, stop the reaction at the aldehyde stage by partially poisoning the palladium catalyst with, for example, quinoline and sulphur. Very good yields of aldehydes are obtainable by this technique, which is known as the *Rosenmund reduction*.

Li[AlH₄] and Na[BH₄]

Acid chlorides can be reduced to primary alcohols by both lithium tetrahydridoaluminate(III) and sodium tetrahydroborate(III). If lithium tetrahydridoaluminate(III) is first treated with ethanol, its reducing power is lessened to such an extent that acid chlorides are reduced only to aldehydes.

RMgX

Like acid anhydrides (but unlike carboxylic acids), acid chlorides react with Grignard reagents to give ketones (§11.5). Reaction may continue, with a second molecule of the Grignard reagent, to give tertiary alcohols.

13.12 Esters

Esters have a $-C\overset{\displaystyle O}{\underset{\displaystyle O-C-}{}}$ grouping and take part in addition–elimination reactions, i.e. nucleophilic addition across the $C{=}O$ bond, followed by the elimination of a molecule of alcohol. The $C{-}O$ bond is unreactive; cf. ethers.

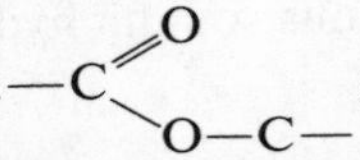

Reactions with Lewis bases

H₂O; H⁺ catalyst (acidic hydrolysis)

On hydrolysis, esters revert to the carboxylic acids and alcohols from which they are made. The reaction with water alone is very slow, partly because the H_2O molecule is a very weak nucleophile and partly because most esters have a low solubility in water, and the reaction may be catalysed by means of a strong acid:

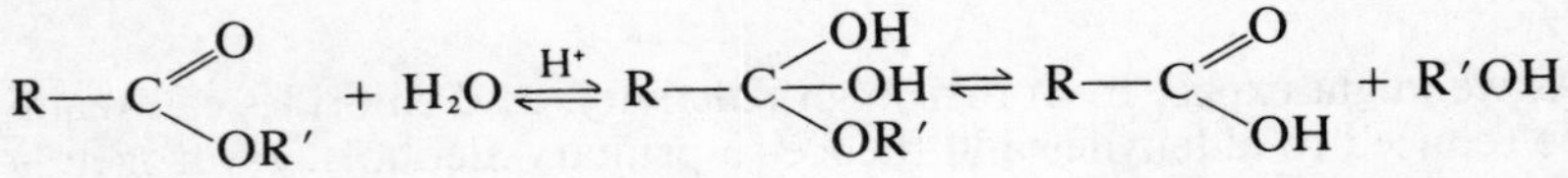

$$R-C\overset{O}{\underset{OR'}{}} + H_2O \underset{}{\overset{H^+}{\rightleftharpoons}} R-C\overset{OH}{\underset{OR'}{\overset{|}{-}OH}} \rightleftharpoons R-C\overset{O}{\underset{OH}{}} + R'OH$$

The acid catalysed addition of water occurs in accordance with mechanism IV; §13.6.

Acid hydrolysis is the exact reverse of acid catalysed esterification, and is of limited use as it leads to the establishment of an equilibrium mixture.

HO⁻ (alkaline hydrolysis)

Alkaline hydrolysis is of much more practical importance, for the reaction goes to completion. Here, the nucleophile is not the H_2O molecule but the much more effective HO^- ion:

$$R-C{\overset{O}{\underset{OR'}{}}} + Na^+ HO^- \rightleftharpoons R-C{\overset{O^{\ominus}Na^+}{\underset{OR'}{-OH}}} \longrightarrow R-C{\overset{O^{\ominus}Na^+}{\underset{O}{}}} + R'OH$$

e.g. $CH_3COOC_2H_5 + NaOH = CH_3COONa + C_2H_5OH$

The reaction goes to completion because the very low reactivity of carboxylate salts in addition reactions causes the final step to be irreversible.

In the laboratory, hydrolysis is conducted by refluxing the ester with aqueous alkali until a homogeneous mixture is obtained. This may well take several hours. Reaction is faster with potassium hydroxide than with sodium hydroxide — KOH is the stronger base — and is assisted by high concentrations of alkali. A mutual solvent for the ester and the aqueous alkali (e.g. ethanol or diethylene glycol) is often included. After hydrolysis is complete, the alcohol is distilled off and the residue then acidified with a strong acid to liberate the carboxylic acid. If the latter is a liquid it may be distilled off or extracted with ether; if solid, it is filtered off.

The alkaline hydrolysis of esters of long chain carboxylic acids (*fatty acids*) is known as *saponification*, for the products — sodium salts of the long chain acids — are soaps (see below).

ROH

An ester and an alcohol can be made to exchange alkyl groups to give another ester and another alcohol:

$$R-C{\overset{O}{\underset{OR'}{}}} + R''OH \rightleftharpoons R-C{\overset{O}{\underset{OR''}{}}} + R'OH$$

Although superficially similar to hydrolysis, the alcoholysis of esters is difficult. It is reversible, takes place only at high temperatures, and requires a base or acid catalyst. The reaction is, however, of considerable importance in the manufacture of polyester *alkyd resins* for the paint industry. Vegetable oils, such as linseed oil, are alcoholysed with 1,2,3-propanetriol (or other polyhydric alcohol) at 230 °C (503 K) in the presence of catalytic amounts of calcium hydroxide. The resulting mixture of mono- and diglycerides (see below) yields an alkyd resin on treatment with 1,2-benzenedicarboxylic anhydride (phthalic anhydride).

RCOOH

The acidolysis of esters, like alcoholysis, has industrial applications:

$$R-C{\overset{O}{\underset{OR'}{}}} + R''-C{\overset{O}{\underset{OH}{}}} \rightleftharpoons R-C{\overset{O}{\underset{OH}{}}} + R''-C{\overset{O}{\underset{OR'}{}}}$$

The temperature needed for this reaction is high; in the region of 270 °C (543 K).

NH_3

The ammonolysis of esters, unlike that of carboxylic acids, occurs readily, for not only are esters more reactive than acids but there is no competing reaction to give an ammonium salt:

$$R-C{\overset{O}{\underset{OR'}{}}} + NH_3 \rightleftharpoons R-C{\overset{OH}{\underset{OR'}{}}}NH_2 \rightleftharpoons R-C{\overset{O}{\underset{NH_2}{}}} + R'OH$$

e.g. $CH_3COOC_2H_5 + NH_3 = CH_3CONH_2 + C_2H_5OH$

This is, perhaps, the best method of preparing amides. The reaction is carried out merely by shaking the ester with aqueous or ethanolic ammonia at room temperature.

RNH_2 and R_2NH

The aminolysis of esters with primary or secondary amines is closely related to the previous reaction and gives rise to substituted amides.

Metals

Aliphatic (but not aromatic) esters can be reduced by sodium and ethanol in the *Bouveault–Blanc reduction*. An addition–elimination reaction to give an aldehyde is at once followed by another addition reaction to give a primary alcohol:

$$R-C{\overset{O}{\underset{OR'}{}}} \xrightarrow{2e^- + 2H^+} R-C{\overset{OH}{\underset{OR'}{}}}H \xrightarrow{-R'OH} R-C{\overset{O}{\underset{H}{}}} \xrightarrow{2e^- + 2H^+} RCH_2OH$$

Two alcohols are thus obtained; one represents the reduction product of the parent acid, while the other is the alcohol which was used in making the ester, e.g.

$$CH_3CH_2C{\overset{O}{\underset{OCH_3}{}}} \xrightarrow{Na + C_2H_5OH} CH_3CH_2CH_2OH + CH_3OH$$

methyl propanoate   1-propanol   methanol

Reactions with C—H acidic compounds

Esters, like aldehydes and ketones, undergo numerous addition–elimination reactions with C—H acidic compounds. A well known example is the *Claisen ester condensation*, in which two molecules of an ester (which must possess an α-hydrogen atom) undergo addition–elimination in the presence of a base to give an ester of a β-keto acid. One molecule of the ester provides the C=O bond, while the second acts as the C—H acidic compound, e.g.

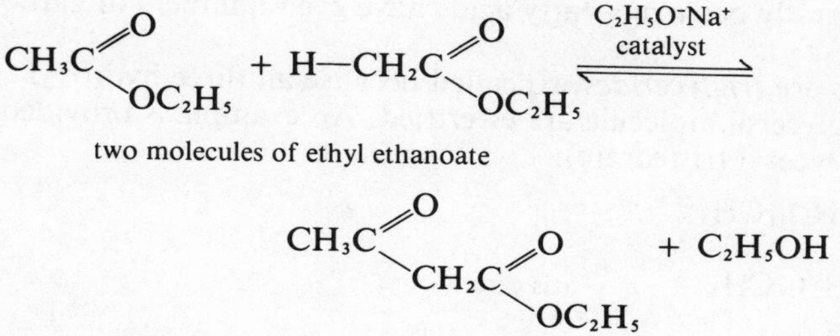

two molecules of ethyl ethanoate

ethyl 3-oxobutanoate
(ethyl acetoacetate)

The mechanism is exactly the same as that of the aldol condensation (§13.7) in the sense that a carbanion, $\overset{\ominus}{C}H_2COOC_2H_5$, is formed initially by the base catalyst, and then combines with the carbonyl carbon atom of a second ester molecule.

Reactions with cryptobases

$Li[AlH_4]$

Lithium tetrahydridoaluminate(III) in dry ether reduces an ester $RCOOR'$ to give a mixture of two alcohols, RCH_2OH and $R'OH$; cf. the Bouveault–Blanc reduction. Although sodium tetrahydroborate(III) will not bring about the reduction, lithium tetrahydroborate(III) is effective.

$RMgX$

Esters, like other derivatives of carboxylic acids, react with Grignard reagents to give ketones, but further reaction, with a second molecule of Grignard reagent, invariably leads to the formation of tertiary alcohols (§11.5). Esters of methanoic acid give aldehydes and, ultimately, secondary alcohols.

Fats, oils and soaps

Animal and vegetable oils and fats are all esters of 1,2,3-propanetriol (glycerol) and long chain carboxylic acids, known as *fatty acids*. Examples of fatty acids are as follows:

Saturated	*Unsaturated*
$CH_3(CH_2)_{14}COOH$	$CH_3(CH_2)_7CH{=}CH(CH_2)_7COOH$
hexadecanoic acid (palmitic acid)	9-octadecenoic acid (oleic acid)
$CH_3(CH_2)_{16}COOH$	$CH_3(CH_2)_4CH{=}CHCH_2CH{=}CH(CH_2)_7COOH$
octadecanoic acid (stearic acid)	9,12-octadecadienoic acid (linoleic acid)

Almost all naturally occurring fatty acids have even numbers of carbon atoms.

Oils and fats are *triglycerides*, so called because all three hydroxyl groups of the glycerol molecule are esterified. An example is provided by tristearin (glyceryl tristearate):

$$CH_2OCO(CH_2)_{16}CH_3$$
$$CHOCO(CH_2)_{16}CH_3 \qquad \text{tristearin}$$
$$CH_2OCO(CH_2)_{16}CH_3$$

Tristearin is a solid and is a constituent of stearin, a fat found in animals and some plants, and the chief constituent of tallow and suet.

Tristearin is a *simple glyceride*, in that all three fatty acid residues are identical. Most natural glycerides, however, are *mixed glycerides*, based on two or even three different fatty acids. In general, a preponderance of saturated fatty acids gives a fat, while a preponderance of unsaturated fatty acids leads to an oil. In the manufacture of margarine, unsaturated oils are partially hydrogenated to give semi-solid fats. Reaction is conducted at 150–200 °C (423–473 K) in the presence of finely divided nickel, which is afterwards recovered by filtration.

Monoglycerides and *diglycerides*, in which the molecules contain respectively one and two fatty acid residues, can be prepared by an alcoholysis reaction between a triglyceride and 1,2,3-propanetriol at 230 °C (503 K) (see above).

$$CH_2OCOR \qquad CH_2OH \qquad\qquad CH_2OCOR \qquad CH_2OCOR$$
$$CHOH \qquad CHOCOR \qquad\qquad CHOCOR \qquad CHOH$$
$$CH_2OH \qquad CH_2OH \qquad\qquad CH_2OH \qquad CH_2OCOR$$

monoglycerides $\qquad\qquad\qquad$ diglycerides

Soaps comprise the sodium and potassium salts of long chain fatty acids. The raw materials used in soap making are low grade oils and fats, such as coconut oil, olive oil, whale oil, and cattle and mutton tallow. In the traditional 'kettle process', the oil or fat is heated with aqueous sodium hydroxide in a large kettle for several hours:

$$CH_2OCOR \qquad\qquad CH_2OH \quad RCOONa$$
$$CHOCOR' \quad + 3NaOH = CHOH \; + \; R'COONa \quad \Big\} \; \text{'soap'}$$
$$CH_2OCOR'' \qquad\qquad CH_2OH \quad R''COONa$$

Potassium hydroxide may be used instead of sodium hydroxide for making liquid soap, soap flakes and some granulated soaps.

In recent years the kettle process has been largely superseded by continuous hydrolysis by steam (not alkali) in a stainless steel column called a hydrolyser. The oil or fat is fed in continuously at the bottom, and hot water at the top. (Some zinc oxide catalyst is also included.) A high temperature (230 °C; 503 K) and a high pressure (40 atm; 4 000 kPa) ensure rapid hydrolysis to fatty acids and 1,2,3-propanetriol. Both products are drawn off continuously, the fatty acids from the top of the column and the 1,2,3-propanetriol, with water, from the bottom. The fatty acids are subsequently neutralised with the exact quantity of alkali to give soap, and the 1,2,3-propanetriol is purified by distillation. (It is a valuable by-product, used mainly for making synthetic resins and explosives.) Among the advantages of this process are a shorter reaction time, greater flexibility, and the need for less equipment.

Soap is a *detergent*, i.e. cleansing agent, and functions as such because one part of the anion — the carboxylate group — is water attracting, while the other part — the hydrocarbon chain — attracts grease and oil. Greasy matter is thus converted into an emulsion with water, and can be removed by rinsing.

13.13 Amides

We might expect amides, with their $-C\begin{smallmatrix}\nearrow O \\ \searrow NH_2\end{smallmatrix}$ grouping, to exhibit reactions of the C=O bond (cf. aldehydes, carboxylic acids, etc) plus reactions of the C—NH_2 bond (cf. primary amines). In practice, the C=O bond is dominant. Except in the reaction with nitrous acid, the C—NH_2 bond contributes little towards the chemistry of amides.

Reactions of the C=O bond

Amides have a relatively low carbonyl activity, comparable with that of carboxylic acids. They take part in a limited number of addition–elimination reactions, i.e. nucleophilic addition across the C=O bond, followed by the elimination of ammonia.

H_2O; H^+ catalyst (acidic hydrolysis)

Amides and substituted amides can be hydrolysed to give the parent carboxylic acid and ammonia or an amine:

$$R-C\underset{NH_2}{\overset{O}{<}} + H_2O \rightleftharpoons R-C\underset{NH_2}{\overset{OH}{-}}OH \rightleftharpoons R-C\underset{OH}{\overset{O}{<}} + NH_3$$

$$R-C\underset{NHR'}{\overset{O}{<}} + H_2O \rightleftharpoons R-C\underset{NHR'}{\overset{OH}{-}}OH \rightleftharpoons R-C\underset{OH}{\overset{O}{<}} + R'NH_2$$

In both theory and practice, the hydrolysis of amides is analogous to that of esters. The reactions are very slow with water alone, but proceed at a reasonable rate in the presence of acid catalysts. Because amides are less reactive than esters, concentrated acid may be needed, e.g.

$$CH_3CONH_2 + H_2O \xrightarrow{H_2SO_4(aq)} CH_3COOH + NH_3$$

Ammonia is retained by the sulphuric acid as ammonium hydrogen-sulphate.

HO^- (alkaline hydrolysis)

Like esters, amides can also be hydrolysed under alkaline conditions, in which case the nucleophile is the HO^- ion, rather than the H_2O molecule, and the organic product is a salt rather than a carboxylic acid:

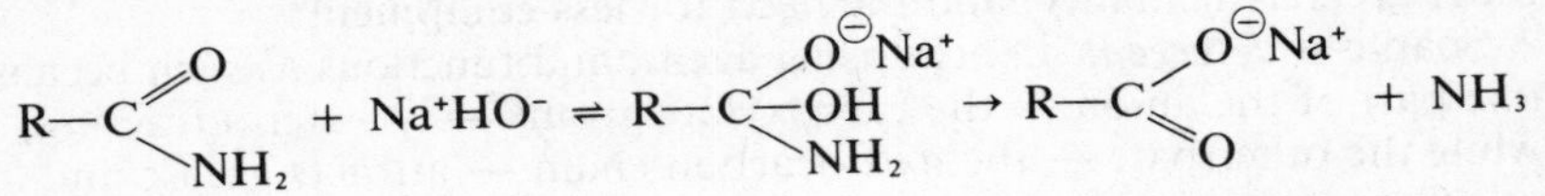

e.g. $CH_3CONH_2 + NaOH \xrightarrow{\text{boil}} CH_3COONa + NH_3$

The hydrolysis of amides under more gentle conditions can be achieved, should this be necessary, by means of nitrous acid at 5 °C (278 K) (see below).

Metals

Amides can be reduced to primary amines by sodium and ethanol:

$$RCONH_2 \xrightarrow{Na + C_2H_5OH} RCH_2NH_2$$

Hydrogenation in the presence of a metallic catalyst is also feasible, but the method requires a high temperature and pressure (typically, 250 °C (523 K) and 300 atm (30 000 kPa)) and is not usually employed.

The reduction of an amide bears a superficial resemblance to the Hofmann reaction ($RCONH_2 \rightarrow RNH_2$), for in both cases a primary amine is formed. The Hofmann reaction, however, is accompanied by a decrease in the number of carbon atoms from n to $n-1$ (§6.4).

$Li[AlH_4]$

Lithium tetrahydridoaluminate(III) is very efficient for converting amides into primary amines, and is often recommended in preference to sodium and ethanol. Sodium tetrahydroborate(III) is ineffective.

$RMgX$

In their reactions with Grignard reagents, amides behave as active hydrogen compounds, thereby converting Grignard reagents into hydrocarbons. In this respect, amides resemble carboxylic acids rather than other acid derivatives.

Elimination of water

Amides can be dehydrated to nitriles by means of various Lewis acids, e.g.

$$CH_3CONH_2 \xrightarrow[\text{heat}]{P_4O_{10}} CH_3CN + H_2O$$

Although phosphorus(V) oxide is the best known dehydrating agent for this purpose, thionyl chloride or boron trifluoride is generally preferable.

The reaction is the exact opposite of those just described in that it is an elimination rather than an addition reaction. It is the reverse of the nucleophilic addition reaction in which water (in the presence of an acid catalyst) converts a nitrile into an amide (§15.4).

Reactions of the C—NH₂ bond (including the N—H bond)

Acidic and basic character

Amides in aqueous solution are neutral to litmus, but have a weakly defined amphoteric character.

Acidity

Amides are slightly acidic for the same reason that carboxylic acids are acidic, i.e. the inductive effect of the carbonyl oxygen atom, which promotes polarisation of the N—H bonds. The acidity of amides is not as great as that of acids because nitrogen is less electronegative than oxygen and polarisation of the N—H bonds is relatively small.

Basicity

Like ammonia and the amines, amides are weakly basic because they can donate a lone pair of electrons from the outer shell of the nitrogen atom to a proton. With hydrogen chloride, for example, they form chlorides which are extensively hydrolysed in aqueous solution, e.g.

$$CH_3C{\overset{\textstyle O}{\underset{\textstyle NH_2}{}}} + HCl \rightleftharpoons CH_3C{\overset{\textstyle O}{\underset{\textstyle \overset{\oplus}{N}H_3Cl^-}{}}} \qquad \text{ethanoylammonium chloride}$$

cf. $CH_3CH_2NH_2 + HCl \rightleftharpoons CH_3CH_2\overset{\oplus}{N}H_3Cl^-$ ethylammonium chloride

The basic character of amides is considerably less than that of ammonia and amines, for two reasons. First, there is the mesomeric effect, whereby the lone pair of electrons on the nitrogen atom is drawn towards the carbonyl carbon atom:

$$R—C{\overset{\textstyle O}{\underset{\textstyle NH_2}{}}}$$

This reduces the ability of the nitrogen atom to coordinate to a proton. Second, when an amide reacts with an acid not all the protons join the NH_2 group; as with carboxylic acids, most of them join the carbonyl oxygen atom:

$$R-C\underset{NH_2}{\overset{O}{<}} + H^+ \rightleftharpoons R-C\underset{NH_2}{\overset{\oplus OH}{<}}$$

Amides as nucleophiles

The chemistry of amines is dominated by their ability to act as nucleophiles (§14.6). With amides, however, because the electron density at the nitrogen atom is reduced by the mesomeric effect, this property is almost entirely absent. There is no reaction between amides and alkyl halides, alcohols, aldehydes, ketones, carboxylic acids or acid derivatives. Only the reaction with nitrous acid proceeds as it does with primary amines:

$$RCONH_2 + HNO_2 \xrightarrow[\text{(273-278 K)}]{\text{0-5 °C}} RCOOH + N_2 + H_2O \text{ (cf. §14.6)}$$

13.14 Syntheses involving diethyl propanedioate and ethyl 3-oxobutanoate

Diethyl propanedioate (diethyl malonate) and ethyl 3-oxobutanoate (ethyl acetoacetate) are two structurally similar carbonyl compounds which are very important in organic syntheses. The preparation of ethyl 3-oxobutanoate by the Claisen ester condensation has already been described (§13.12), while diethyl propanedioate is obtained by the simultaneous hydrolysis and esterification of potassium cyanoethanoate:

$$ClCH_2COOK \xrightarrow{KCN} CNCH_2COOK \xrightarrow[\text{100 °C (373 K)}]{C_2H_5OH + \text{conc. HCl}} CH_2(COOC_2H_5)_2$$

They are colourless liquids. Diethyl propanedioate boils at 199 °C (472 K), and ethyl 3-oxobutanoate at 181 °C (454 K).

Both substances are C—H acidic compounds of unusually high acidity. (Diethyl propanedioate has a pK_a value of ~15, while that of ethyl 3-oxobutanoate is 10.1.) In each case the acidity is due to the same structural feature, namely the presence of a C—H bond which is in an α-position to *two* carbonyl groups, and subject to the electron withdrawing effects of both.

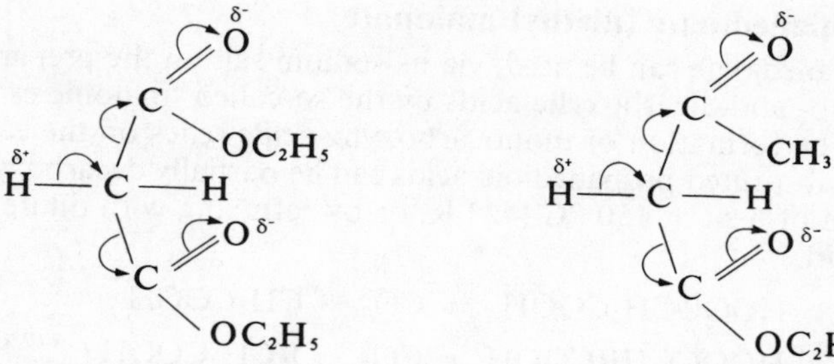

Both compounds, on treatment with sodium ethoxide, are largely converted to salts which consist of sodium ions and carbanions. (To avoid the risk of ester hydrolysis, sodium hydroxide must not be used for these reactions.)

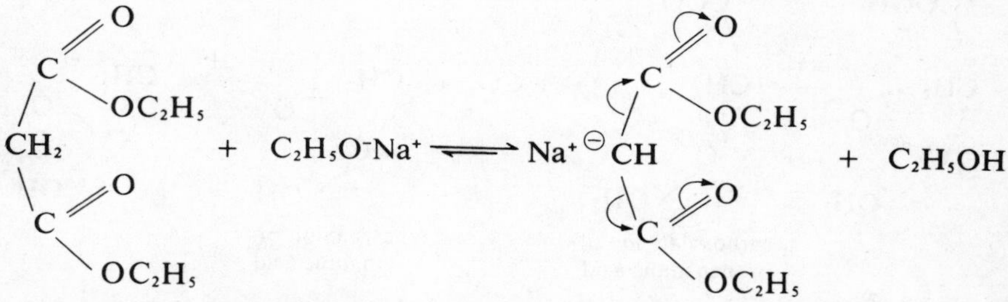

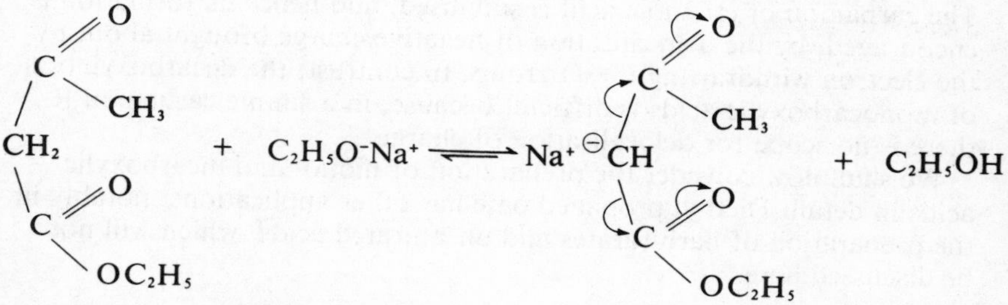

$$CH_2(CO_2C_2H_5)_2 + C_2H_5O^-Na^+ \rightleftharpoons Na^+{}^{\ominus}CH(CO_2C_2H_5)_2 + C_2H_5OH$$

$$CH_2(COCH_3)(CO_2C_2H_5) + C_2H_5O^-Na^+ \rightleftharpoons Na^+{}^{\ominus}CH(COCH_3)(CO_2C_2H_5) + C_2H_5OH$$

In both cases equilibrium lies to the right-hand side. The electron withdrawing carbonyl groups cause the charge on each carbanion to be delocalised to an unusual extent, and these carbanions are considerably more stable than those derived from other acids.

Carbanions are highly effective nucleophilic reagents, and because of this the sodium salts will readily take part in substitution reactions with many organic halides and with the halogens themselves. The products can subsequently be modified by hydrolysis and decarboxylation. In this way diethyl propanedioate can be used to produce a wide range of carboxylic acids, while ethyl 3-oxobutanoate is a reagent for the preparation of a great many carboxylic acids and ketones.

Diethyl propanedioate (diethyl malonate)

Diethyl propanedioate can be used, via its sodium salt, in the preparation of both mono- and dicarboxylic acids by the so-called 'malonic ester synthesis'. The formation of monocarboxylic acids relies on the ease with which substituted propanedioic acids can be partially decarboxylated by the action of heat at 150 °C (423 K) or by refluxing with dilute sulphuric acid.

$$HOOCCH_2COOH = CO_2 + CH_3COOH$$

substituted propanedioic acids $\begin{cases} HOOCCHRCOOH = CO_2 + RCH_2COOH \\ HOOCCR_2COOH = CO_2 + R_2CHCOOH \end{cases}$ substituted ethanoic acids

Decarboxylation proceeds by the following mechanism:

carboxylate ion of propanedioic acid

carbanion of ethanoic acid

The carbanion of ethanoic acid is stabilised, and hence its formation is encouraged, by the delocalisation of negative charge brought about by the electron withdrawing $C=O$ group. In contrast, the decarboxylation of monocarboxylic acids is difficult because, in a simple carbanion R^-, there is no scope for delocalisation of charge.

We shall now consider the preparation of mono- and dicarboxylic acids in detail. Diethyl propanedioate has other applications, notably in the preparation of barbiturates and unsaturated acids, which will not be discussed here.

Monocarboxylic acids

Any monosubstituted ethanoic acid, RCH_2COOH, can be obtained by reacting the sodium salt of diethyl propanedioate with an alkyl halide, RX. The diethyl alkylpropanedioate produced in this way is hydrolysed by refluxing with aqueous potassium hydroxide, and the hydrolysate is then acidified to liberate the free alkylpropanedioic acid from its potassium salt. Finally, partial decarboxylation as described above gives the monosubstituted ethanoic acid.

$$Na^+\overset{\ominus}{C}H(COOC_2H_5)_2 \xrightarrow{RX} RCH(COOC_2H_5)_2 \xrightarrow{KOH(aq)} RCH(COOK)_2$$

$$\xrightarrow{HCl} RCH(COOH)_2 \xrightarrow[\text{at 150 °C (423 K)}]{\text{heat, dry,}} RCH_2COOH$$

Suppose, for example, that we have to prepare 3-methylbutanoic acid, $(CH_3)_2CHCH_2COOH$. This compound can be regarded as isopropylethanoic acid, i.e. $iPrCH_2COOH$. We therefore need to react the sodium salt of diethyl propanedioate with an isopropyl halide, such as 2-bromopropane, $CH_3CHBrCH_3$.

A disubstituted ethanoic acid, $RR'CHCOOH$, can be synthesised by first introducing one alkyl group with an alkyl halide RX, and then repeating the procedure with an alkyl halide $R'X$ to introduce the second alkyl group:

$$Na^+ \overset{\ominus}{C}H(COOC_2H_5)_2 \xrightarrow{RX} RCH(COOC_2H_5)_2 \xrightarrow{C_2H_5O^-Na^+} Na^+ R\overset{\ominus}{C}(COOC_2H_5)_2$$

$$\xrightarrow{R'X} RR'C(COOC_2H_5)_2 \xrightarrow[\text{acidification}]{\text{hydrolysis and}} RR'C(COOH)_2 \xrightarrow[\text{decarboxylation}]{\text{partial}}$$

$$RR'CHCOOH$$

Dicarboxylic acids

Syntheses, as indicated below, can be devised for most dicarboxylic acids.

$$\begin{array}{c} CH_2COOH \\ | \\ \end{array}$$

Butanedioic acid (succinic acid), $\begin{array}{c} CH_2COOH \\ | \\ CH_2COOH \end{array}$

Two routes are available. We can, if we wish, adopt the previous approach and regard butanedioic acid as an $HOOCCH_2$-substituted ethanoic acid, in which case we must react the sodium salt of diethyl propanedioate with, in principle, a haloethanoic acid, $HOOCCH_2X$. In practice, however, an *ester* of the haloacid is employed:

$$Na^+ \overset{\ominus}{C}H(COOC_2H_5)_2 \xrightarrow[\text{(ethyl chloroethanoate)}]{C_2H_5OOCCH_2Cl} C_2H_5OOCCH_2CH(COOC_2H_5)_2$$

$$\xrightarrow[\text{acidification}]{\text{hydrolysis and}} HOOCCH_2CH(COOH)_2 \xrightarrow[\text{decarboxylation}]{\text{partial}} HOOCCH_2CH_2COOH$$

The haloacid itself cannot be used, for it undergoes double decomposition with the sodium salt of diethyl propanedioate:

$$CH_2ClCOOH + Na^+ \overset{\ominus}{C}H(COOC_2H_5)_2 = CH_2ClCOONa$$
$$+ CH_2(COOC_2H_5)_2$$

Alternatively, both —CH_2COOH groups of butanedioic acid can be supplied by diethyl propanedioate. This approach requires the use of 2 mol of the sodium salt of diethyl propanedioate, joined together by means of iodine:

$$\begin{array}{l} \overset{\ominus}{Na^+CH(COOC_2H_5)_2} \\ \overset{\ominus}{Na^+CH(COOC_2H_5)_2} \end{array} + I_2 \rightarrow \begin{array}{l} CH(COOC_2H_5)_2 \\ | \\ CH(COOC_2H_5)_2 \end{array} + 2NaI$$

$$\xrightarrow[\text{acidification}]{\text{hydrolysis and}} \begin{array}{l} CH(COOH)_2 \\ | \\ CH(COOH)_2 \end{array} \xrightarrow[\text{decarboxylation}]{\text{partial}} \begin{array}{l} CH_2COOH \\ | \\ CH_2COOH \end{array}$$

Monosubstituted butanedioic acids, $\begin{array}{l} RCHCOOH \\ | \\ CH_2COOH \end{array}$

The first of the above methods for butanedioic acid can be adapted by selecting the correct α-haloester. Again, we argue that the diethyl propanedioate supplies the—CH_2COOH group; we must therefore introduce HOOCCHR— with, for example, $C_2H_5OOCCHRBr$.

$$\overset{\ominus}{Na^+CH(COOC_2H_5)_2} \xrightarrow{C_2H_5OOCCHRBr} C_2H_5OOCCHRCH(COOC_2H_5)_2$$

$$\xrightarrow[\text{and decarboxylation}]{\text{hydrolysis, acidification}} HOOCCHRCH_2COOH$$

This synthesis is particularly useful because α-haloacids and their esters are easy to prepare.

Disubstituted butanedioic acids, $\begin{array}{l} RCHCOOH \\ | \\ RCHCOOH \end{array}$

Such compounds can be obtained by using iodine to unite 2 mol of the sodium salt of the appropriate diethyl alkylpropanedioate:

$$\begin{array}{l} \overset{\ominus}{Na^+RC(COOC_2H_5)_2} \\ \overset{\ominus}{Na^+RC(COOC_2H_5)_2} \end{array} + I_2 \rightarrow \begin{array}{l} RC(COOC_2H_5)_2 \\ | \\ RC(COOC_2H_5)_2 \end{array} + 2NaI$$

$$\xrightarrow[\text{and decarboxylation}]{\text{hydrolysis, acidification}} \begin{array}{l} RCHCOOH \\ | \\ RCHCOOH \end{array}$$

Pentanedioic acid (glutaric acid), $(CH_2)_3 \overset{\displaystyle COOH}{\underset{\displaystyle COOH}{<}}$, and substituted pentanedioic acids

Pentanedioic acid itself can be prepared by uniting 2 mol of the sodium salt of diethyl propanedioate by means of diiodomethane:

$$\begin{array}{l} Na^+\overset{\ominus}{C}H(COOC_2H_5)_2 \\ Na^+\overset{\ominus}{C}H(COOC_2H_5)_2 \end{array} + CH_2I_2 \rightarrow \begin{array}{c} CH(COOC_2H_5)_2 \\ | \\ CH_2 \\ | \\ CH(COOC_2H_5)_2 \end{array} + 2NaI$$

$$\xrightarrow[\text{and decarboxylation}]{\text{hydrolysis, acidification}} \begin{array}{c} CH_2COOH \\ | \\ CH_2 \\ | \\ CH_2COOH \end{array}$$

Pentanedioic acids which are substituted on the central carbon atom can be made by using higher 1,1-dihalides:

$$\begin{array}{l} Na^+\overset{\ominus}{C}H(COOC_2H_5)_2 \\ Na^+\overset{\ominus}{C}H(COOC_2H_5)_2 \end{array} + RCHI_2 \rightarrow \begin{array}{c} CH(COOC_2H_5)_2 \\ | \\ RCH \\ | \\ CH(COOC_2H_5)_2 \end{array} \rightarrow \begin{array}{c} CH_2COOH \\ | \\ RCH \\ | \\ CH_2COOH \end{array}$$

To prepare a pentanedioic acid which is substituted on the other carbon atoms it is necessary to use a diethyl alkylpropanedioate:

$$\begin{array}{l} Na^+\overset{\ominus}{R}C(COOC_2H_5)_2 \\ Na^+\overset{\ominus}{R}C(COOC_2H_5)_2 \end{array} + CH_2I_2 \rightarrow \begin{array}{c} RC(COOC_2H_5)_2 \\ | \\ CH_2 \\ | \\ RC(COOC_2H_5)_2 \end{array} \rightarrow \begin{array}{c} RCHCOOH \\ | \\ CH_2 \\ | \\ RCHCOOH \end{array}$$

Higher dicarboxylic acids, $\begin{array}{c} COOH \\ | \\ (CH_2)_n \\ | \\ COOH \end{array}$

For an unsubstituted acid, diethyl propanedioate is used with an α,ω-dihalide. (An alpha, omega-dihalide is one in which the two halogen atoms reside on the carbon atoms at the beginning and end of the chain.)

$$\begin{array}{l} Na^+\overset{\ominus}{C}H(COOC_2H_5)_2 \\ Na^+\overset{\ominus}{C}H(COOC_2H_5)_2 \end{array} + \begin{array}{c} Br \\ | \\ (CH_2)_{n-2} \\ | \\ Br \end{array} \rightarrow \begin{array}{c} CH(COOC_2H_5)_2 \\ | \\ (CH_2)_{n-2} \\ | \\ CH(COOC_2H_5)_2 \end{array} \rightarrow \begin{array}{c} CH_2COOH \\ | \\ (CH_2)_{n-2} \\ | \\ CH_2COOH \end{array}$$

The preparation of a substituted acid requires the use of a substituted diethyl propanedioate, a substituted α,ω-dihalide, or both:

$$\begin{matrix} Na^+\overset{\ominus}{R}C(COOC_2H_5)_2 \\ Na^+\overset{\ominus}{R}C(COOC_2H_5)_2 \end{matrix} + \begin{matrix} Br \\ / \\ (R'CH)_n \\ \backslash \\ Br \end{matrix} \rightarrow \begin{matrix} RC(COOC_2H_5)_2 \\ / \\ (R'CH)_n \\ \backslash \\ RC(COOC_2H_5)_2 \end{matrix} \rightarrow \begin{matrix} RCHCOOH \\ / \\ (R'CH)_n \\ \backslash \\ RCHCOOH \end{matrix}$$

Alicyclic carboxylic acids

These compounds can be prepared from the sodium salt of diethyl propanedioate and α,ω-dihalides. The route to alicyclic monocarboxylic acids is as follows:

$$Na^+\overset{\ominus}{C}H(COOC_2H_5)_2 + \begin{matrix} Br \\ / \\ (CH_2)_n \\ \backslash \\ Br \end{matrix} \rightarrow \begin{matrix} CH(COOC_2H_5)_2 \\ / \\ (CH_2)_n \\ \backslash \\ Br \end{matrix} \xrightarrow{C_2H_5O^-Na^+}$$

$$\begin{matrix} Na^+\overset{\ominus}{C}(COOC_2H_5)_2 \\ | \\ (CH_2)_n \\ | \\ Br \end{matrix} \rightarrow (\widehat{CH_2)_n}C(COOC_2H_5)_2 \rightarrow (\widehat{CH_2)_n}CHCOOH$$

The preparation of alicyclic dicarboxylic acids requires the use of two α,ω-dihalides (or one dihalide, and iodine). The method is flexible. Various combinations of reagents can be used, and it is possible to prepare a range of dicarboxylic acids with up to seven carbon atoms in the ring.

$$\begin{matrix} Na^+\overset{\ominus}{C}H(COOC_2H_5)_2 \\ Na^+\overset{\ominus}{C}H(COOC_2H_5)_2 \end{matrix} + \begin{matrix} Br \\ / \\ (CH_2)_n \\ \backslash \\ Br \end{matrix} \rightarrow \begin{matrix} CH(COOC_2H_5)_2 \\ / \\ (CH_2)_n \\ \backslash \\ CH(COOC_2H_5)_2 \end{matrix} \xrightarrow{2(C_2H_5O^-Na^+)}$$

$$\begin{matrix} Na^+\overset{\ominus}{C}(COOC_2H_5)_2 \\ | \\ (CH_2)_n \\ | \\ Na^+\overset{\ominus}{C}(COOC_2H_5)_2 \end{matrix} \begin{matrix} Br \\ | \\ (CH_2)_m \\ | \\ Br \end{matrix} \xrightarrow{} \begin{matrix} C(COOC_2H_5)_2 \\ / \quad \backslash \\ (CH_2)_n \ (CH_2)_m \\ \backslash \quad / \\ C(COOC_2H_5)_2 \end{matrix} \longrightarrow \begin{matrix} CHCOOH \\ / \quad \backslash \\ (CH_2)_n \ (CH_2)_m \\ \backslash \quad / \\ CHCOOH \end{matrix}$$

Ethyl 3-oxobutanoate (ethyl acetoacetate or acetoacetic ester)

The great versatility of ethyl 3-oxobutanoate in organic syntheses is due partly to its high acidity, and partly to the fact that it has three modes of hydrolysis, as follows.

(i) With dilute aqueous sodium hydroxide at room temperature, the ester is hydrolysed to the sodium salt of the acid in the usual way:

$$CH_3COCH_2COOC_2H_5 + NaOH = CH_3COCH_2COONa + C_2H_5OH$$

(ii) On boiling with 85 per cent phosphoric acid, or with a dilute solution of potassium hydroxide in water or ethanol, *ketonic hydrolysis* occurs to give a ketone. Essentially,

$$CH_3COCH_2|COO|C_2H_5 \rightarrow CH_3COCH_3 + CO_2 + C_2H_5OH$$

The reaction should be regarded as hydrolysis, to give CH_3COCH_2COOH, followed by decarboxylation to give CH_3COCH_3. 3-Oxobutanoic acid, in common with all β-ketoacids, undergoes decarboxylation even more readily than propanedioic acid, and by a similar mechanism.

The so-called 'acetoacetic ester synthesis of ketones' relies on the fact that alkyl (and acyl) substituted versions of ethyl 3-oxobutanoate undergo ketonic hydrolysis in the same way.

(iii) With boiling concentrated ethanolic potassium hydroxide, ethyl 3-oxobutanoate undergoes *acidic hydrolysis* to produce the potassium salt of ethanoic acid:

$$CH_3CO|CH_2COO|C_2H_5 + 2KOH = 2CH_3COOK + C_2H_5OH$$

The reaction is a reversed Claisen ester condensation to give $CH_3COOC_2H_5$, followed by hydrolysis to give CH_3COOK and C_2H_5OH.

Alkyl (and acyl) substituted versions of ethyl 3-oxobutanoate undergo acidic hydrolysis in a similar manner to give a mixture of potassium salts:

$$CH_3CO|CHRCOO|C_2H_5 + 2KOH$$
$$= CH_3COOK + CH_2RCOOK + C_2H_5OH$$

One salt is always potassium ethanoate, while the other is derived from a higher carboxylic acid. The so-called 'acetoacetic ester synthesis of acids' is based on this principle.

We shall now consider the applications of ethyl 3-oxobutanoate in detail.

Ketones

Any monosubstituted propanone, CH_3COCH_2R, can be obtained by reacting the sodium salt of ethyl 3-oxobutanoate with an alkyl halide, RX. Suppose, for example, that we have to prepare 2-pentanone, $CH_3COCH_2CH_2CH_3$. This compound can be regarded as ethylpropanone, i.e. CH_3COCH_2Et; therefore the required reagent is an ethyl halide:

$$Na^+ \; CH_3CO\overset{\ominus}{C}HCOOC_2H_5 \xrightarrow{\;CH_3CH_2Br\;} CH_3COC\overset{\displaystyle CH_2CH_3}{\overset{|}{H}}|COO|C_2H_5 + NaBr$$

$$\downarrow \text{ketonic hydrolysis}$$

$$CH_3COCH_2CH_2CH_3 + CO_2 + C_2H_5OH$$

A disubstituted propanone, $CH_3COCHRR'$, can be synthesised by introducing, first, an alkyl group R with a reagent RX, and then, after retreatment with sodium ethoxide, a second alkyl group R' with a reagent $R'X$:

$$Na^+CH_3CO\overset{\ominus}{C}HCOOC_2H_5 \xrightarrow{RX} CH_3COCHRCOOC_2H_5 \xrightarrow{C_2H_5O^-Na^+}$$

$$Na^+CH_3CO\overset{\ominus}{C}RCOOC_2H_5 \xrightarrow{R'X} CH_3COCRR' \vert COO \vert C_2H_5 \xrightarrow[\text{hydrolysis}]{\text{ketonic}}$$

$$CH_3COCHRR' + CO_2 + C_2H_5OH$$

Diketones

Ethyl 3-oxobutanoate provides a simple means of preparing 1,3- and 1,4-diketones. In a 1,3-diketone, the carbonyl groups are at positions 1 and 3 *with respect to each other* (not carbon atoms 1 and 3 of the chain). In other words, the carbonyl groups are next but one to each other. The simplest example is 2,4-pentanedione, $CH_3COCH_2COCH_3$. This can be regarded as a CH_3CO-substituted propanone; therefore the reagent required in its preparation is CH_3COCl.

In a 1,4-diketone, the carbonyl groups are separated by two other carbon atoms, e.g. $CH_3COCH_2CH_2COCH_3$. There are two synthetic routes available. We can, if we wish, adopt the previous approach and regard the compound as a CH_3COCH_2-substituted propanone, in which case we need to select CH_3COCH_2Cl as the reagent. Alternatively, we can allow ethyl 3-oxobutanoate to supply both CH_3COCH_2— groups, in which case we have the task of joining together 2 mol of the sodium salt of ethyl 3-oxobutanoate by means of iodine. (This is a similar technique to that used with diethyl propanedioate in making butanedioic acid.)

$$\begin{array}{c}
Na^+CH_3CO\overset{\ominus}{C}HCOOC_2H_5 \\
Na^+CH_3CO\overset{\ominus}{C}HCOOC_2H_5
\end{array} + I_2 \rightarrow
\begin{array}{c}
CH_3COCH \vert COO \vert C_2H_5 \\
\vert \\
CH_3COCH \vert COO \vert C_2H_5
\end{array}$$

$$\downarrow \text{ketonic hydrolysis}$$

$$\begin{array}{c}
CH_3COCH_2 \\
\vert \\
CH_3COCH_2
\end{array} + 2CO_2 + 2C_2H_5OH$$

By using CH_2I_2 instead of iodine we can make a 1,5-diketone, and by selecting suitable α,ω-dihalides we can obtain a range of higher diketones.

Monocarboxylic acids

Ethyl 3-oxobutanoate, like diethyl propanedioate, can be used to prepare any monosubstituted ethanoic acid, RCH_2COOH. An alkyl halide RX supplies the group R.

$$Na^+CH_3COCH\overset{\ominus}{C}OOC_2H_5 \xrightarrow{RX} CH_3CO|CHRCOO|C_2H_5$$

$$\downarrow \text{ acidic hydrolysis}$$

$$CH_3COOK + CH_2RCOOK + C_2H_5OH$$

Acidification of the potassium salts with, for example, hydrochloric acid gives the free carboxylic acids. The method is thus exactly the same as that for ketones, except that acidic hydrolysis is adopted for the final stage.

The preparation of a disubstituted ethanoic acid, $RR'CHCOOH$, requires the usual double treatment, first with an alkyl halide RX, and then (after more $C_2H_5O^-Na^+$) with an alkyl halide $R'X$. In other words, we follow the procedure described above for disubstituted propanones, but employ acidic hydrolysis:

$$CH_3CO|CRR'COO|C_2H_5 \xrightarrow[\text{hydrolysis}]{\text{acidic}} CH_3COOK + CHRR'COOK$$

$$+ C_2H_5OH$$

Dicarboxylic acids

As in the diethyl propanedioate syntheses of dicarboxylic acids, two routes are open to us. Consider, for example, the preparation of

butanedioic acid, $\overset{\displaystyle CH_2COOH}{\underset{\displaystyle CH_2COOH}{|}}$. We can, if we wish, argue that this is an $HOOCCH_2$-substituted ethanoic acid, in which case we react the sodium salt of ethyl 3-oxobutanoate with an ester of $CH_2ClCOOH$. Alternatively, we allow both $-CH_2COOH$ groups of butanedioic acid to be supplied by 2 mol of ethyl 3-oxobutanoate, linked together by the use of iodine (see above, under 'diketones').

$$\begin{array}{l} CH_3CO|CHCOO|C_2H_5 \\ \quad\quad | \\ CH_3CO|CHCOO|C_2H_5 \end{array} \xrightarrow[\text{hydrolysis}]{\text{acidic}} 2CH_3COOK + \begin{array}{l} CH_2COOK \\ | \\ CH_2COOK \end{array} + 2C_2H_5OH$$

Higher dicarboxylic acids can be obtained by adapting either of these routes. For example, to prepare pentanedioic acid, $HOOC(CH_2)_3COOH$, we could use 1 mol of the sodium salt of ethyl 3-oxobutanoate to supply $-CH_2COOH$, and a reagent such as an ester of CH_2ClCH_2COOH to supply the missing $-CH_2CH_2COOH$ group. Alternatively, we could use 2 mol of the sodium salt of ethyl 3-oxobutanoate to supply two $-CH_2COOH$ groups, and employ CH_2I_2 as the linking agent.

294

γ-Keto acids

Two routes are available. We can either use a haloacid ester as the reagent, followed by ketonic hydrolysis, or use a haloketone as the reagent, followed by acidic hydrolysis. For example,

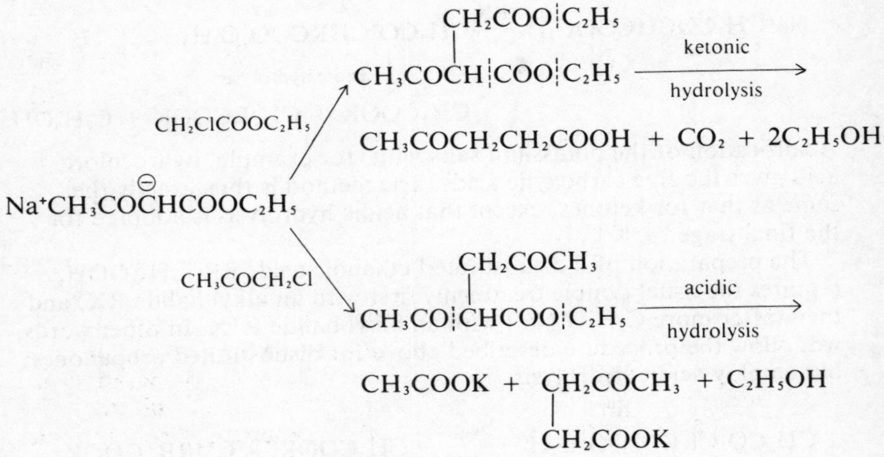

$$CH_2ClCOOC_2H_5 \nearrow \begin{array}{l} CH_2COO\vert C_2H_5 \\ \vert \\ CH_3COCH\vert COO\vert C_2H_5 \end{array} \xrightarrow[\text{hydrolysis}]{\text{ketonic}}$$

$$CH_3COCH_2CH_2COOH + CO_2 + 2C_2H_5OH$$

$$Na^+CH_3CO\overset{\ominus}{C}HCOOC_2H_5$$

$$CH_3COCH_2Cl \searrow \begin{array}{l} CH_2COCH_3 \\ \vert \\ CH_3CO\vert CHCOO\vert C_2H_5 \end{array} \xrightarrow[\text{hydrolysis}]{\text{acidic}}$$

$$CH_3COOK + \begin{array}{l} CH_2COCH_3 \\ \vert \\ CH_2COOK \end{array} + C_2H_5OH$$

Alicyclic ketones and carboxylic acids

Ethyl 3-oxobutanoate is a recognised reagent for the preparation of alicyclic (carbocyclic) compounds with 3, 5, 6 or 7 carbon atoms in the ring. (Cyclobutane derivatives cannot be made in this way.) There are two methods, one for monoketones and monocarboxylic acids, and the other for diketones and dicarboxylic acids. Both require the use of an α,ω-dihalide. We shall illustrate the methods by considering the preparation of cyclohexane derivatives.

Monoketones and monocarboxylic acids

In this synthesis the ethyl 3-oxobutanoate supplies one carbon atom of the ring. The dihalide supplies the other five; therefore we need a 1,5-dihalopentane:

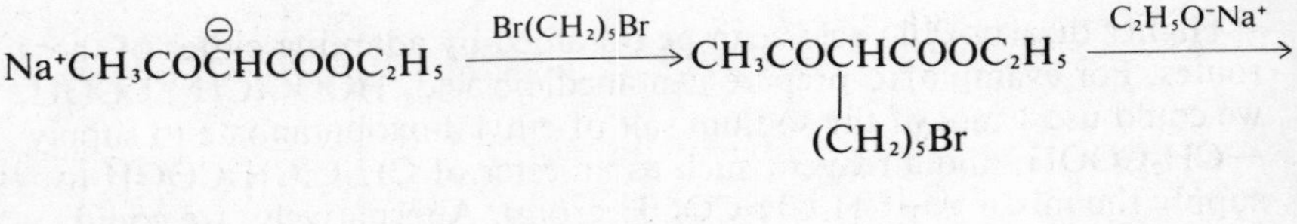

$$Na^+CH_3CO\overset{\ominus}{C}HCOOC_2H_5 \xrightarrow{Br(CH_2)_5Br} \begin{array}{l} CH_3COCHCOOC_2H_5 \\ \vert \\ (CH_2)_5Br \end{array} \xrightarrow{C_2H_5O^-Na^+}$$

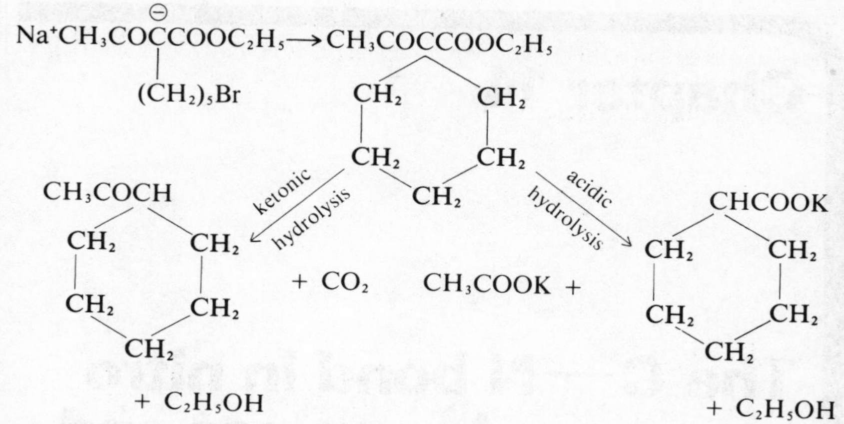

$$Na^+CH_3CO\overset{\ominus}{C}COOC_2H_5 \longrightarrow CH_3COCCOOC_2H_5$$

$$(CH_2)_5Br$$

CH_3COCH CH_2 CH_2

ketonic hydrolysis CH_2 CH_2 *acidic hydrolysis*

CH_3COCH

CH_2 CH_2 CH_2

CH_2 CH_2 $+ CO_2$ $CH_3COOK +$ $CHCOOK$

CH_2 CH_2 CH_2

$+ C_2H_5OH$ CH_2

$+ C_2H_5OH$

Diketones and dicarboxylic acids

By means of iodine 2 mol of the sodium salt of ethyl 3-oxobutanoate are first linked together, as described above under the preparation of diketones. The sodium salt of this dimer is then treated with the dihalide. By this method the dimerised ethyl 3-oxobutanoate contributes two carbon atoms to the final product; therefore the dihalide must supply the remaining four, i.e. a 1,4-dihalobutane is needed:

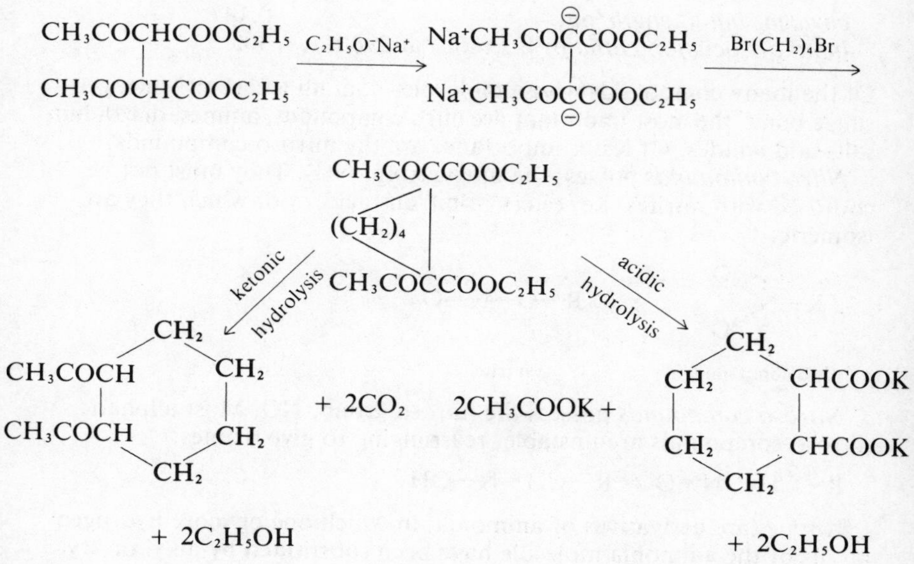

$$CH_3COCHCOOC_2H_5 \xrightarrow{C_2H_5O^-Na^+} Na^+CH_3CO\overset{\ominus}{C}COOC_2H_5 \xrightarrow{Br(CH_2)_4Br}$$
$$CH_3COCHCOOC_2H_5 \qquad\qquad Na^+CH_3CO\underset{\ominus}{C}COOC_2H_5$$

$$CH_3COCCOOC_2H_5$$
$$(CH_2)_4$$

ketonic hydrolysis $CH_3COCCOOC_2H_5$ *acidic hydrolysis*

CH_2 CH_2

CH_3COCH CH_2 CH_2 $CHCOOK$

CH_3COCH CH_2 $+ 2CO_2$ $2CH_3COOK +$ CH_2 $CHCOOK$

CH_2 CH_2

$+ 2C_2H_5OH$ $+ 2C_2H_5OH$

Chapter 14

The C—N bond in nitro compounds, amines and diazonium salts

Data on the C—N bond

covalent bond length/nm   0.147

bond dissociation enthalpy/kJ mol^{-1} at 298 K   305

Of the many compounds whose molecules contain a carbon–nitrogen single bond, the most important are nitro compounds, amines, diazonium salts and amides. Of lesser importance are the nitroso compounds.

Nitro compounds possess the nitro group, NO_2. They must not be confused with nitrites, i.e. esters of nitrous acid, with which they are isomeric.

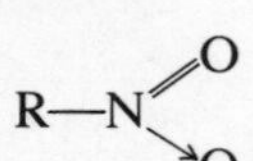

$$R-N{\overset{\displaystyle O}{\underset{\displaystyle O}{}}} \qquad\qquad R-O-N=O$$

nitro compound   nitrite

Nitroso compounds possess the nitroso group, NO. Most aliphatic nitroso compounds are unstable, rearranging to give oximes:

$$R-CH_2-N=O \rightleftharpoons R-CH=N-OH$$

Amines are derivatives of ammonia, in which one or more hydrogen atoms of the ammonia molecule have been substituted by alkyl or aryl groups. They are classified as primary, secondary and tertiary amines, according to the number of such groups.

Diazonium salts are electrovalent compounds consisting of a

diazonium ion, Ar$\overset{\oplus}{N}$≡N, in association with an anion, usually Cl⁻. The best known example is benzenediazonium chloride, $C_6H_5\overset{\oplus}{N}$≡N Cl⁻.
Aliphatic diazonium salts are highly unstable and decompose rapidly, even in cold dilute aqueous solution.

Amides, $RCONH_2$ or $ArCONH_2$, are derivatives of carboxylic acids. They possess, in addition to the C—N bond, a C=O bond which dominates their chemistry (§13.13).

14.1 Nomenclature of compounds with a C—N bond

Nitro compounds

All nitro compounds have substitutive names which are formed from the prefix 'nitro-' and the name of the appropriate alkane or arene, e.g.

CH_3NO_2

$C_6H_5NO_2$

nitromethane

nitrobenzene

Primary amines

Simple primary amines have radicofunctional names which, unlike other such names (e.g. methyl alcohol and methyl chloride), are written as a single word. Examples are as follows:

$CH_3CH_2CH_2NH_2$

$(CH_3)_2CHNH_2$

$CH_3CH_2CH(NH_2)CH_3$

propylamine

isopropylamine

sec-butylamine

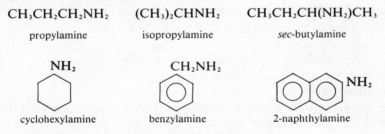

cyclohexylamine

benzylamine

2-naphthylamine

A few trivial names are retained by IUPAC, e.g.

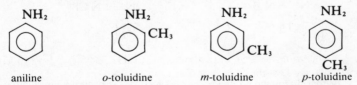

aniline

o-toluidine

m-toluidine

p-toluidine

Primary diamines are given substitutive names by addition of the suffix '-diamine' to the name of the parent hydrocarbon, e.g.

$NH_2CH_2CH_2NH_2$ 1,2-ethanediamine

Additive names, e.g. ethylenediamine for $NH_2CH_2CH_2NH_2$, are also allowed.

Secondary and tertiary amines

Symmetrical secondary and tertiary amines are named by adding the prefix 'di-' or 'tri-' to the name of the radical, e.g.

$(C_2H_5)_2NH$ $(C_2H_5)_3N$

diethylamine   triethylamine

Unsymmetrical secondary and tertiary amines are named as *N*-substituted primary amines. The most complex radical is regarded as belonging to the parent primary amine, and the other radicals are cited in alphabetical order, e.g.

$CH_3CH_2N(CH_3)_2$ $CH_3CH_2CH_2NCH_2CH_3$
                                        CH_3

N,N-dimethylethylamine   *N*-ethyl-*N*-methylpropylamine

$C_6H_5NH(CH_3)$ $C_6H_5N(CH_3)_2$

N-methylaniline or   *N,N*-dimethylaniline or
N-methylphenylamine   *N,N*-dimethylphenylamine

Of the trivial names retained by IUPAC, the following are worthy of note:

pyridine   quinoline

Diazonium salts

These compounds are named by adding the suffix '-diazonium' and the name of the anion to that of the parent arene, e.g.

benzenediazonium bromide   4-methylbenzenediazonium chloride

14.2 Formation of the C—N bond

Both substitution and addition reactions are available. Substitution may be employed to establish the C—N bond in the form of nitro compounds and primary amines, while the addition of hydrogen across the C=N or C≡N bonds gives the C—N bond in primary or secondary amines.

There is no direct way of forming the C—N bond in tertiary amines or diazonium salts; such compounds must be prepared from others that already possess the bond.

Substitution methods

From C—H; the substitution of H by NO₂
Both alkanes and arenes react with nitric acid to give nitro compounds, e.g.

$$CH_3CH_2CH_3 + HNO_3 = CH_3\underset{\underset{NO_2}{|}}{C}HCH_3 + H_2O$$

$$C_6H_6 + HNO_3 = C_6H_5NO_2 + H_2O$$

Although these reactions are superficially similar, they proceed by different mechanisms and require different conditions. The nitration of arenes is an electrophilic substitution in which concentrated sulphuric acid serves as a catalyst, but that of alkanes is a free radical substitution that requires a high temperature (§10.2). Sulphuric acid is valueless as a catalyst.

From C—Cl, C—Br or C—I; the substitution of halogen by NH₂ (§11.5)
Alkyl halides undergo nucleophilic substitution with ammonia to give primary amines. Further reactions invariably occur because amines, like ammonia, act as nucleophiles and attack the remaining alkyl halide. The final product is thus a mixture of primary, secondary and tertiary amines and a quaternary ammonium salt. The Gabriel synthesis, from alkyl halides and potassium phthalimide, gives primary amines only.

Aryl halides, because of the non-lability of their halogen atoms, participate in such reactions only under drastic conditions.

From C—OH; the substitution of OH by NH₂ (§12.5)
Alcohols resemble alkyl halides in their ability to undergo nucleophilic substitution with ammonia to give amines. Phenols behave in a similar fashion.

Addition methods

From C=N
The principal compounds to possess a carbon—nitrogen double bond are derived from aldehydes and ketones by addition—elimination reactions with nitrogen-containing reagents. Examples are oximes, phenylhydrazones and Schiff's bases, made with hydroxylamine, phenylhydrazine and primary amines respectively.

C=N compounds can readily be reduced, either by catalytic hydrogenation or by means of a metal and a protic solvent, the product being a primary or secondary amine depending on the type of atom to which the doubly bonded nitrogen atom is attached. The N—C bond in

Schiff's bases remains intact so that the reduction products are secondary amines, but the N—O bond in oximes and the N—N bond in phenyl-hydrazones do not survive and the reduction products are *primary* amines. A comparison of the bond dissociation enthalpies for the N—C, N—O and N—N bonds will aid understanding; the values are 305, 222 and 163 kJ mol^{-1} respectively. Examples are as follows.

$$CH_3CHO \xrightarrow{NH_2OH} CH_3CH\!\!=\!\!NOH \xrightarrow{Na + C_2H_5OH} CH_3CH_2NH_2 + H_2O$$

ethanal   ethanal oxime   ethylamine — a *primary* amine

$$CH_3CHO \xrightarrow{C_6H_5NHNH_2} CH_3CH\!\!=\!\!NNHC_6H_5 \xrightarrow{Zn + CH_3COOH} CH_3CH_2NH_2$$

ethanal phenylhydrazone

$$+ \; C_6H_5NH_2$$

$$CH_3CHO \xrightarrow{CH_3NH_2} CH_3CH\!\!=\!\!NCH_3 \xrightarrow[\text{Ni catalyst}]{H_2} CH_3CH_2NHCH_3$$

N-ethylidenemethylamine   *N*-methylethylamine — a *secondary* amine

From C≡N

This bond is present in nitriles, RCN. On reduction, either by catalytic hydrogenation or by means of a metal plus protic solvent, nitriles give primary amines as the principal product. Because of a side reaction, however, significant quantities of secondary amines are also formed, e.g.

$$CH_3CN \xrightarrow[\text{or Na + C}_2\text{H}_5\text{OH}]{\text{H}_2 \text{ with Ni catalyst}} CH_3CH_2NH_2 \text{ and } (CH_3CH_2)_2NH$$

The use of lithium tetrahydridoaluminate(III) avoids the formation of secondary amines.

14.3 Interconversion of C—N compounds

The preparation of amines by the ammonolysis of halides or alcohols is of limited use, because the reactions are difficult to carry out and lead to a mixture of products whose separation may present problems. Consequently, it is common practice to establish the C—N bond in nitro compounds or amides and then prepare other C—N compounds from these starting materials.

Primary amines

In both the laboratory and the chemical industry aromatic primary amines are usually prepared by the reduction of nitro compounds, e.g.

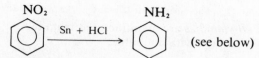

(see below)

The method is not well suited to the preparation of aliphatic primary amines because of the difficulty of making aliphatic nitro compounds.

Amides can be converted into primary amines either by reduction (§13.13) or by treatment with bromine and alkali (§6.4), e.g.

$$CH_3CONH_2 \begin{array}{c} \xrightarrow[\text{reduction}]{Li[AlH_4]} CH_3CH_2NH_2 \\ \xrightarrow[\text{Hofmann reaction}]{Br_2 + KOH} CH_3NH_2 \end{array}$$

Secondary and tertiary amines

Secondary and tertiary amines, both aliphatic and aromatic, are usually made by the alkylation (or arylation) of primary amines. Halides or alcohols can be used as alkylating agents (§11.5 and §12.5), e.g.

NH₂ → (heat with CH₃I or CH₃OH and H₂SO₄) → NHCH₃ → N(CH₃)₂

Diazonium salts

The diazotisation of aromatic primary amines is achieved by treatment with sodium nitrite and dilute hydrochloric acid (or sulphuric acid) at a temperature of 0–5 °C (273–278 K) (§14.6), e.g.

NH₂ + NaNO₂ + 2HCl = N≡N Cl⁻ + NaCl + 2H₂O

14.4 Manufacture of nitro compounds and amines

Nitro compounds

The only common nitroalkanes are nitromethane, nitroethane, and 1-and 2-nitropropane, all made by the nitration of propane (§10.2). They are used as solvents.

Nitrobenzene is made in the chemical industry, as it is in the laboratory, by the nitration of benzene by means of a mixture of nitric acid and sulphuric acid at 50–55 °C (323–328 K). It is used mainly for the preparation of phenylamine and other dyestuff intermediates.

Amines

The *methylamines* and the *ethylamines* are manufactured from the corresponding alcohols by treatment with ammonia:

$$CH_3OH \xrightarrow[\text{Al}_2\text{O}_3 \text{ catalyst}]{\text{NH}_3,\ 400\ °C\ (673\ K),\ 60\ atm\ (6\ 000\ kPa)} CH_3NH_2,\ (CH_3)_2NH \text{ and } (CH_3)_3N$$

$$C_2H_5OH \xrightarrow[\text{Ni catalyst}]{\text{NH}_3 + \text{H}_2,\ 200\ °C\ (473\ K),\ \text{under pressure}} C_2H_5NH_2,\ (C_2H_5)_2NH \text{ and } (C_2H_5)_3N$$

The most important of these compounds is dimethylamine. On reaction with methyl methanoate it gives the useful solvent, *N,N*-dimethylmethanamide (dimethylformamide). It is also used for making rocket fuels, agricultural and rubber chemicals, and certain detergents.

Phenylamine is obtained by the reduction of nitrobenzene. The Béchamp process, which employs iron and dilute hydrochloric acid as the reducing agent, is now obsolescent. Instead, nitrobenzene is catalytically hydrogenated in the vapour phase:

$$C_6H_5NO_2 + 3H_2 \xrightarrow[\text{or NiS on Al}_2\text{O}_3]{\text{Cu on silica gel}} C_6H_5NH_2 + 2H_2O$$

The reaction is exothermic, and efficient removal of heat is essential to prevent the further reduction of phenylamine to benzene.

Phenylamine is required principally for the preparation of antioxidants and vulcanisation catalysts for the rubber industry. Its other uses include the manufacture of dyestuffs, rigid polyurethane foams and sulphonamide drugs.

14.5 General properties of compounds with a C—N bond

Nitro compounds

Pure nitroalkanes are colourless liquids with boiling temperatures that are substantially higher than those of the isomeric alkyl nitrites; e.g. CH_3NO_2 101 °C (374 K); CH_3ONO — 12 °C (261 K). The relatively high boiling temperatures are due to the intermolecular attraction that results from polarisation of the nitrogen–oxygen bonds, $\overset{\delta+}{N}\!-\!\overset{\delta-}{O}$. (Oxygen is more electronegative than nitrogen.)

Nitroalkanes are stable and non-toxic. They have a limited solubility in water as a result of weak hydrogen bonding between the oxygen atoms of nitro groups and the hydrogen atoms of water molecules.

Most nitroarenes are crystalline solids, although nitrobenzene and 1-methyl-2-nitrobenzene (*o*-nitrotoluene) are oily liquids. All are yellow. They all have a density greater than 1 g cm^{-3} (1 kg dm^{-3}), and a low solubility in water. **Except for simple mononitro compounds, they are potentially explosive and should be distilled only in steam or, with care, under reduced pressure.**

Nitrobenzene has a smell reminiscent of almonds. It is a particularly toxic nitro compound and it should not be allowed to come into contact with the skin; neither should its vapour be inhaled.

Amines

Aliphatic amines are colourless compounds with a fishy, ammoniacal smell. (Dimethylamine and trimethylamine are actually present in herrings.) Because the charge separation in the N—H bond is less than in the O—H bond, amines are more volatile than alcohols of comparable relative molecular mass, e.g.

	Boiling temperature/			Boiling temperature/	
	°C	(K)		°C	(K)
CH_3NH_2	−6.3	(266.9)	CH_3OH	64.5	(337.7)
$CH_3CH_2NH_2$	16.6	(289.8)	CH_3CH_2OH	78.5	(351.7)
$CH_3CH_2CH_2NH_2$	48.6	(321.8)	$CH_3CH_2CH_2OH$	97.2	(370.4)

So volatile are the methylamines and the ethylamines that (like ammonia) they irritate the sensitive tissues of the nose and eyes.

Hydrogen bonding between amine and water molecules is responsible for the lower amines having a high solubility in water.

Most aromatic amines are crystalline solids, although the lower ones, notably phenylamine, are oily liquids with a density greater than 1 g cm^{-3} (1 kg dm^{-3}). They have a sweet, sickly smell and are very poisonous. The lower members can be steam distilled. This is because they have a low solubility in water, are steam-volatile, and are not hydrolysed by hot water.

Diazonium salts

In general, aromatic diazonium salts are only marginally more stable than their aliphatic counterparts. They are normally prepared in cold aqueous solution and used at once, for they decompose slowly even at 0 °C (273 K).

If, like benzenediazonium chloride, they are derived from strong acids, they can be isolated from aqueous solution by precipitation with ether. Obtained in this way, they are colourless crystalline solids with a high solubility in water. On shock or gentle heating, the dry crystals detonate violently.

14.6 Chemical properties of compounds with a C—N bond

The electronegativity difference between carbon and nitrogen leads to a permanent charge separation of the carbon–nitrogen bond, $\overset{\delta+}{C}$ $\overset{\delta-}{N}$. A similar effect in the carbon–halogen bond causes alkyl halides to take part in many nucleophilic substitution reactions, and we must consider the possibility of such reactions with amines, nitro compounds and diazonium salts:

$$Nu:^- + \underset{|}{\overset{|}{C}}\!-NH_2 = Nu\!-\!\underset{\diagdown}{\overset{|}{C}} + NH_2^-$$

$$Nu:^- + \underset{\diagdown}{\overset{|}{C}}\!-NO_2 = Nu\!-\!\underset{\diagdown}{\overset{|}{C}} + NO_2^-$$

$$Nu:^- + \underset{\cdot\cdot}{\overset{\diagup}{C}}\!-\!\overset{\oplus}{N}\!\equiv\!N = \underset{\cdot\cdot}{\overset{\diagup}{C}}\!-\!Nu + N_2$$

However, the nitrogen atom is small and the C—N bond is not readily polarised any further. As a result, NH_2^- and NO_2^- ions are very poor leaving groups, and **amines and nitro compounds do not undergo subsitution by nucleophilic reagents.**

A similar difficulty among alcohols is overcome by the use of an acid catalyst, which encourages polarisation of the C—OH bond and leads to the loss of H_2O (a very stable molecule) rather than the HO^- ion (§12.5). With amines, however, acid catalysts are ineffective, partly because nitrogen is less electronegative than oxygen so that the C—N bond is less polarised than the C—O bond, and partly because ammonia, which would be formed, is a less stable molecule than water.

The only C—N compounds which do react in this way are diazonium salts. Reactions are energetically favoured because of the instability of the diazonium ion — there is a *positive* charge on a nitrogen atom — coupled with the relative stability of the nitrogen molecule that is being formed. Even so, aromatic diazonium salts do not react readily except in the presence of catalysts.

Nitro compounds

Nitro compounds may be described as *hetero-analogous carbonyl compounds,* because the N=O bond is similar in its chemistry to the C=O bond in carbonyl compounds. Nevertheless, the carbonyl activity of the N=O bond is low. It is comparable with that of the C=O bond in the carboxylate ion, and for the same reason, namely the presence of a delocalised π bond covering all three atoms.

$$-N\!\!\underset{\searrow O}{\overset{\diagup O}{}} \quad I \qquad\qquad -N\!\!\underset{\diagdown O}{\overset{\diagup O}{}} \quad II$$

Although the nitro group is often represented by formula I, a more accurate representation is formula II. The group has the same number of orbitals and electrons as the carboxylate ion and therefore possesses the same type of structure (§1.4). This is reflected by a remarkable correspondence between bond lengths and bond angles in nitro compounds and carboxylate ions, e.g.

$$C_6H_5\!-\!N\!\!\underset{O}{\overset{O}{<}}\,124° \qquad\qquad \left[H\!-\!C\!\!\underset{O}{\overset{O}{<}}\,124°\right]^-$$

0.121 nm   ~0.125 nm

The delocalised π bond largely compensates for the partial positive charge on the nitrogen atom, and precludes reaction by most of the reagents which attack the C=O bond in aldehydes and ketones.

Reduction of nitro compounds

Nitro compounds are readily attacked only by reducing agents. Reduction is often accomplished by a non-noble metal (e.g. Fe, Zn or Sn) in acidic solution, but can also be achieved electrolytically or by catalytic hydrogenation. The mechanism is similar to that of the reduction of carbonyl compounds (§13.7), except that it is an addition—elimination rather than a straightforward addition reaction. When, for example, metal plus acid is used, electrons from the metal attack first, and addition is completed by protons from the acid. Elimination of water then leads to the formation of a nitroso compound:

$$R-N{\overset{O}{\underset{O}{}}} + 2e^- \rightarrow R-N{\overset{O^\ominus}{\underset{O^\ominus}{}}} \overset{2H^+}{\rightarrow} R-N{\overset{OH}{\underset{OH}{}}} \rightarrow R-N=O + H_2O$$

The reduction cannot be stopped at this point, for nitroso compounds, also, are hetero-analogous carbonyl compounds and have a higher carbonyl activity than nitro compounds. They in turn become reduced in a similar manner to give substituted hydroxylamines which, when suitable reducing agents are used, can be obtained in good yield, e.g.

nitrosobenzene   phenylhydroxylamine
(not isolated)

In strongly acidic solution, the substituted hydroxylamine is reduced still further to give a primary amine, e.g.

$$C_6H_5NO_2 \xrightarrow{Sn + HCl} C_6H_5NO \rightarrow C_6H_5NHOH \rightarrow C_6H_5NH_2$$

Amines

The chemistry of amines is dominated by the lone pair of electrons in the outer shell of the nitrogen atom. Donation of this lone pair to a proton accounts for the basic properties of amines, while donation to another species, usually a carbon atom, is responsible for amines acting as nucleophiles in their reactions with alkyl halides, alcohols, carbonyl compounds, nitrous acid, and trichloromethane.

Amines as bases

All amines, aliphatic and aromatic, resemble ammonia in that their molecules are able to coordinate to a proton, e.g.

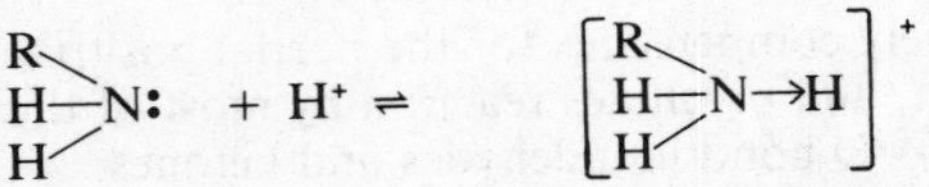

$$R-H-N: \quad + H^+ \rightleftharpoons \left[\begin{array}{c} R \\ H-N \rightarrow H \\ H \end{array} \right]^+$$

primary amine   monosubstituted ammonium ion

Secondary and tertiary amines, in a similar manner, give rise to di- and trisubstituted ammonium ions respectively. (Nitro compounds, in contrast, have no basic properties because there is no lone pair of electrons on the nitrogen atom.)

Ionisation occurs in water, with the result that aqueous solutions of amines are alkaline:

$$RNH_2 + H_2O \rightleftharpoons RNH_3^+ + HO^-$$
cf. $$NH_3 + H_2O \rightleftharpoons NH_4^+ + HO^-$$

Amines, like ammonia, are weak bases, for their ionisation in water is far from complete. Because α (the degree of ionisation) is low, pK_b values are relatively high (Table 3.2).

Neutralisation reactions occur with acids at room temperature to give amine salts. If, for example, a bottle of aqueous methylamine is held close to one of concentrated hydrochloric acid, a dense white smoke of methylammonium chloride is observed:

$$CH_3NH_2 + HCl = CH_3\overset{\oplus}{N}H_3 \ Cl^-$$
cf. $$NH_3 + HCl = NH_4^+ \ Cl^-$$

If hydrochloric acid is added to phenylamine, and the resulting mixture is cooled, crystals of phenylammonium chloride are deposited:

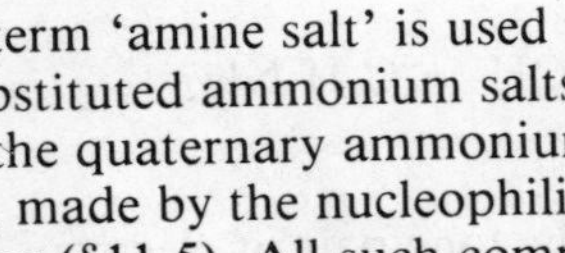

Amine salts

The term 'amine salt' is used to describe the mono-, di- and trisubstituted ammonium salts made by dissolving amines in acids, and also the quaternary ammonium salts, i.e. tetrasubstituted ammonium salts, made by the nucleophilic attack of tertiary amines on alkyl halides (§11.5). All such compounds are named as substituted ammonium salts, e.g.

$[(CH_3)_2NH_2]^+ \ Cl^-$ dimethylammonium chloride
$[(CH_3)_4N]^+ \ Cl^-$ tetramethylammonium chloride

If different hydrocarbon radicals are present, they are cited in alphabetical order, e.g.

$$\left[CH_3CH_2CH_2-\underset{\underset{H}{|}}{\overset{\overset{CH_2CH_3}{|}}{N}}-CH_3 \right]^+ \ Cl^-$$

ethylmethylpropylammonium chloride

Should the name of the parent base end in '-ine', rather than '-amine', the cation name is formed by adding the suffix '-ium' to that of the base, with elision of the final 'e', e.g.

$[C_6H_5NH_3]^+ \ Cl^-$ anilinium chloride (from 'aniline')
or phenylammonium chloride (from 'phenylamine')

As may be expected from their ionic structures, amine salts are crystalline solids with a moderate solubility in water. They are extensively hydrolysed in aqueous solution and give an acidic reaction if they are derived from strong acids.

On treatment with moist silver oxide or ethanolic potassium hydroxide, quaternary ammonium halides form quaternary ammonium hydroxides, $[NR_4]^+ \ HO^-$. Such compounds can be isolated as deliquescent crystalline solids. Because of their ionic structures they are as strongly basic as sodium hydroxide or potassium hydroxide.

A quaternary ammonium hydroxide decomposes at a temperature of about 130 °C (403 K) into a tertiary amine, an alkene and water. The reaction, called the *Hofmann degradation,* should be regarded as a base (HO⁻) induced elimination of a proton and a tertiary amine from a quaternary ammonium ion. It is essentially similar to the base induced elimination of HX from an alkyl halide (§11.5). In both cases a proton is eliminated from the β-carbon atom, while the other species (tertiary amine or halide ion) is lost from the α-position, e.g.

$$HO^- \left[\begin{array}{c} CH_3 \\ | \\ CH_2 \\ \overset{\delta^-}{H}-C^\beta-H \\ | \\ H-C^\alpha \overset{\delta^-}{\rightarrow} N(CH_3)_3 \\ | \\ H \end{array} \right]^+ \xrightarrow{\text{heat}} HOH + \begin{array}{c} CH_3 \\ | \\ CH_2 \\ C-H \\ \| \\ C \\ / \ \backslash \\ H \quad H \end{array} + N(CH_3)_3$$

butyltrimethylammonium hydroxide   water   2-butene   trimethylamine

If more than one alkene can be formed, ethene (from ethyl radicals) is always formed preferentially. Failing this, *Hofmann's rule* states that there will be a predominance of the least substituted ethene. Methyl groups cannot form an alkene and are always retained on the nitrogen atom. If methyl groups are the only ones present, the quaternary ammonium ion decomposes by substitution rather than elimination:

$$[N(CH_3)_4]^+ + HO^- = CH_3OH + N(CH_3)_3$$

Amines as nucleophiles

Amines are able to react with a considerable variety of Lewis acids. Often, coordination of the amine is followed by the elimination of another species. In some of these reactions, e.g. with nitrous acid and with carbonyl compounds, amines resemble alcohols. They are, however, better than alcohols at donating a lone pair of electrons, and will react with certain compounds (e.g. alkyl halides) which are not attacked by alcohols.

Alkyl halides

In nucleophilic substitution reactions with alkyl halides, primary amines give secondary amines, secondary amines give tertiary amines, and tertiary amines give quaternary ammonium salts (§11.5).

Alcohols

In similar reactions with alcohols and phenols, primary amines give secondary amines, while secondary amines give tertiary amines. Quaternary ammonium compounds are not formed (§12.5).

Carbonyl compounds

Primary and secondary amines undergo nucleophilic addition—elimination reactions with carbonyl compounds. Tertiary amines do not react at all; because they lack an 'active' hydrogen atom on the nitrogen atom, water (or other molecule) cannot be eliminated.

With aldehydes and ketones, primary amines give Schiff's bases while secondary amines give enamines (§13.7). With carboxylic acids and their derivatives, primary and secondary amines give N-substituted amides. The analytical importance of these reactions has been stressed (§13.11).

Nitrous acid

Nitrous acid is an unstable compound and must be generated *in situ* from sodium nitrite and hydrochloric acid:

$$NaNO_2 + HCl = HNO_2 + NaCl$$

The amine under test is first dissolved in excess dilute hydrochloric acid. The solution is cooled to 5 °C (278 K) by means of ice, and sodium nitrite is then added in small portions with stirring and continued cooling.

An excess of hydrochloric acid is needed in order to form the nitrosyl cation, NO^+:

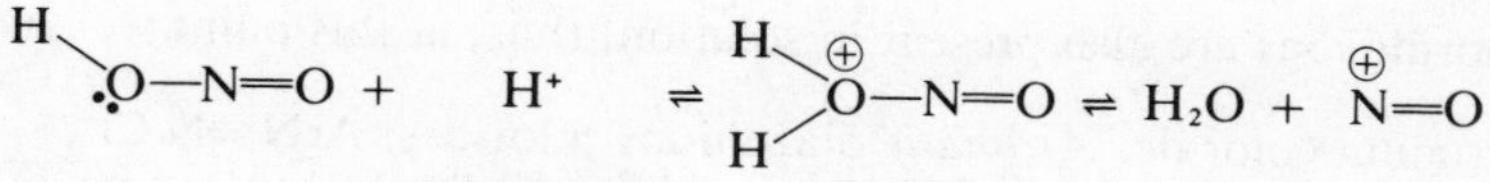

$$\overset{\bullet\bullet}{\text{O}}\text{—N}=\text{O} + \text{H}^+ \rightleftharpoons \overset{H}{\underset{H}{}}\text{O}\text{—N}=\text{O} \rightleftharpoons \text{H}_2\text{O} + \overset{\oplus}{\text{N}}=\text{O}$$

(from HCl)   protonated   nitrosyl
      nitrous acid   cation

This is very similar to the protonation of nitric acid and the formation of the nitryl cation, NO_2^+, that precedes the nitration of benzene.

Primary and secondary amines attack the nitrosyl cation in addition–elimination reactions; nucleophilic addition of the amine to NO^+ is followed by the elimination of a proton. (Alternatively, the reactions may be regarded as the electrophilic attack of the nitrosyl cation on an amine.) Tertiary amines behave differently.

Secondary amines Coordination from the nitrogen atom of the amine to that of the nitrosyl cation is followed by the elimination, as a proton, of the hydrogen atom from the amine nitrogen atom. The product is a yellow, oily *nitrosoamine,* e.g.

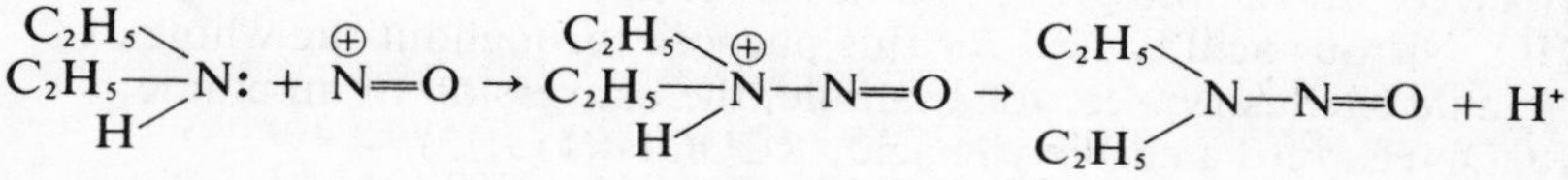

N-nitrosodiethylamine

Likewise,

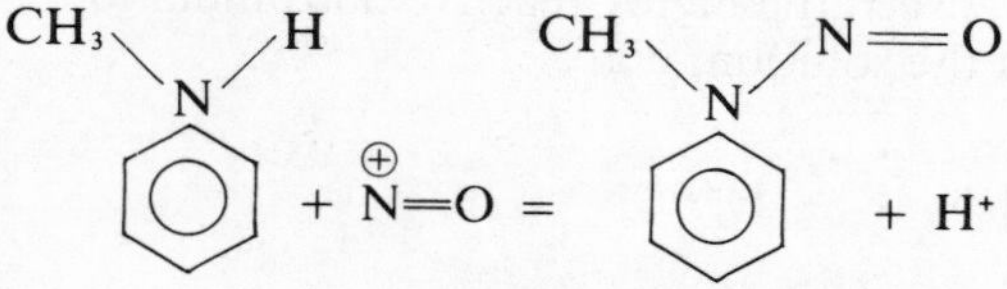

N-nitroso-N-methylphenylamine

Primary amines Several reactions occur in rapid succession. Initially, a nitrosoamine is formed, exactly as before:

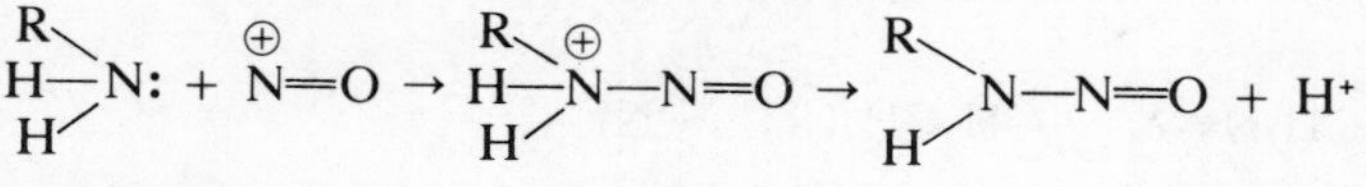

However, the nitrosoamine is unstable, and in the presence of acid immediately changes into a diazonium ion, by a series of reactions which can be summarised by the following equation:

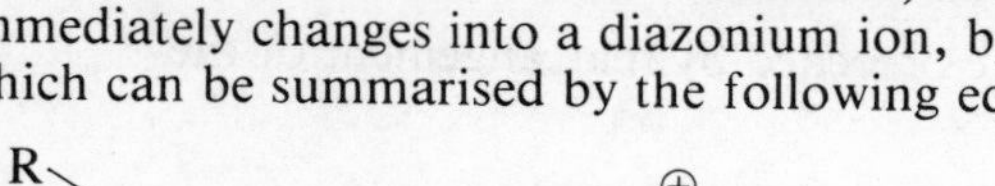

Chloride ions are also present in solution; thus, at this point we have a diazonium chloride. *Aromatic* diazonium chlorides, $Ar\overset{\oplus}{N}{\equiv}N\ Cl^-$, remain in solution at 5 °C (278 K), and the overall reaction can be summarised as follows:

$$ArNH_2 + NaNO_2 + 2HCl = Ar\overset{\oplus}{N}{\equiv}N\ Cl^- + NaCl + 2H_2O$$

Hydrolysis, to phenols, occurs at an appreciable rate only on warming (see below).

Aliphatic diazonium salts, in contrast, are highly unstable and cannot be isolated. They are immediately hydrolysed to give alcohols:

$$\overset{\oplus}{R}N{\equiv}N\ Cl^- + H_2O = ROH + N_2 +^- HCl$$

The overall reaction can therefore be represented as follows:

$$RNH_2 + HNO_2 = ROH + N_2 + H_2O$$

In effect, the amino group, NH_2, is replaced by the hydroxyl group, HO. Nitrous acid is used for this purpose throughout the whole of organic chemistry; see, for example, the conversion of an amide, $RCONH_2$, into a carboxylic acid, $RCOOH$ (§13.13).

Alcohols are not the only compounds to be formed when aliphatic primary amines are treated with nitrous acid. The product, even at 0 °C (273 K), is likely to be a mixture of an alcohol, alkene, alkyl nitrite, alkyl chloride and other compounds. These substances arise mainly through reactions between the highly reactive diazonium ion and various nucleophiles in the solution, e.g.

$$R\overset{\oplus}{N}{\equiv}N + H_2O = ROH + N_2 + H^+$$
$$R\overset{\oplus}{N}{\equiv}N + NO_2^- = RONO + N_2$$
$$R\overset{\oplus}{N}{\equiv}N + Cl^- = RCl + N_2$$

Alkenes are formed from carbonium ions, which in turn arise through the decomposition of diazonium ions, e.g.

$$CH_3CH_2CH_2\overset{\oplus}{N}{\equiv}N \rightarrow CH_3CH_2\overset{\oplus}{C}H_2 + N_2$$
$$CH_3CH_2\overset{\oplus}{C}H_2 \rightleftharpoons CH_3CH{=}CH_2 + H^+$$

The loss of a proton may be preceded by rearrangement of the carbonium ion:

$$CH_3CH_2\overset{\oplus}{C}H_2 \rightleftharpoons CH_3\overset{\oplus}{C}HCH_3 \rightleftharpoons CH_3CH{=}CH_2 + H^+$$

Tertiary amines Aliphatic tertiary amines undergo no immediate reaction with nitrous acid, except for neutralisation. There is therefore no visible change.

Aromatic tertiary amines undergo nitrosation of the ring, mainly at the 4-position (§10.2). For example, N,N-dimethylphenylamine, when treated with sodium nitrite and hydrochloric acid, gives the chloride of 4-nitroso-N,N-dimethylphenylamine, $NOC_6H_4\overset{\oplus}{N}H(CH_3)_2$ Cl^-. A dark red solution is formed, from which yellow crystals separate on cooling. On treatment with alkali, a green precipitate of the free base is obtained.

Summary Nitrous acid is an excellent reagent for distinguishing between primary, secondary and tertiary amines. If the solution remains clear and almost colourless, and if bubbles of nitrogen are evolved, the amine is primary. The presence of a diazonium salt derived from an aromatic primary amine may be confirmed by coupling with alkaline 2-naphthol to produce a red azo dye (see below). If an emulsion appears, followed by the separation of a yellow, oily nitrosoamine, the amine is secondary; the presence of a nitrosoamine can be confirmed by *Liebermann's nitroso reaction*, in which a series of distinctive colour changes follows the addition of phenol and concentrated sulphuric acid. If nothing at all appears to happen, the amine is tertiary aliphatic. If a red solution appears, giving rise to yellow crystals, the amine is tertiary aromatic; confirmation is provided by a colour change to green on the addition of aqueous sodium hydroxide.

Trichloromethane and alkali; the *carbylamine test*

A confirmatory test for primary amines is provided by the *isocyanide reaction*, in which the amine is warmed with trichloromethane and ethanolic alkali. A foul smelling isocyanide or 'carbylamine' is evolved, e.g.

phenyl isocyanide

In this reaction the primary amine is believed to coordinate to dichlorocarbene, CCl_2, an electron deficient intermediate formed from trichloromethane by the action of alkali:

$$CHCl_3 + HO^- \rightarrow CCl_2 + H_2O + Cl^-$$
$$CCl_2 + RNH_2 \rightarrow RN{\equiv}C + 2Cl^- + 2H^+$$

Isocyanides are toxic, and it is preferable to test for the presence of aromatic primary amines by diazotisation. Whenever the carbylamine test is performed, the isocyanide should be destroyed before disposal by the addition of concentrated hydrochloric acid:

$$RN{\equiv}C + HCl + 2H_2O = R\overset{\oplus}{N}H_3\ Cl^- + HCOOH$$

Aromatic diazonium salts

Although the formula of the aryldiazonium ion is usually written

$Ar\overset{\oplus}{-}N\equiv N$, the positive charge is not located entirely on the penultimate nitrogen atom but is shared among both nitrogen atoms and the aromatic

ring. (The formula $\overset{\oplus}{\overbrace{Ar-N\equiv N}}$ may be considered more appropriate.) This accounts for the relative stability of aromatic diazonium ions; in aliphatic ones there is less scope for delocalisation of charge.

Most of the reactions of diazonium salts are nucleophilic substitutions. They may take place with or without the loss of nitrogen.

I. The sharing of positive charge with the adjoining carbon atom leads to nucleophilic substitution with the loss of molecular nitrogen:

$$Nu\!:^- + \;\;\overset{\oplus}{\overbrace{>\!\!C-N\equiv N}} \;=\; >\!\!C-Nu \;+\; N_2$$

II. The sharing of positive charge with the terminal nitrogen atom leads to nucleophilic substitution with the retention of nitrogen:

$$\overset{\oplus}{\overbrace{>\!\!C-N\equiv N}} \;+\; Nu\!:^- \;=\; >\!\!C-N\!=\!N-Nu$$

Although a few reactions of type I take place satisfactorily without a catalyst, most need to be catalysed by a copper(I) compound. Reactions of type II are known as *coupling reactions*. They occur with phenols and aromatic amines, and give rise to brightly coloured azo dyes.

Diazonium salts can also undergo the nucleophilic addition of hydrogen across the $N\equiv N$ bond. The reactions are brought about by certain reducing agents, and the products are substituted hydrazines.

Uncatalysed nucleophilic substitution with loss of nitrogen
Kinetic experiments indicate that these reactions proceed by an S_N1 mechanism. First, the diazonium ion decomposes, in the presence of the nucleophile, into an arylcarbonium ion and nitrogen:

$$Ar\overset{\oplus}{N}\!\equiv\!N \rightarrow Ar^+ + N_2$$

The simplest arylcarbonium ion is the phenyl cation, $C_6H_5^+$.

The carbonium ion is then rapidly attacked by the nucleophile to give the final product:

$$Nu\!:^- + Ar^+ \rightarrow Nu\!\longrightarrow\!Ar$$

The reactions are comparable with the S_N1 reactions of organic halides

(§11.5). Although aryldiazonium salts undergo these reactions more readily than aryl halides, the only nucleophiles which react without a catalyst are H_2O, ROH, $[BF_4]^-$ and reducing agents. Details are as follows.

H_2O

Diazonium salts react with water to give phenols; slowly at room temperature, but more rapidly on warming, e.g.

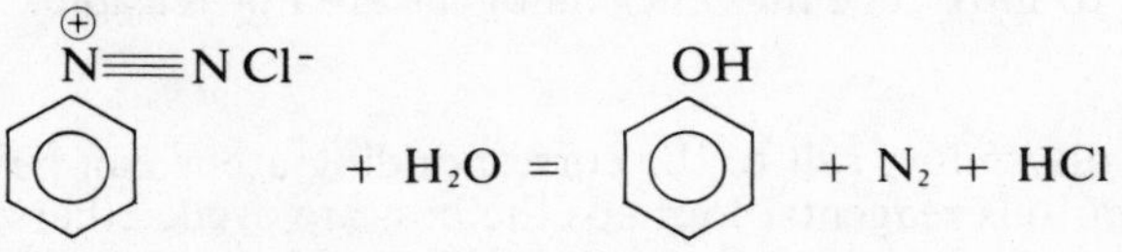

$$\overset{\overset{\oplus}{N}\equiv N\ Cl^-}{\bigcirc} + H_2O = \overset{OH}{\bigcirc} + N_2 + HCl$$

Competing reactions occur (principally the formation of an aryl chloride), the yield of phenol is low, and the method is employed only in special circumstances, e.g. when it is impracticable to make the phenol by other means.

The use of the hydroxide ion as nucleophile, instead of the water molecule, is ruled out because under alkaline conditions a diazonium ion is converted into a *diazotate ion* which, with its negative charge, is not susceptible to nucleophilic attack:

$$Ar—\overset{\oplus}{N}\equiv N + HO^- \rightleftharpoons Ar—N\equiv N—OH \rightleftharpoons Ar—N\equiv N—O^{\ominus} + H^+$$

diazonium ion   diazotate ion

ROH

On boiling with methanol or ethanol, diazonium salts react in the same way as they do with water, ethers being formed instead of phenols, e.g.

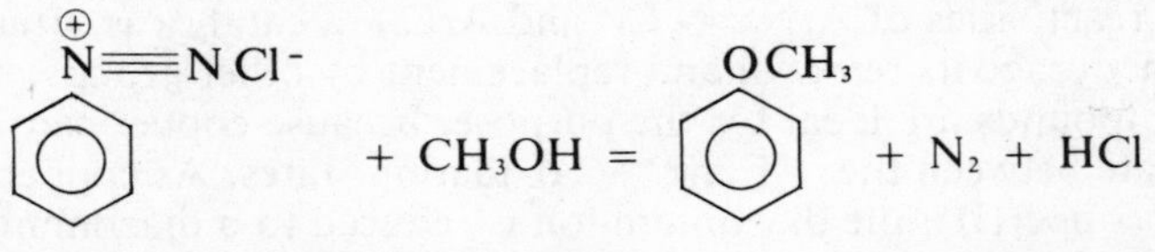

$$\overset{\overset{\oplus}{N}\equiv N\ Cl^-}{\bigcirc} + CH_3OH = \overset{OCH_3}{\bigcirc} + N_2 + HCl$$

The formation of methoxybenzene is accompanied by a small amount of benzene, due to reduction of the diazonium salt.

$[BF_4]^-$

Diazonium salts do not react with simple fluorides. In the *Schiemann reaction* the fluoride ion is introduced as a complex, by means of tetrafluoroboric(III) acid:

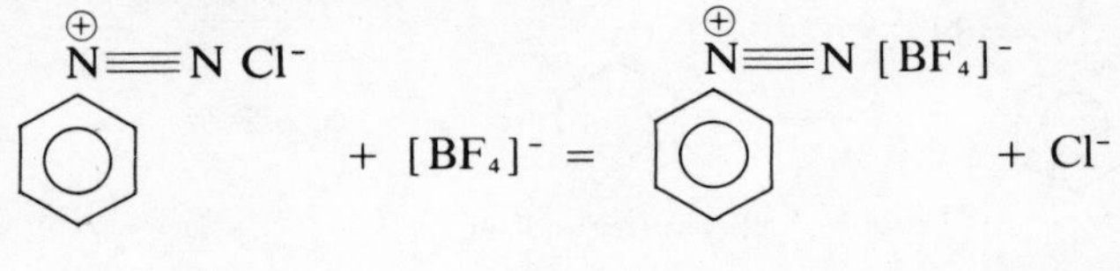

$$\overset{\overset{\oplus}{N}\equiv N\ Cl^-}{\bigcirc} + [BF_4]^- = \overset{\overset{\oplus}{N}\equiv N\ [BF_4]^-}{\bigcirc} + Cl^-$$

The insoluble tetrafluoroborate(III) is filtered off, dried, and then decomposed by gentle heat:

$$\underset{\bigcirc}{\overset{\oplus}{N}\!\equiv\!N\;[BF_4]^-} \quad\overset{F}{\underset{\bigcirc}{}} \quad=\quad \underset{\bigcirc}{} \;+\; N_2 \;+\; BF_3$$

Similar reactions to introduce the other halogens are not feasible.

Reducing agents

The reduction of a diazonium salt to the corresponding arene can be brought about by various reagents. Perhaps the best are cyclic ethers, notably 1,4-dioxan and tetrahydrofuran, although in student laboratories it is common practice to use phosphinic acid. (A copper catalyst may be included.) For example

$$\underset{\bigcirc}{\overset{\oplus}{N}\!\equiv\!N\;Cl^-} \;+\; HPH_2O_2 \;+\; H_2O \xrightarrow[\text{Cu catalyst}]{0\ °C\ (273\ K)} \underset{\bigcirc}{} \;+\; H_2PHO_3 \;+\; N_2 \;+\; HCl$$

In all such reactions a hydride ion, H^-, is effectively transferred from the reducing agent to the arylcarbonium ion (or to the aryl radical, if a copper catalyst is used; see below).

Catalysed nucleophilic substitution with loss of nitrogen (Sandmeyer reactions)

Although N_2 is a much better leaving group than, say, Cl^- or HO^- (compare the reactivities of $ArN\!\overset{\oplus}{\equiv}\!N\;Cl^-$ and $ArCl$), a catalyst is often required to encourage its removal and replacement by other groups. Copper(I) compounds are ideal for the purpose, because copper can readily alternate between the +1 and +2 oxidation states. As copper(I) is oxidised to copper(II), the diazonium ion is reduced to a diazonium free radical, which at once decomposes into an aryl radical and nitrogen. Free radicals, like carbonium ions, are readily attacked by nucleophiles:

$$\underset{\bigcirc}{\overset{\oplus}{N}\!\equiv\!N} \;+\; Cu^+ \;\longrightarrow\; \underset{\bigcirc}{N\!=\!\overset{\bullet}{N}} \;+\; Cu^{2+}$$

$$\underset{\bigcirc}{N\!=\!\overset{\bullet}{N}} \;\longrightarrow\; \underset{\bigcirc}{\overset{\bullet}{}} \;+\; N_2$$

phenyl free radical

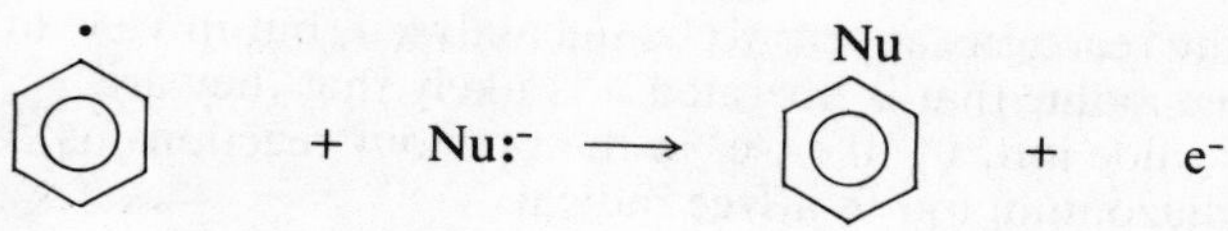

$$\text{C}_6\text{H}_5^{\bullet} \quad + \quad \text{Nu:}^- \quad \longrightarrow \quad \text{C}_6\text{H}_5\text{-Nu} \quad + \quad e^-$$

$$\text{Cu}^{2+} \quad + \quad e^- \quad \longrightarrow \quad \text{Cu}^+$$

Support for this mechanism is provided by the formation of by-products such as biphenyl, which must result from the joining together of free radicals, e.g.

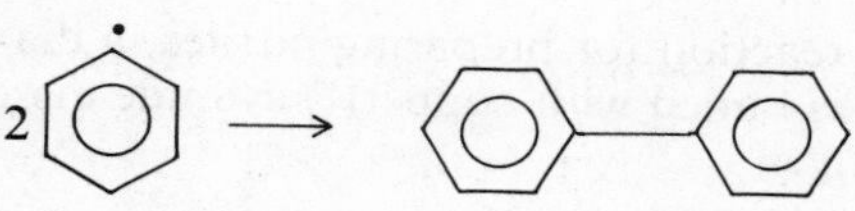

$$2\,\text{C}_6\text{H}_5^{\bullet} \quad \longrightarrow \quad \text{C}_6\text{H}_5\text{-C}_6\text{H}_5$$

Nucleophiles which can be introduced by means of Sandmeyer reactions include Cl^-, Br^- and CN^- (catalysed by Cu^{I}), and I^- (self-catalysed).

Cl^-

In the Sandmeyer reaction for the introduction of Cl^-, the diazonium chloride, at 0 °C (273 K), is added to a solution of copper(I) chloride in concentrated hydrochloric acid, and the mixture is then boiled until there is no further evolution of nitrogen, e.g.

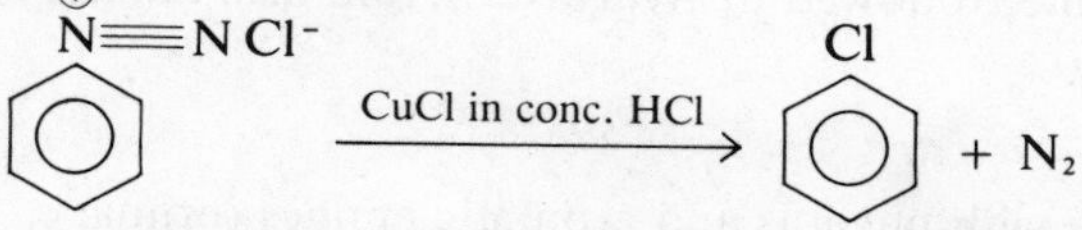

$$\text{C}_6\text{H}_5\text{-}\overset{\oplus}{\text{N}}{\equiv}\text{N}\ \text{Cl}^- \quad \xrightarrow{\text{CuCl in conc. HCl}} \quad \text{C}_6\text{H}_5\text{-Cl} \quad + \quad \text{N}_2$$

Metallic copper, too, is suitable as a catalyst, probably because of the thin film of copper(I) oxide that exists on its surface.

Br^-

The bromide ion can be introduced in a similar fashion to the chloride ion, by treating the diazonium bromide with copper(I) bromide dissolved in hydrobromic acid.

I^-

For the preparation of aryl iodides it is necessary only to boil the diazonium chloride solution with aqueous potassium iodide, e.g.

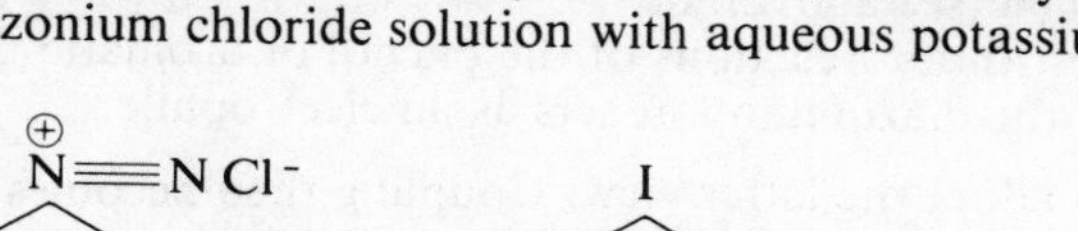

$$\text{C}_6\text{H}_5\text{-}\overset{\oplus}{\text{N}}{\equiv}\text{N}\ \text{Cl}^- \quad + \quad \text{KI} \quad = \quad \text{C}_6\text{H}_5\text{-I} \quad + \quad \text{N}_2 \quad + \quad \text{KCl}$$

Superficially, the reactions appear to be uncatalysed, but in view of the amount of free iodine that is liberated it is likely that they are catalysed by the iodide ion. I^-, like Cu^+ in the previous reactions, is able to reduce a diazonium ion to a free radical:

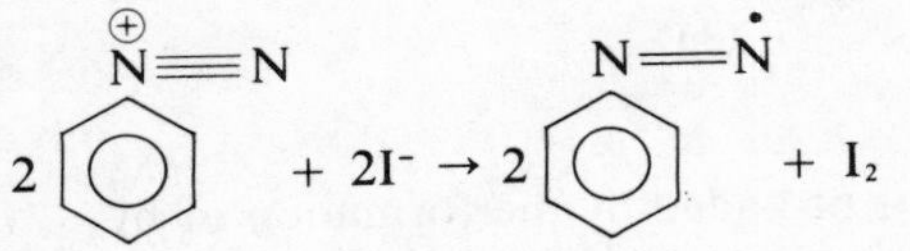

CN^-

In the Sandmeyer reaction for preparing nitriles, a diazonium hydrogensulphate is boiled with copper(I) cyanide dissolved in aqueous sodium cyanide, e.g.

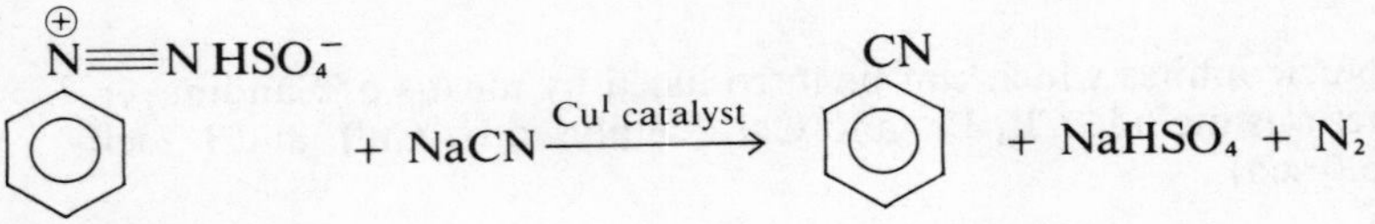

benzonitrile

The reaction should not be performed in student laboratories because of the toxicity of cyanides.

From nitriles can be made carboxylic acids (by hydrolysis), aldehydes (by reduction to an imine, followed by hydrolysis), and also amines (by the Mendius reduction).

Coupling reactions

Diazonium salts couple with phenols and aromatic amines (primary, secondary and tertiary) to give brightly coloured *azo dyes*. Their colour is usually at the red end of the spectrum; i.e. it may be red, orange or yellow, but seldom blue or green. The colour is due to the presence of the azo group, $N{=}N$, referred to as a *chromophoric* (literally 'colour bearing') group. It has the effect of absorbing the blue components of white light, so that only the red and yellow ones are transmitted or reflected.

Coupling reactions can equally well be regarded as:

(i)   nucleophilic substitution reactions of diazonium salts, in which nitrogen is retained (see above); or

(ii) electrophilic substitution reactions of the phenol or aromatic amine, in which the diazonium ion acts as an electrophile.

It is conventional to adopt the latter view. Coupling then becomes a typical electrophilic substitution reaction of the aromatic ring (cf. nitration, chlorination and Friedel–Crafts reactions) and not a special type of reaction.

Diazonium ions are relatively weak electrophiles because of the low concentration of positive charge on the terminal nitrogen atom. Benzene itself does not react with diazonium ions because the electron density of the ring is insufficient. However, in phenols and amines the electron density is increased by the mesomeric effect (§1.4). As a result, phenols and amines are much more reactive than benzene in all electrophilic substitutions.

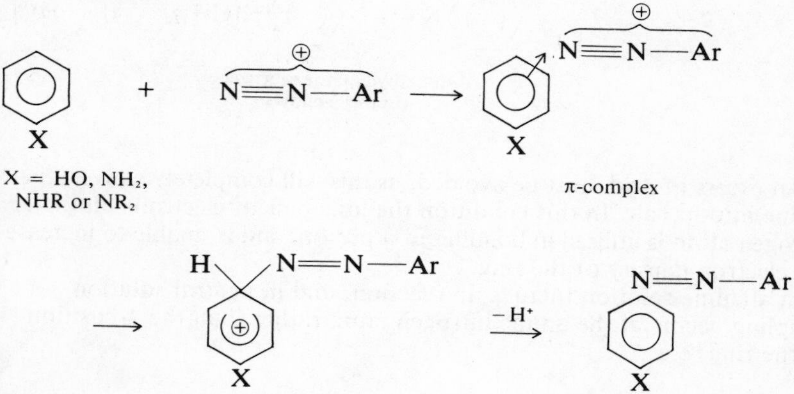

$$X = HO, NH_2, NHR \text{ or } NR_2$$

π-complex

σ-complex   azo dye

Coupling occurs almost entirely at the 4-position, unless this is blocked, in which case coupling is observed at the 2-position.

The pH of the solution is critical. Phenols couple most rapidly in alkaline solution at a pH of 9−10, because the negative charge of the phenoxide ion, which is present under these conditions, helps to increase the electron density of the ring still further. (Strongly alkaline solution should be avoided, for this converts the diazonium ion into the diazotate ion (see above), which does not couple.) For example

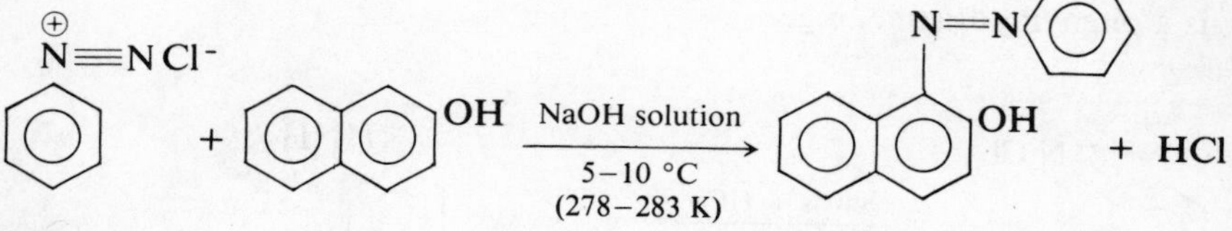

2-naphthol   1-phenylazo-2-naphthol (red)

Best results with amines are obtained in weakly acidic solution. The optimum pH is ~5, achieved by the use of a buffer solution of ethanoic acid and sodium ethanoate, e.g.

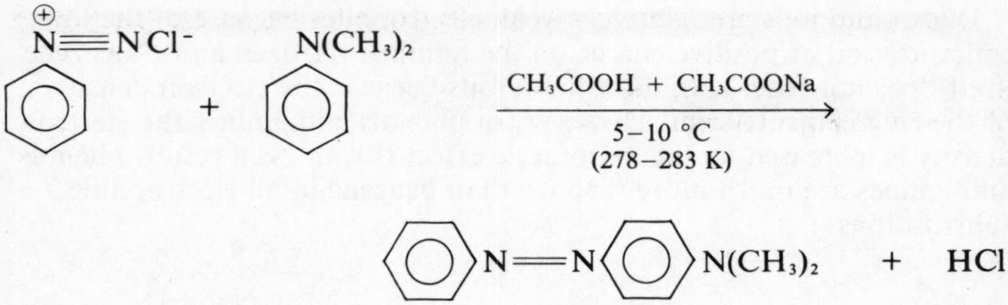

$$\overset{\oplus}{N}\!\!\equiv\!\!N\,Cl^- \quad + \quad N(CH_3)_2 \quad \xrightarrow[\substack{5-10\ °C \\ (278-283\ K)}]{CH_3COOH\ +\ CH_3COONa}$$

$$\langle O \rangle N\!=\!N \langle O \rangle N(CH_3)_2 \quad + \quad HCl$$

4-dimethylaminoazobenzene
(Butter Yellow)

An excess of acid must be avoided, as this will completely convert the amine into its salt. In this condition the lone pair of electrons on the nitrogen atom is utilised in bonding to a proton, and is unable to increase the electron density of the ring.

In alkaline solution there is no reaction, and in neutral solution coupling occurs at the amine nitrogen atom rather than the 4-position of the ring, e.g.

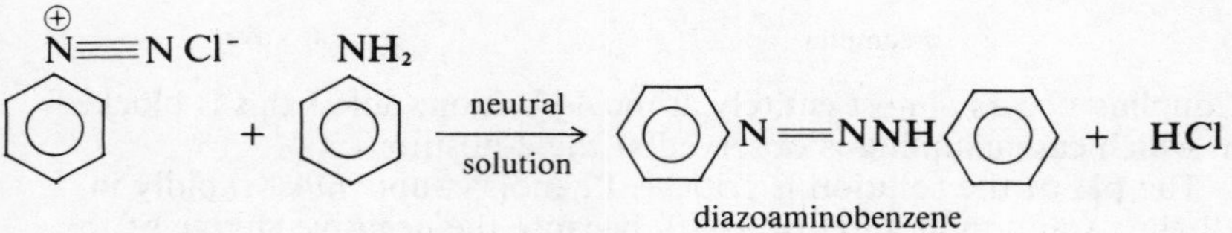

$$\overset{\oplus}{N}\!\!\equiv\!\!N\,Cl^- \quad + \quad NH_2 \quad \xrightarrow[\text{solution}]{\text{neutral}} \quad \langle O \rangle N\!=\!NNH \langle O \rangle \quad + \quad HCl$$

diazoaminobenzene

Nucleophilic addition

We have already seen that when a diazonium salt is reduced with phosphinic acid or a cyclic ether, nitrogen is eliminated and an arene is formed. However, when certain other reductants are used, notably tin(II) chloride or sodium sulphite, nitrogen is retained and the product is a phenylhydrazine, e.g.

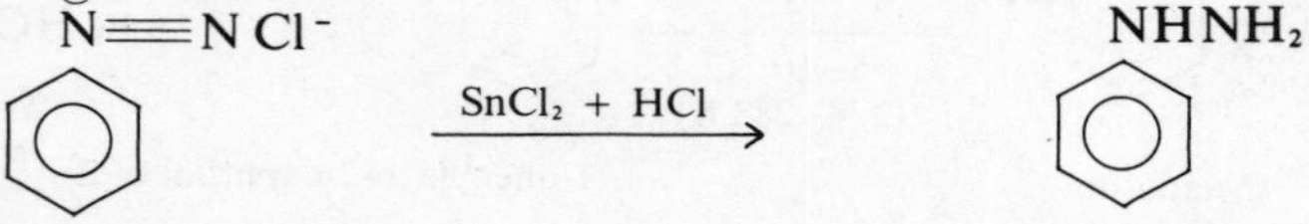

$$\overset{\oplus}{N}\!\!\equiv\!\!N\,Cl^- \quad \xrightarrow{SnCl_2\ +\ HCl} \quad NHNH_2$$

phenylhydrazine (as the chloride)

Chapter 15

The C≡N bond in nitriles and isocyanides

Data on the C≡N bond

covalent bond length/nm	0.116
bond dissociation enthalpy/kJ mol^{-1} at 298 K	890

The principal compounds to possess a carbon–nitrogen triple bond are known as *nitriles* or *cyanides*. Isomeric with them are the *isocyanides*, in which the *nitrogen* atom is joined to the remainder of the molecule.

R—C≡N   R—N≡C   (often written as R—N⁺≡C⁻)

  alkyl nitrile   alkyl isocyanide

15.1 Nomenclature of nitriles and isocyanides

Nitriles

IUPAC recognises the following three systems for the naming of nitriles.

(i) Substitutive names, usually preferred for aliphatic compounds, are formed by adding the suffix '-nitrile', '-dinitrile', etc to the name of the parent hydrocarbon, e.g.

    CH$_3$CN   NCCH$_2$CH$_2$CN

      ethanenitrile   butanedinitrile

(ii) According to another system, nitriles are named after the carboxylic acids to which they are hydrolysed. The ending '-ic acid' or '-oic acid' of the name of the carboxylic acid is substituted by the suffix '-onitrile', e.g.

CH₃CN   C₆H₅CN

acetonitrile   benzonitrile

(iii) Radicofunctional names can be formed by writing the word 'cyanide' after the name of the appropriate radical, e.g.

CH₃CH₂CN   ethyl/cyanide

Like the radicofunctional names of alcohols and halides, they are now little used.

Isocyanides

Such compounds are given radicofunctional names, e.g.

C₆H₅NC   phenyl isocyanide

IUPAC does not recommend naming them as isonitriles or carbylamines.

15.2 Formation of the C≡N bond

Analogy with carbon–carbon multiple bonds suggests the use of elimination methods for establishing the C≡N bond. The principal route available — the dehydration of amides — is an elimination reaction.

Nitriles and isocyanides can also be made by substitution reactions between alkyl halides and metallic cyanides, notably potassium cyanide and sodium cyanide. In such compounds the C≡N bond already exists in the form of the cyanide ion, CN⁻.

Elimination methods

Water can be eliminated from amides at elevated temperatures with the help of dehydrating agents, namely Lewis acid catalysts such as phosphorus(V) oxide or phosphoryl chloride, e.g.

$$CH_3CONH_2 \xrightarrow{P_4O_{10}} CH_3CN + H_2O$$

Substitution methods

The cyanide ion, from sodium cyanide or potassium cyanide, has a lone pair of electrons on both the carbon and nitrogen atoms. In substitution reactions with alkyl halides or arylalkyl halides (such as (chloromethyl)benzene) the ion may coordinate either through the carbon atom, to give a nitrile, or via nitrogen, to give an isocyanide:

$$:N \equiv C: + \overset{|}{\underset{/}{C}} - X \xrightarrow[\text{or aqueous ethanol}]{\text{reflux in ethanol}} N \equiv C \rightarrow \overset{|}{\underset{/}{C}} + X^-$$

nitrile

or

$$:C \equiv N: + \overset{|}{\underset{/}{C}} - X \longrightarrow C \equiv N \rightarrow \overset{|}{\underset{/}{C}} + X^-$$

isocyanide

The use of silver cyanide, rather than sodium cyanide or potassium cyanide, leads mainly to the formation of an isocyanide, because silver cyanide has a covalent structure and behaves as $Ag-C \equiv N$. It is therefore unable to coordinate via carbon.

When a primary alkyl halide is treated with an ionic cyanide the product is almost entirely a nitrile, but when a tertiary alkyl halide is used the product is predominantly an isocyanide. A secondary alkyl halide gives a mixture of a nitrile and an isocyanide.

These facts are explained by the theory that a primary alkyl halide reacts by an S_N2 mechanism, via a transition state, whereas a tertiary alkyl halide reacts by an S_N1 mechanism, via a carbonium ion (§11.5). (With a secondary alkyl halide the reaction is partly S_N1 and partly S_N2.)

$$:C \equiv N: + \overset{|}{\underset{/}{C}} - X$$

S_N2 route

S_N1 route

$$\overset{\delta^-}{NC} \cdots \overset{|}{\underset{/}{\overset{\delta^+}{C}}} \cdots \overset{\delta^-}{X} \longrightarrow NC - \overset{|}{\underset{/}{C}} + X^-$$

transition state

nitrile

$$CN^- \quad \overset{|}{\underset{/}{\overset{\oplus}{C}}} \quad X^- \longrightarrow CN - \overset{|}{\underset{/}{C}} + X^-$$

carbonium ion

isocyanide

In the formation of the transition state (S_N2 route) the cyanide ion coordinates to the alkyl halide through carbon rather than nitrogen. Carbon is said to have the 'greater nucleophilicity'. The reason is not entirely clear, but it may be that the nucleophilicity of the nitrogen atom is reduced by hydrogen bonding with the solvent. The product is thus a nitrile.

S_N1 reactions are less selective, for the cyanide ion is strongly attracted to the carbonium ion and is able to donate an electron pair from either its carbon or nitrogen atom. The product is thus a mixture of nitrile and isocyanide. The isocyanide tends to predominate because, in linking to a carbonium ion, CN^- coordinates preferentially through nitrogen, the more electronegative atom.

15.3 General properties of nitriles and isocyanides

Nitriles

The electronegativity difference between carbon and nitrogen leads to the π electrons of the $C\equiv N$ bond being concentrated in the vicinity of the nitrogen atom: $R-C\equiv\overset{\frown}{N}$ (cf. carbonyl compounds). Intermolecular attraction therefore occurs, causing the boiling temperatures of nitriles to be substantially higher than those of hydrocarbons of similar relative molecular mass, e.g.

	Relative molecular mass	Boiling temperature/°C (K)
CH_3CN	41	81.6 (354.8)
$CH_3CH_2CH_3$	44	-42.2 (231.0)
C_6H_5CN	103	191 (464)
$C_6H_5CH_2CH_3$	106	136 (409)

Nitriles have reasonably pleasant odours, and, although toxic, are less poisonous than inorganic cyanides. Because hydrogen bonding can occur, the lower nitriles are soluble in water. Solutions are neutral.

Isocyanides

Isocyanides are poisonous, foul smelling liquids with boiling temperatures below those of the isomeric nitriles. Hydrogen bonding with water is impossible, because the nitrogen atom carries a partial *positive* charge (see below), and the water solubility of isocyanides is low.

15.4 Chemical properties of nitriles and isocyanides

Nitriles and isocyanides have very different chemical properties and cannot be studied together. (We shall concentrate on nitriles.) The difference stems from the fact that the charge separation displayed by the $C\equiv N$ bond is fundamentally different in the two types of compound:

$$R-\overset{\delta^+}{C}\equiv\overset{\delta^-}{N} \qquad R-\overset{\delta^+}{N}\equiv\overset{\delta^-}{C}$$

nitrile   isocyanide

In nitriles, charge separation is similar to that in carbonyl compounds, and this leads to a similarity in chemistry. Nitriles may be described as *nitrogen-analogous carbonyl compounds*. Their 'carbonyl activity', however, is low, and the only reactions of importance are hydrolysis and reduction.

In isocyanides, because of the donation of a lone pair of electrons from nitrogen to carbon, the carbon atom is *negatively* charged. Nucleophilic attack on this atom is therefore impossible, as shown by the fact that isocyanides (unlike nitriles) are not attacked by bases.

Hydrolysis of nitriles

Hydrolysis may be conducted in either acidic or alkaline solution, e.g. with concentrated hydrochloric acid or 30 per cent aqueous sodium hydroxide. With water alone the reaction is very slow. The initial product is an *imidic acid*, which immediately rearranges into an amide. Further hydrolysis, of the amide, leads to the formation of a carboxylic acid and ammonia:

$$R—C\equiv N + H_2O \rightarrow R—C\overset{\textstyle OH}{\underset{\textstyle NH}{<}} \rightarrow R—C\overset{\textstyle O}{\underset{\textstyle NH_2}{<}} \xrightarrow{H_2O} R—C\overset{\textstyle O}{\underset{\textstyle OH}{<}} + NH_3$$

| nitrile | imidic acid | amide | carboxylic acid |

The hydrolysis of nitriles involves nucleophilic addition across the $C\equiv N$ bond, and corresponds closely with the hydrolysis of esters and amides in which there is a similar addition to the $C=O$ bond. Thus, in acidic solution the reagent is the H_2O molecule; protons, from the acid, serve as a catalyst. Under alkaline conditions the reagent is not the H_2O molecule but the HO^- ion.

Acidic hydrolysis

The addition of a proton to the nitrile increases the positive charge on the carbon atom so that it can more easily be attacked by a molecule of water. Loss of the proton catalyst completes the addition. (Cf. acid catalysed addition of molecular nucleophiles to carbonyl compounds; mechanism IV, §13.6.)

$$R—\overset{\delta^+}{C}\equiv\overset{\delta^-}{N} + H^+ \rightleftharpoons R—\overset{\oplus}{C}=NH \overset{H_2O}{\rightleftharpoons} R—C\overset{\textstyle \overset{\oplus}{O}H_2}{\underset{\textstyle NH}{<}} \overset{-H^+}{\rightleftharpoons} R—C\overset{\textstyle OH}{\underset{\textstyle NH}{<}}$$

$$\rightleftharpoons R—C\overset{\textstyle O}{\underset{\textstyle NH_2}{<}}$$

Alkaline hydrolysis

HO^- coordinates to the carbon atom of the nitrile to give the anion of an imidic acid. A proton, from the water, then completes the addition to give the imidic acid itself:

$$R—\overset{\delta^+}{C}\equiv\overset{\delta^-}{N} + HO^- \rightleftharpoons R—C\overset{\textstyle OH}{\underset{\textstyle N^\ominus}{<}} \overset{H^+}{\rightleftharpoons} R—C\overset{\textstyle OH}{\underset{\textstyle NH}{<}} \rightleftharpoons R—C\overset{\textstyle O}{\underset{\textstyle NH_2}{<}}$$

Reduction of nitriles

Nitriles can be reduced; first to aldimines, and then to primary amines:

$$RC\equiv N \xrightarrow{H_2} RCH=NH \xrightarrow{H_2} RCH_2NH_2$$

nitrile   aldimine   primary amine

In the presence of water aldimines are immediately hydrolysed to give aldehydes, so that in practice the reduction product of a nitrile is either an aldehyde or an amine.

Aldehydes are formed in the *Stephen reaction*, in which the nitrile is reduced with tin(II) chloride and hydrochloric acid.

With hydrogen and a catalyst of nickel, platinum or palladium, or with a non-noble metal and a protic solvent, or with lithium tetrahydridoaluminate(III) in ether, the product is a primary amine. The reaction is known as the *Mendius reduction*. Except when lithium tetrahydridoaluminate(III) is the reductant, a certain amount of secondary amine is also produced by a reaction between the aldimine and the primary amine already formed.

Chapter 16

The C—S bond in sulphonic acids

Data on the C—S bond

covalent bond length/nm	0.181
bond dissociation enthalpy/kJ mol⁻¹ at 298 K	272

Sulphur in organic compounds can have valencies of 2, 4 or 6. Divalent sulphur takes the place of oxygen in compounds such as thiols and sulphides. Thiols (literally 'thio-alcohols') have a chemical as well as a structural similarity to alcohols, while sulphides are relatively inert compounds related to ethers. Tetravalent sulphur serves to replace carbon in such compounds as sulphoxides and sulphinic acids, which bear a structural resemblance to ketones and carboxylic acids respectively (see below). The hexavalent state is the most important, and is encountered in sulphonic acids and their derivatives. Although there is no precise structural correspondence, sulphonic acids bear a strong chemical resemblance to carboxylic acids.

Organic sulphur compounds fall into the two oxidation sequences shown on p. 326.

Sulphinic acids, sulphonic acids, sulphoxides and sulphones all contain at least one sulphur–oxygen double bond, in which there is charge

separation of the type $\overset{\delta^+}{S}=\overset{\delta^-}{O}$; cf. $\overset{\delta^+}{C}=\overset{\delta^-}{O}$. Such compounds therefore take part in nucleophilic addition reactions, and may be regarded as hetero-analogous carbonyl compounds.

We shall restrict further discussion to sulphonic acids and their derivatives.

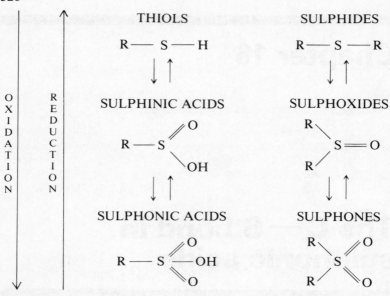

16.1 Nomenclature of sulphonic acids

Sulphonic acids, both aliphatic and aromatic, are given substitutive names, e.g.

CH₃CH(SO₃H)CH₂CH₂CH₂CH₃   2-hexanesulphonic acid

C₆H₅SO₃H   benzenesulphonic acid

16.2 Formation of the C—S bond in sulphonic acids

Sulphonic acids are most simply prepared by the uncatalysed sulphonation of the appropriate hydrocarbon. Arenes readily undergo electrophilic substitution with concentrated sulphuric acid, e.g.

$$C_6H_6 \ + \ H_2SO_4 \xrightarrow[\text{(353 K)}]{80\ °C} C_6H_5SO_3H \ + \ H_2O$$

The reaction is reversible, and desulphonation can be effected by means of superheated steam (§10.2).

    Alkanes are not subject to electrophilic substitution, but only to free radical substitution under more drastic conditions, and sulphonation requires the use of fuming sulphuric acid containing at least 30 per cent of sulphur trioxide. Attack is concentrated at tertiary and secondary hydrogen atoms, e.g.

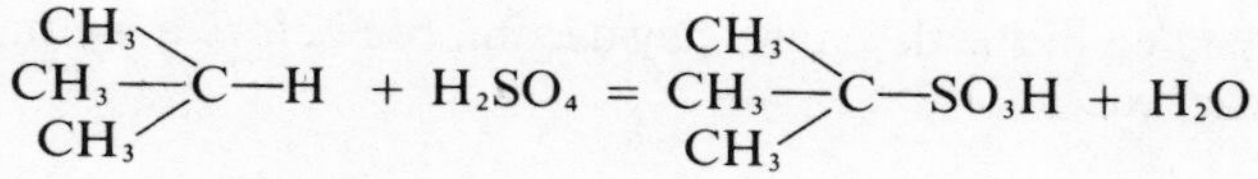

$$CH_3 \diagdown$$
$$CH_3 \!-\!\!-\!C\!-\!H + H_2SO_4 = CH_3 \!-\!\!-\!C\!-\!SO_3H + H_2O$$
$$CH_3 \diagup \qquad\qquad\qquad CH_3 \diagup$$

2-methylpropane   2-methyl-2-propanesulphonic acid

16.3 General properties of sulphonic acids

Sulphonic acids are either syrupy liquids or crystalline solids of low
melting temperature. (Benzenesulphonic acid has a melting temperature
of 50−51 °C (323−324 K).) They have a very high solubility in water.
Because the sulphonic acid group is so strongly water attracting,
sulphonation of drugs and dyes is often carried out to make them
water-soluble and hence more efficacious.

16.4 Chemical properties of sulphonic acids

To obtain a complete understanding of the chemistry of sulphonic
acids, we need to consider three covalent bonds, namely S—OH, S=O
and C—S.

Reactions of the S—OH bond (including the O—H bond)

As with alcohols and carboxylic acids, the hydroxyl group is responsible
for acidic and basic properties.

Acidity

Sulphonic acids are extremely strong acids — about as strong as
sulphuric acid — on account of the inductive effect of *two* S=O bonds:

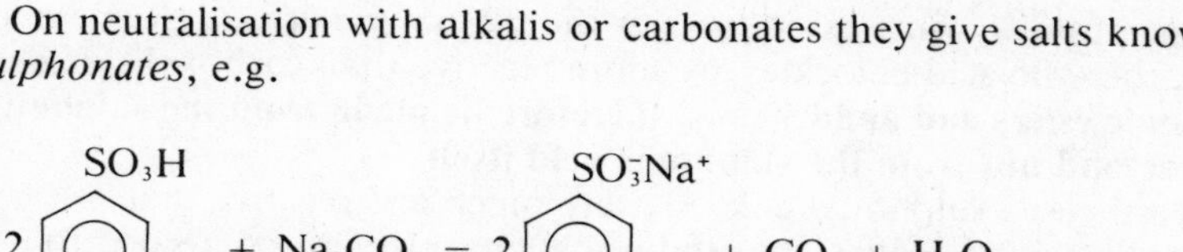

$$Ar\!-\!S\!\!\underset{O}{\overset{O}{\lessgtr}}\!\!O\!-\!H \;\rightleftharpoons\; \left[Ar\!-\!S\!\!\underset{O}{\overset{O}{\lessgtr}}\!\!O \right]^{-} + H^{+}$$

On neutralisation with alkalis or carbonates they give salts known as
sulphonates, e.g.

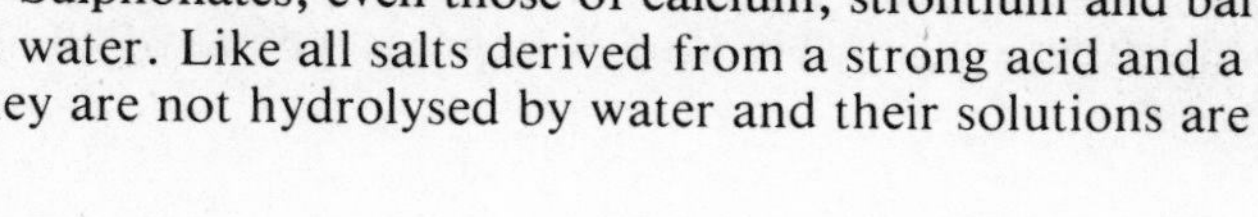

$$2\;\underset{}{\bigcirc}^{SO_3H} + Na_2CO_3 = 2\;\underset{}{\bigcirc}^{SO_3^{-}Na^{+}} + CO_2 + H_2O$$

sodium benzenesulphonate

Sulphonates, even those of calcium, strontium and barium, are soluble
in water. Like all salts derived from a strong acid and a strong base,
they are not hydrolysed by water and their solutions are neutral. When

sulphonates are heated in the dry state they decompose before a melting temperature is reached.

Basicity

Sulphonic acids, like carboxylic acids, can coordinate to phosphorus pentachloride via a lone pair of electrons on the hydroxyl oxygen atom (§13.8). As a result, the HO group is replaced by an atom of chlorine, so that the sulphonic acid is converted into its acid chloride, e.g.

benzenesulphonyl chloride

Reactions of the S=O bond

The electronegativity of sulphur is the same as that of carbon (2.5 on the Pauling scale), so that there is a similar charge separation in the S=O and C=O bonds. From our knowledge of carboxylic acids, we might expect sulphonic acids to take part in nucleophilic addition reactions across the S=O bond, followed by the elimination of water, but this does not happen. The carbonyl activity of the S=O bond is very low. The reason is that the mesomeric effect (§1.4) is more pronounced in sulphonic acids than in carboxylic acids. There is more scope for delocalisation of π electrons because more atoms are involved, namely one sulphur and three oxygen. A tripod shaped delocalised π orbital is set up, which encompasses these four atoms and almost completely shields the partial positive charge on the sulphur atom.

Sulphonic acids do not react with alcohols, ammonia or amines (except to give salts), and are not susceptible to reduction. Sulphonyl chlorides, however, are able to react, because the −I effect of the chlorine atom to some extent offsets the mesomeric effect. (Remember that carboxylic acid chlorides are more reactive than carboxylic acids.) **Sulphonic esters and amides must therefore be made from the sulphonyl chloride and not from the sulphonic acid itself.**

Nevertheless, sulphonyl chlorides are much less reactive than carboxylic acid chlorides, and sodium hydroxide is often needed to assist in the elimination of HCl.

H_2O

Sulphonyl chlorides are relatively resistant to hydrolysis, and some can even be recrystallised from water. Hydrolysis requires the use of alkali, e.g.

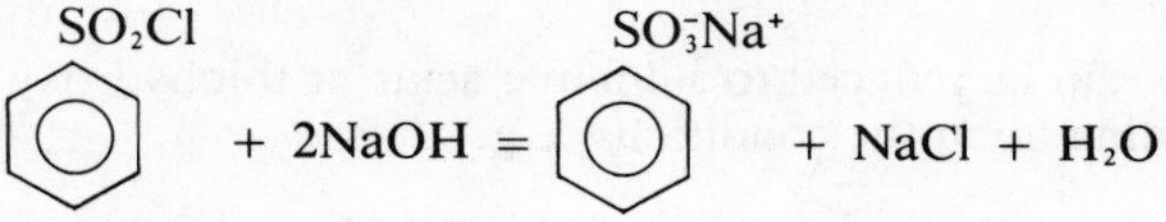

$$\underset{\text{SO}_2\text{Cl}}{\bigcirc} + 2\text{NaOH} = \underset{\text{SO}_3^-\text{Na}^+}{\bigcirc} + \text{NaCl} + \text{H}_2\text{O}$$

ROH or ArOH

Sulphonyl chlorides react with alcohols or phenols to give *sulphonic esters*, e.g.

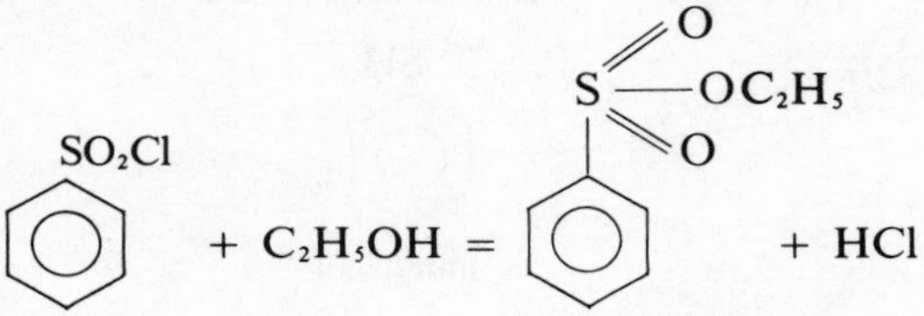

$$\underset{\text{SO}_2\text{Cl}}{\bigcirc} + \text{C}_2\text{H}_5\text{OH} = \overset{\displaystyle S \overset{O}{\underset{O}{\diagup}} -\text{OC}_2\text{H}_5}{\bigcirc} + \text{HCl}$$

ethyl benzenesulphonate

The reaction is best accomplished by shaking the reactants with aqueous sodium hydroxide at room temperature. (This is an example of a Schotten–Baumann reaction, §13.11.)

Sulphonic esters, because they are crystalline solids with well defined melting temperatures, are useful derivatives for assisting the identification of alcohols and phenols. Predictably, they are hydrolysed to sulphonates and alcohols on refluxing with alkali.

NH₃ or amines

Sulphonamides are formed when sulphonyl chlorides react with concentrated aqueous ammonia or primary or secondary amines, e.g.

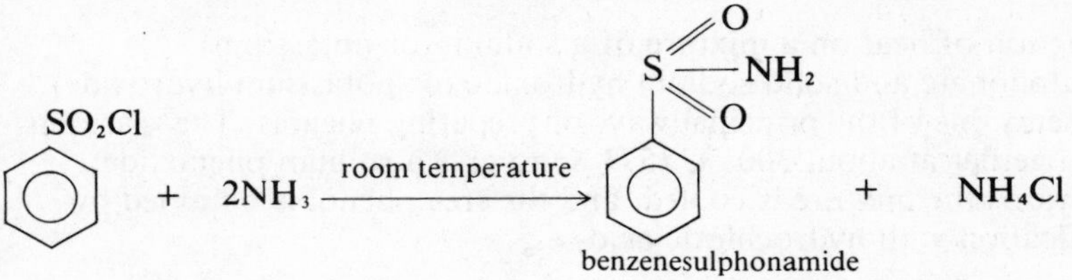

$$\underset{\text{SO}_2\text{Cl}}{\bigcirc} + 2\text{NH}_3 \xrightarrow{\text{room temperature}} \overset{\displaystyle S \overset{O}{\underset{O}{\diagup}} -\text{NH}_2}{\bigcirc} + \text{NH}_4\text{Cl}$$

benzenesulphonamide

$$\underset{\text{SO}_2\text{Cl}}{\bigcirc} + \underset{\text{NH}_2}{\bigcirc} \xrightarrow[\substack{\text{shake at room} \\ \text{temperature}}]{\text{aqueous NaOH}} \overset{\displaystyle S \overset{O}{\underset{O}{\diagup}} -\text{NH}\bigcirc}{\bigcirc} + \text{HCl} \\ \text{(absorbed by NaOH)}$$

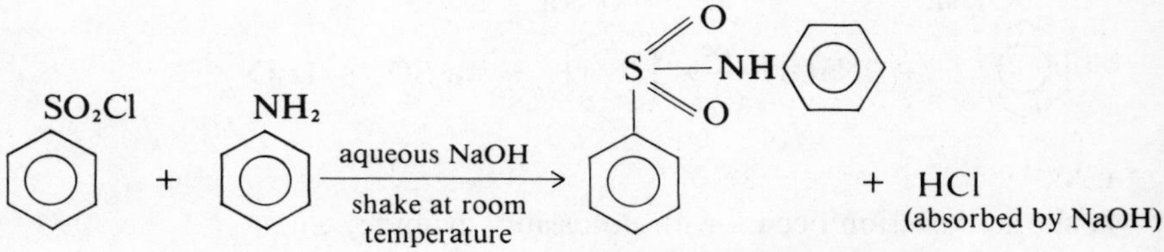

N-phenylbenzenesulphonamide

As may be expected, all sulphonamides can be hydrolysed to give sulphonic acids and ammonia (or amines) by refluxing with acid or alkali.

Reducing agents

Sulphonyl chlorides can be reduced to sulphinic acids or thiohydroxy compounds, depending upon the conditions, e.g.

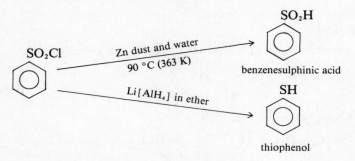

Reactions of the C—S bond

Arylsulphonates, $ArSO_3^-M^+$ (where M = Na or K), take part in several nucleophilic substitution reactions:

$$Nu:^- + \ \overset{\diagup}{\underset{\diagdown}{C}}-SO_3^- \rightarrow \ \overset{\diagup}{\underset{\diagdown}{C}}-Nu + SO_3^{2-}$$

Although the reactions require a high temperature, they occur more readily than the corresponding reactions of aryl halides. The C—S bond is relatively weak (see data), and the sulphite ion is fairly easily detached from the aromatic ring.

HO^-

The action of heat on a mixture of a sodium (or potassium) arylsulphonate and solid sodium hydroxide (or potassium hydroxide) represents one of the principal ways of preparing phenols. The reactants fuse together at about 300 °C (573 K) to give a sodium phenoxide; afterwards the mixture is cooled, and the free phenol is obtained by acidification with hydrochloric acid, e.g.

$$\text{(} SO_3Na \text{)} + 2NaOH \xrightarrow{\text{fuse}} \text{(} O^-Na^+ \text{)} + Na_2SO_3 + H_2O$$

CN^-

A similar reaction occurs with potassium cyanide, e.g.

$$\text{(} SO_3K \text{)} + KCN \xrightarrow{\text{strong heat}} \text{(} CN \text{)} + K_2SO_3$$

benzonitrile

In this case, however, the reactants do not fuse together and the yield of nitrile is relatively low.

NH₂

With sodium amide or potassium amide, a primary amine is formed in moderately good yield, e.g.

16.5 Soapless detergents

The word *detergent* means 'cleansing agent', and includes both soap and soapless detergent. All such materials function in essentially the same way, their molecules or ions having one or more polar portions which are water attracting, and one or more non-polar portions which attract grease.

Soaps are by no means ideal, for in hard water they are inefficient and form an insoluble scum of calcium and magnesium salts. Soapless detergents do not suffer from this disadvantage, because their calcium and magnesium salts are water-soluble. They fall into three categories.

Non-ionic detergents

Many liquid detergents are of this type. They are prepared from a long chain alcohol and oxirane according to the following equation:

$$ROH + n\overset{O}{CH_2-CH_2} = R(OCH_2CH_2)_nOH$$

The ether linkages and the terminal hydroxyl group cause a certain polarity, while the long chain alkyl radical (which may include an aromatic ring) is non-polar.

Anionic detergents

Such materials are so-called because they owe their detergent action to an anion. They are sodium salts of acids with a high relative molecular mass. Sodium salts of carboxylic acids are, of course, the familiar soaps: sodium salts of sulphonic acids constitute most of the solid soapless detergents in use today.

The hydrocarbon portion of the sulphonic acid needs to be fairly large to be sufficiently grease attracting, as does the hydrocarbon portion of a soap. It needs to be both aliphatic, for the best detergent action, and aromatic, for ease of sulphonation. For these reasons such detergents are usually sodium alkylarylsulphonates. A typical hydrocarbon used in their manufacture is 5-phenyldodecane. On reaction with sulphuric acid, this compound undergoes sulphonation at the 4-position on the ring:

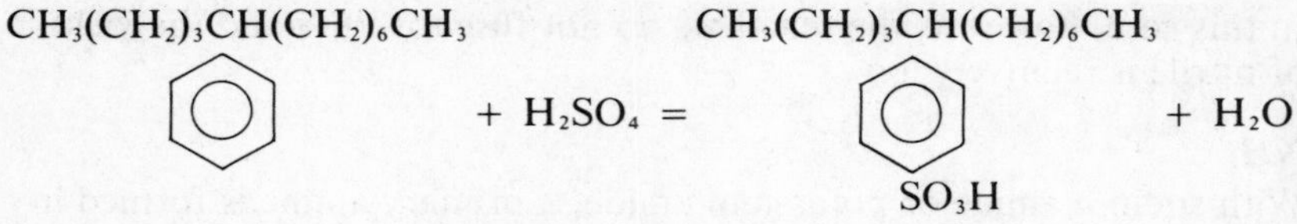

$$CH_3(CH_2)_3CH(CH_2)_6CH_3 \qquad\qquad CH_3(CH_2)_3CH(CH_2)_6CH_3$$

$$\bigcirc \qquad + \ H_2SO_4 \ = \qquad \bigcirc\!\!-\!\!SO_3H \qquad + \ H_2O$$

The sodium salt of this sulphonic acid is an anionic detergent which has the advantage of being *biodegradable*, i.e. it can be broken down by bacteria at sewage treatment works. Earlier anionic detergents, made by sulphonating tetrapropylenebenzene, were not biodegradable and have now been withdrawn from the UK market.

Another group of anionic detergents are the sodium alkylsulphates, which are closely related to conventional soaps. Instead of consisting of sodium salts of long chain carboxylic acids, they are the sodium salts of long chain sulphuric acids:

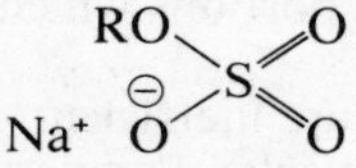

$$Na^+ \ \ {}^{\ominus}O\!\!-\!\!\underset{\displaystyle O}{\overset{\displaystyle O}{\underset{\displaystyle \|}{\overset{\displaystyle \|}{S}}}}\!\!-\!\!OR$$

They are made from the same materials as soaps, e.g. olive oil, by treatment with concentrated sulphuric acid, followed by sodium hydroxide. 'Turkey-red oil', made by sulphating castor oil, is particularly well known, but in general these materials have been superseded by the alkylarylsulphonates.

Cationic detergents

These substances are quaternary ammonium salts of high relative molecular mass. They are little used as detergents, but find some application in germicidal products.

Appendix
IUPAC nomenclature — rules of priority

In substitutive nomenclature, the only common groups which are cited only as prefixes are Cl (chloro), Br (bromo), I (iodo), NO_2 (nitro) and NO (nitroso). Other groups can be cited as a prefix or a suffix, but if only one is present it must be represented by a suffix. Thus, for CH_3OH the name methanol is correct; not hydroxymethane.

If more than one group is present, only one, termed the *principal group*, can be cited as a suffix. All the others appear in the name as prefixes. When applying this rule to difunctional compounds, difficulty often arises in deciding which group to cite as the suffix and which as the prefix. For example, the compound $CH_3CH(OH)NH_2$ could conceivably be called 1-aminoethanol or 1-hydroxyethylamine. To resolve such problems, IUPAC has published a list of functional groups in decreasing order of priority for citation as the principal group and hence the suffix in the name. An abridged version of the list is as follows.

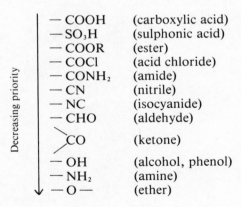

Decreasing priority

— COOH	(carboxylic acid)
— SO_3H	(sulphonic acid)
— COOR	(ester)
— COCl	(acid chloride)
— $CONH_2$	(amide)
— CN	(nitrile)
— NC	(isocyanide)
— CHO	(aldehyde)
$>$CO	(ketone)
— OH	(alcohol, phenol)
— NH_2	(amine)
— O —	(ether)

Because OH lies above NH_2 in this list, the correct name for the compound $CH_3CH(OH)NH_2$ is 1-aminoethanol. By similar reasoning, the compound $CH_3CH(OH)CN$ is called 2-hydroxypropanenitrile; not 1-cyanoethanol.

Index